NEW DEVELOPMENTS IN COMBUSTION RESEARCH

NEW DEVELOPMENTS IN COMBUSTION RESEARCH

WILLIAM J. CAREY
EDITOR

Nova Science Publishers, Inc.
New York

For permission to use material from this book please contact us:
Telephone 631-231-7269; Fax 631-231-8175
Web Site: http://www.novapublishers.com

NOTICE TO THE READER

LIBRARY OF CONGRESS CATALOGING-IN-PUBLICATION DATA
New developments in combustion research / William J. Carey, editor.
p. cm.
Includes index.
ISBN 1-59454-826-9
1. Combustion--Research. I. Carey, William J.
QD516.N49 2006
621.402'3--dc22 2005033130

Published by Nova Science Publishers, Inc. ✣ New York

CONTENTS

PREFACE

Combustion or burning is an exothermic reaction between a substance and a gas to release heat. Combustion normally occurs in oxygen (often in the form of gaseous O2) to form oxides, However, combustion can also take place in other gases like chlorine. The products of such reactions usually include water (H2 O) as well as carbon monoxide (CO) or carbon dioxide (CO2), or both. Other byproducts, such as partially reacted fuel and elemental carbon (C), may generate visible smoke and soot. This new book presents leading research from around the world in this frontal field.

Chapter 1 propounds an approach to modeling of turbulent premixed combustion at strong turbulence and fast chemistry based on generalize for this case fundamental for turbulence modeling Kolmogorov's ideas. This approach gives an opportunity to overcome assuming small-scale equilibrium the fundamental difficulty of modeling (so called "challenge of turbulent combustion") caused by the fact that combustion is controlled by small-scale coupling of chemistry, and turbulence, and develop a first approximation but justified theory and model equations, which conceptually close to the Kolmogorov one for the small-scale turbulence and his two parametric "$k-\omega$" model.

The basic principles of this approach are the following:

1. Thin and strongly wrinkled laminar or microturbulent flamelet sheet is interpreted as a random surface with continuous spectrum of the wrinkle sizes, and in main the small-scale wrinkles control the sheet area and hence the flame speed U_t while the large-scale ones control the brush width δ_t.
2. We distinguish in our analysis three successive stage of the premixed flame:

   (i) short initial stage when both small- and large-scale wrinkles are nonequlibrium that results in growing U_t and δ_t.

   (ii) dominated in experiments intermediate stage with quasi-equilibrium small-scale wrinkles and practically not equilibrium large-scale wrinkles that results in nearly constant U_t and increasing δ_t;

   (iii) practically unattainable final stage with equilibrium wrinkles of all sizes.

3. Relevant effects of small-scale coupling are expressed in terms of u', L and τ_{ch}, it permits to express by exponential algebraic expressions in these terms the thickened flamelet parameters, and U_t and δ_t for all three stages.
4. The counter-gradient flux is treated as a pressure-driven phenomenon and developed gasdynamics model for $\overline{u}_u$ and $\overline{u}_b$ does not bring additional constants.
5. The velocity of the steady sate flame is identified with the speed of the concentration front in turbulent diffusion and the theoretical result for the letter is used for defining the speed of the steady state flame.

The main results are the following:

1. Theoretical expressions for the thickened flamelet width (including influence of the flamelet stretch) and U_t and δ_t for all combustion stages;
2. Differential relaxation equations describing all three stages in the case $\rho = const$.
3. Gasdynamics equations for the counter-gradient component of the scalar flux.
4. Model equations for ISP flame $(\rho \neq const)$ in context of RANS, LES and original joint RANS/LES approaches.
5. Results of validation and numerical RANS, LES and Joint RANS/LES simulations of the premixed flames in different situations (stirred bomb, open and channel flames, impingent flames) with description the field of parameters and counter-gradient transport phenomenon.

In Chapter 2 we derive an original unclosed ensemble averaged equation for the Favre averaged chemical source $\widetilde{W}$ in the progress variable balance equation ($\widetilde{c}-$ equation) describing premixed turbulent combustion. This derived $\widetilde{W}-$equation directly describing the source at the flamelet combustion mechanism can be an alternative one for designing of combustion model to existing unclosed equation for the average surface density function (SDF) Σ which uses the intermediate model presentation $\overline{\rho}\widetilde{W} = \rho_u S_L \Sigma$. The $\widetilde{W}-$ equation is of the same type as any other equation used in CFD for combustion analysis and simulations. However, it has fundamental advantage as it admits to avoid completely a mismatch of a surface averaged SDF $\Sigma-$equation and others Favre averaged CFD equations that inevitably provokes a principal difficulty in designing model equations.

For completeness of analysis we also investigated an intermediate way to describe the chemical source. It which consists in derivation of an unclosed equation for ensemble averaged SDF $\overline{\Sigma}$ which is not used in the classical SDF approach. We show that, in spite of equivalence of deduced $\widetilde{W}-$ and $\overline{\Sigma}-$ equations, the former is preferable as it has logically more simple structure typical for others equations and contains only Favre averaged variables while the latter contains both Favre and Reynolds averaged values. We also address the problem of modeling of the chemical source for premixed combustion in flamelet regime and compare capabilities of the deduced ensemble averaged $\overline{\Sigma}-$ and $\widetilde{W}-$ equations with those of the usually used $\Sigma-$equation.

In conclusion of the chapter we show that that the approaches based on the scalar G and the progress c variables are equivalents. We deduce defined on one particular isosurface kinematical $c-$equation that is completely equivalent to the classical $G-$equation and show that only infinite numbers of such equations both in terms of G or c are sufficient for deducing the Favre average moment equations. Known in the literature speculations about advantage of the scalar G over the progress variable c for resolving of the challenging problems of the turbulent premixed combustion modeling have no any theoretical justification.

Recent developments in Reynolds-Averaged Navier-Stokes (RANS) turbulence models are discussed in Chapter 3 in the context of applications in numerical simulations of turbulent non-premixed flames. Both two-equation (k–ε type) models and second moment closure models are considered. Some results are discussed with both the pre-assumed PDF (probability density function) approach and the transported PDF methodology to model interaction between turbulence and chemistry. In the pre-assumed PDF simulations the conserved scalar approach is followed, with mixture fraction as conserved scalar. Combustion is modelled by assumption of (constrained) chemical equilibrium or by means of a steady laminar flamelet model in the results discussed. In the transported PDF approach detailed chemistry with finite rate kinetics is applied in the form of skeletal schemes or augmented reduced mechanisms. Results are discussed for different turbulent non-premixed jet type flames: piloted jet flames ('Sandia Flame D-F' and 'Delft Flame III') and a bluff-body stabilized flame ('Sydney Flame HM1-3').

Chapter 4 undertakes an analysis of parameters defining the rate of chemical reactions in turbulent premixed combustion with particular emphasis on the strong turbulence case characterised by a break-down of a simple relationship between the flame speed and turbulent rms velocity. It is argued that, in addition to commonly considered flow properties such as the laminar flame speed and rms turbulent velocity, both the details of chemical kinetics and the molecular transport coefficients are important in this case. Mathematical description of this combustion regime should be based on use of the probability density functions (pdf's), or similar probabilistic concepts, and possible frameworks for doing so are investigated. Two novel approaches are presented: one is based on consideration of a transport equation for two-point probability density function and the second is an extension of the reference scalar field (rsf) model with an improved representation of the small-scale mixing. While the former method is self-consistent and self-contained, nevertheless it is expressed in terms of a multidimensional function and is rather complicated. The latter method is based on a use of several reference fields possessing only one extra dimension and the small-scale mixing is modelled as a diffusion process in this extra dimension. The proposed model is capable of predicting the small-scale mixing in agreement with the DNS results for homogeneous turbulence. When applied to a reacting case it predicts very strong effects of molecular transport coefficients, consistently with the analysed experiments.

Chapter 5 addresses detailed chemical kinetics which was implemented into a state-of-the-art 3-D CFD code to simulate the complex diesel spray atomization, combustion and emission processes. Spray atomization is simulated by considering instability of the surface wave that results in fuel drops breakup from the fuel jet. The chemistry model uses a reaction mechanism to simulate the fuel oxidation and NOx formation processes. Soot emission is modeled by using a competing formation and oxidation rate formulation. The model is used to study the diesel lift-off flame structure under high ambient temperature and pressure conditions and results are compared with experimental imaging. Good levels of agreement are obtained in lift-off length, local equivalence ratio at the lift-off location, flame structure and soot distribution within the reaction zone. Results show that low-temperature cool flame heat release occurs prior to the lift-off location. As the cool flame mixture moves downstream of the spray plume, second-stage ignition occurs such that the flame is stabilized at the lift-off location. It is also found that the amount of soot emission is related to the local conditions at the lift-off location, which, in turn, is affected by ambient conditions. A decrease in ambient gas temperature results in a longer lift-off length (i.e., ignition delay) and less soot emission. Such a low-temperature combustion technology can be used to reduce soot and NOx emissions in diesel engine operations. The present model is also applied to simulate diesel engine combustion under conditions with highly premixed burn. Computed cylinder pressure, heat release rate, soot and NOx emissions are compared with experimental measurements with good agreement. Results show that NOx emission decreases monotonically as the start of injection (SOI) is retarded. However, soot emission increases as start of injection is retarded approaching top-dead-center and then decreases at further retarded timings, as also observed from engine-out measurement. The model indicates that low soot emission at early SOI is due to better oxidation while low soot emission at late SOI is due to less formation. Results show that emission reduction strategies such as late injection and exhaust gas recirculation (EGR) can be used such that combustion occurs at a relatively low temperature environment.

In Chapter 6 a model is presented to predict turbulent combustion of premixed CH_4 flames at gas turbine conditions. Mixing, combustion and heat loss are described with five scalar variables. These are the mixture fraction, enthalpy variable and three reaction progress variables for combustion of CH_4, H_2 and CO. The effect of turbulence on combustion is taken into account by weighting with an assumed shape probability density function. In a gas turbine the flame is stabilized in an internal flow recirculation area. In this configuration not only the mean but also the variance of the methane progress variable has to be taken into account. The model is tested on a premixed lean CH_4 flame stabilized by a coflow of burnt stoichiometric CH_4 flow. Essential in this configuration is the turbulent mixing of areas burnt and unburnt. The calculated velocity field and CH_4 concentration field have been compared with experiments done by P. Moreau and A. Boutier on this geometry. The results show good agreement concerning the velocity field. The reaction progress of CH_4 to H_2 and CO has to be assessed for good accuracy. Good results can be obtained when the derived thermochemical database is considered in detail for both source terms and calculated species concentrations.

In Chapter 7 we present a detailed strategy for performing quantitative spatially-resolved multiscalar measurements of major species concentration and temperature in high pressure H_2-air combustion systems using spontaneous Raman scattering (SRS). This strategy can also be extended to include CO, CO_2 and hydrocarbons (such as CH_4) in both high pressure and atmospheric pressure hydrocarbon-air flames. Central to this strategy is the Raman calibration matrix formalism which is often used to account for spectral interferences resulting from inter-species spectral crosstalk. We describe in detail, the theory, instrumentation, facilities, and data processing techniques used to generate the Raman calibration matrix, and its implementation to obtain quantitative multiscalar measurements. The Raman calibration matrix is realized with the aid of a novel high-pressure fully-premixed calibration burner, an optically accessible high-pressure combustion facility, and a state-of-art 532 nm wavelength laser SRS system. This system and data processing methods, collectively, provide an empirically based, but analytic formulation of the temperature-dependent Raman calibration matrix elements based on the assumption of chemical equilibrium at the measured temperature in the post-flame products of a premixed H_2-air calibration burner operated at a pressure of 10 atm. At pressures up to 30 atm, we find the inclusion of a pressure-dependent term in the calibration matrix unnecessary for the relatively low, but practical, spectral resolution of the optical setup, mainly because the pressure broadening (or narrowing, if any) effect is minor relative to the instrumental resolution (approx. 1.5 nm) . The high signal-to-noise ratio SRS signals yielded accurate (< 2% uncertainty) temperature measurements via the rotational H_2 and vibrational N_2 spectra, and together, enable the generation of a high quality Raman calibration matrix. The Raman calibration matrix determined in this study can be used as a transferable SRS standard for use with other 532 nm wavelength laser-Raman setups to probe H_2-air flames since the signal response was calibrated using traceable wavelength and intensity standards.

It is shown in chapter 8 that the study of wildland fire is a growing field bringing important improvements to our understanding of heat and mass transfer and combustion. Indeed, forest fire is a complex phenomenon in which the levels of description cover a huge range, from the details of the kinetics of gaseous combustion and thermal degradation of fuels, up to the physical-chemistry characterization of the flame and the vegetation cover as a fuel. This paper addresses experiments, modelling and simulation of surface forest fires. Experimental techniques and results are first displayed to investigate fire behaviour. Then modelling tools and simulation results devoted to understand the mechanisms involved in such fires are presented. Different methods and measuring structures build in order to improve the knowledge on the fundamental physical mechanisms that control the propagation of wildland fire are detailed. The basic properties of flames encountered in forest fire are presented along with the major tools developed for measuring their characteristics at both laboratory and field scales. Indeed, although the modelling of forest fire spread has made considerable progress over the last ten years, however there remains a lack of reliable experimental results concerning fire front properties for the validation of the different existing models. Then, a theoretical model for the spreading of a surface fire is proposed incorporating the effects of wind and slope. The aim of our ongoing research is to propose a model for the spreading of a forest fire to be integrated in an operational management tool. To proceed, we carry out the reduction of a detailed multiphase set of equations that describe the finest mechanisms involved in forest fire. This reduction allows us to develop a simplified model by considering mainly the buoyancy effect induced by combustion in the flaming zone. The contributions of both convection and radiation impinging on the top of the fuel-bed ahead of the fire-front are included. Finally, model predictions are compared to laboratory-scale experiments. To this end, dedicated fires were carried out for horizontal spreading in still air across beds of pine needles, in order to measure flame and fire fronts properties, thermocouples and heat flux sensors. Experiments conducted under wind and slope conditions are also considered for the validation of the model.

In: New Devel. in Combustion Research
Editor: William J. Carey, pp. 1-93
ISBN: 1-59454-826-9

Chapter 1

KOLMOGOROV'S LEGACY AND TURBULENT PREMIXED COMBUSTION MODELING

Vladimir L. Zimont *
CRS4 Research Center, Polaris, 09010 Pula(CA), Italy
N.N.Semenov Institute of Chemical Physics, Moscow, 117977, Russia

Abstract

The chapter propound an approach to modeling of turbulent premixed combustion at strong turbulence and fast chemistry based on generalize for this case fundamental for turbulence modeling Kolmogorov's ideas. This approach gives an opportunity to overcome assuming small-scale equilibrium the fundamental difficulty of modeling (so called "challenge of turbulent combustion") caused by the fact that combustion is controlled by small-scale coupling of chemistry, and turbulence, and develop a first approximation but justified theory and model equations, which conceptually close to the Kolmogorov one for the small-scale turbulence and his two parametric "$k-\omega$" model.
The basic principles of this approach are the following:

1. Thin and strongly wrinkled laminar or microturbulent flamelet sheet is interpreted as a random surface with continuous spectrum of the wrinkle sizes, and in main the small-scale wrinkles control the sheet area and hence the flame speed U_t while the large-scale ones control the brush width δ_t.
2. We distinguish in our analysis three successive stage of the premixed flame:

   (i) short initial stage when both small- and large-scale wrinkles are nonequlibrium that results in growing U_t and δ_t.
   (ii) dominated in experiments intermediate stage with quasi-equilibrium small-scale wrinkles and practically not equilibrium large-scale wrinkles that results in nearly constant U_t and increasing δ_t;
   (iii) practically unattainable final stage with equilibrium wrinkles of all sizes.

---

* E-mail address: zimont@crs4.it

3. Relevant effects of small-scale coupling are expressed in terms of u', L and τ_{ch}, it permits to express by exponential algebraic expressions in these terms the thickened flamelet parameters, and U_t and δ_t for all three stages.
4. The counter-gradient flux is treated as a pressure-driven phenomenon and developed gasdynamics model for $\overline{u}_u$ and $\overline{u}_b$ does not bring additional constants.
5. The velocity of the steady sate flame is identified with the speed of the concentration front in turbulent diffusion and the theoretical result for the letter is used for defining the speed of the steady state flame.

The main results are the following:

1. Theoretical expressions for the thickened flamelet width (including influence of the flamelet stretch) and U_t and δ_t for all combustion stages;
2. Differential relaxation equations describing all three stages in the case $\rho = const$.
3. Gasdynamics equations for the counter-gradient component of the scalar flux.
4. Model equations for ISP flame $(\rho \neq const)$ in context of RANS, LES and original joint RANS/LES approaches.
5. Results of validation and numerical RANS, LES and Joint RANS/LES simulations of the premixed flames in different situations (stirred bomb, open and channel flames, impingent flames) with description the field of parameters and counter-gradient transport phenomenon.

Keywords: Turbulent premixed combustion; Combustion modeling; Kolmogorov ideas. Counter-gradient transport phenomenon.

Nomenclature

c	= progress variable
Y_i	= concentration of the i spice
ρ_u	= unburned gas (reactants) density
ρ_b	= burned gas (products) density
$\Theta = \rho_u / \rho_b$	= densities ratio
$\overline{\vec{u}}_u$	= reactants conditional average speed
$\overline{\vec{u}}_b$	= products conditional average speed
$u' = (\overline{u'^2})^{1/2}$	= root-mean-square velocity pulsation
$\kappa = (3/2)u'^2$	= turbulent energy
L	= Eulerian integral scale of turbulence
$\tau_t = L/u'$	= turbulent time
$\omega = u'/L$	= Kolmogorov turbulent frequency
$T_L = \int_0^\infty R_L(t)dt$	= Lagrange integral time scale

$R_L(t) = \overline{u_1'(0)u_1'(t)} / u'^2$	= Lagrange correlation coefficient
$l_L = u'T_L$	= Lagrange integral scale
$D_t = u'l_L \approx u'L$	= turbulent diffusion coefficient
S_L	= normal laminar speed
ν	= kinematical viscosity coefficient
$\mathrm{Re}_t = u'L / \nu$	= turbulent Reynolds number
$\eta = L\,\mathrm{Re}_t^{-3/4}$	= Kolmogorov microscale
D	= molecular diffusion coefficient
$\chi(\approx D \approx \nu)$	= molecular heat transfer coefficient
$\tau_{ch} = \chi / S_L^2$	= chemical combustion time
$Da = \tau_t / \tau_{ch}$	= Damköhler number
$Ka = (\delta_L / \eta)^2$	= Karlovitz number
$E(k)$	= turbulence spectrum
$F(k)$	= spectrum of the random flamelet sheet
k	= wave number
$\lambda = \int F(k)dk / \int k^2 F(k)dk$	=Taylor mocroscale of a random surface
u_f'	= turbulent pulsation inside the flamelet
L_f	= scale of turbulence inside the flamelet
Re	= global Reynolds number
Re_{cr}	= critical Reynolds number
$(\overline{A} / A_0)$	= dimensional flamelet sheet area
u_f'	= microturbulent velocity inside the flamelet
u_η'	= velocity pulsation in the dissipative eddies
U_f	= thickened flamelet speed
W	= chemical source
δ_f	= thickened flamelet width
δ_L	= normal laminar flame width
δ_t	= turbulent flame brush width
χ	= molecular heat transfer coefficient
$\chi_f(\approx D_f)$	= heat transfer coefficient in the flamelet
$\chi_t(\approx D_t)$	= turbulent heat transfer coefficient
ε	= mean turbulent dissipation rate
ν	= kinematical molecular viscosity

$\nu_t (\approx D_t)$	= turbulent viscosity
Φ	= equivalence ratio
$\alpha = 1/\Phi$	= air excess coefficient
u	= index for reactants (unburned mixture)
b	= index for products (burned mixture)

1 Introduction

Modeling of turbulent premixed combustion is more difficult problem in comparison with modeling of non-premixed combustion. One of main reason of it is that the latter is predominantly defined by mixing of a fuel and oxidizer and coupling of turbulence, chemistry and molecular transport has relatively small influence at least on the flame aerodynamics. Namely due to it the diffusion models (based on assumption of infinite reaction rates) in many cases could be a good first approximation. At the same time premixed combustion is completely controlled by this coupling and assumption of infinite reaction rates has no any physical meaning. So-called "challenge of turbulent combustion" relates mainly to turbulent combustion at strong turbulence and fast chemistry.

S. Pope [1] and K. Bray [2] used this term in scientific papers, for example, to emphasize that there is a fundamental difficulty on the way of developing of turbulent premixed combustion modeling on the theoretically justified base. The reason of it is connected with the fact that at real turbulence and chemical kinetics the combustion rates are controlled by coupling of chemistry, micro-turbulent mixing and molecular diffusion at such small scales that cannot be resolved in the context of RANS or LES balance model equations, so "the accurate prediction of mean reaction rates, which can be influenced strongly by molecular diffusion caused by small-scale turbulent mixing, represents the central problem and challenge of turbulent combustion"[2]. In other words, the challenge of turbulent combustion implies a fundamental inability of moment and PDF turbulent combustion models to describe correctly basic physico-chemical mechanisms controlling combustion rates, i. e. to predict accurately the dependence of combustion processes in turbulent flows on turbulence, chemistry and molecular properties of mixtures.

Before describing of our approach to practical resolving this problem we point out that observations demonstrate in fact paradoxical properties of the turbulent premixed flame. We attract attention to four apparent paradoxes of the turbulent premixed combustion:

1 Dependence of the Speed U_t on Chemistry at Strong Turbulence

Plausible reasoning of the pioneers of the turbulent combustion theory (that it seems nobody even try to refute) result in that at strong turbulence $u' >> S_L$ the turbulent flame speed $U_t \sim u'$ [3], [4]. Their argument is that this speed is controlled by the large-scale eddies with the characteristic velocity u' which throw volumes of the products in the reactants so at strong turbulence the flame speed does not depend directly on chemistry (but obviously can depend on it indirectly through influence of the gas expansion due to combustion on turbulence). At the same time a large body of evidence demonstrate such direct dependence and many experimental and theoretical investigation are devoted to it.

The first question: why U_t depends on chemistry at strong turbulence?

2 Contradiction between Increasing Brush Width and Constant Speed

Many observations show that the turbulent premixed flame is not a steady state flame (the classical example in the 1D stationary flame analyzed theoretically and numerically in many papers) but strange, unusual transient flames with increasing width δ_t and at the same time nearly constant speed U_t. Visualization of the flame clearly demonstrates this increasing of the width and practically constant inclination to the flow both for the open flame [5] and for combustion behind the stabilizer in the uniform turbulent flow [6]. At the same time by analogy with laminar premixed combustion we can expect intuitively that combustion take place in the steady state flame with constant speed and width, i.e. in 1D stationary flame analyzed theoretically and numerically in many papers. We notice that in transient laminar flame (for example, at transition from the step distribution to the normal 1D stationary flame) increasing of the width inevitably is accompanied by increasing of the flame speed. Besides the transient time for actual chemistry is so small that usual approximation in the combustion aerodynamics that the reacting sheet is the running wave with practically constant S_L and δ_L is theoretically justified and it is confirmed by observations.

The second question: Why in contrast to laminar premixed combustion turbulent premixed combustion takes place not in the form of the running turbulent combustion front with constant speed $U_t = U_t^{st}$ and width $\delta_t = \delta_t^{st}$ but in the form of moving turbulent mixing layer with increasing width δ_t and nearly constant speed U_t?

3 Contradiction between Increasing Brush Width and the Counter-Gradient Transport

Experimental data demonstrate that in transient premixed flames with increasing brush width the scalar transport is predominantly counter-gradient [7] (the mean heat flux across the flame is directed from cold reactants to hot products, mean reactants flux is directed from products to reactants and mean products flux is directed from reactants to products), but it could be also gradient or neutral [8]. At the same time increasing of the width in the laminar transient flames is connected exclusively with the normal gradient molecular diffusion as well as increasing of the width of the turbulent nonreacting mixing layer takes place due to "normal" direction of the scalar flux and in the hypothetical case of negative molecular or turbulent diffusion coefficient we would have decreasing of the width.

The third question: How to resolve the contradiction between increasing brush width and the counter-gradient transport observed in the experiments?

4 Contradiction between Width of the Instantaneous Flame and the Continuous Turbulent Spectrum

Recent direct measurements demonstrate that the width of the instantaneous reaction zone (so called “flamelet”) in the turbulent flame with developed turbulence is wider than the laminar flame as in the case of weak turbulence. This result seems natural as at strong turbulence with $\eta < S_L$ (η is the minimal size of the turbulent eddies, the Kolmogorove microscale) the small eddies fall inside the flamelet and intensify the transport that results in the flamelet thickening. Paradoxical here is the fact that this broadening is relatively small. For experimental condition in [9] flamelet width is only three-four times as large as the normal laminar flame width, i.e. the thickened flamelet width is small in comparison with the turbulence integral scale L and the instantaneous flamelet sheet remains thin and strongly wrinkled. The paradox here is following: everybody can expect that the first stage of the laminar flame broadening will be followed by the second and further steps when more and more large eddies fall inside more and more broadened flamelet so step by step all eddies of the continuous spectrum will be involved in the reaction zone.

The fourth question: Why at strong turbulence with continuous spectrum of eddies and developed small-scale structure premixed combustion takes place in relatively thin and strongly wrinkled reacting sheets instead of intuitively expected distributed stirred-reactor type combustion mechanism due to successive involving of more and more large eddies in more and more thickened by these eddies reaction zone?

Our aim is not only to give plausible qualitative explanation of these apparent paradoxes but also to present a combustion model that describes quantitatively the turbulent premixed combustion with all these paradoxical peculiarities.

Turning back to the turbulent combustion challenge our practical resolving of it is based on the Kolmogorov ideas of turbulence modeling modified to our case of turbulent premixed combustion. We assume not only equilibrium small-scale turbulence but also an equilibrium small-scale structure of the reaction zone (including the equilibrium thickened flamelet and statistically equilibrium small-scale wrinkles of the sheet) controlled by the small-scale eddies. We first express theoretically in the analytical exponential form relevant effects of the small-scale coupling of turbulence and chemistry in terms of large-scale turbulent parameters and the integral chemical combustion time, and second use them directly for designing model equations, i. e. methodologically we follow the Kolmogorov pattern.

To gain a better understanding of our approach we remind the Kolmogorov approach and point out that many years ago conceptually similar practically resolved by Kolmogorov challenge arose at modeling of the mean turbulent dissipation rate. In this case small-scale coupling of turbulence and viscous stresses resulting in the dissipation rate takes place in unresolved "viscous" eddies. This "challenge of turbulence" was resolved by A. Kolmogorov[10],[11] by assuming (reasoning from the cascade mechanism) the equilibrium structure of small-scale turbulence. In the case of large turbulent Reynolds number Re_t the non-linear cascade flux of turbulent energy to the small-scale dissipative eddies is controlled by the large-scale turbulence and this is equal to the dissipation rate. It let him to derive in [10] simple formula for the averaged dissipation rate $\varepsilon \sim u'^3 L^{-1}$. A year later[11] Kolmogorov inserted this expression (we stress that it is a theoretical one, i.e. the exponents 3 and -1 are not empirical) in his original "$k-\omega$" model equations describing in general case nonequilibrium turbulence. We stress that it was first-ever two-parametric model which was analyzed in the West literature in [12]. It is recognized that widely used now "$k-\varepsilon$" model occurred in fact a modification of the Kolmogorov model[13]).

A logical consequence of Kolmogorov analysis is that for strong turbulence every attempt to model the dissipation rate ε in terms of the molecular viscosity coefficient ν is a misconception in spite of existing of well known exact expression for the dissipation rate

$$\varepsilon = (\rho\nu/2)\sum_{i,j}\overline{(\partial u_i'/\partial x_j + \partial u_j'/\partial x_i)^2}\,.$$

The physical meaning of this expression, which can be presented as $\varepsilon \sim \rho\cdot\nu\cdot g^2$, is the following: the mean dissipation rate is proportional to the molecular viscosity coefficient and to the quadrate of the characterization velocity gradient g controlling by the small-scale turbulent structure. Independence of ε on ν means that in the developed turbulence there is a self-compensation mechanism: increasing of ν decreasing g (make small-scale structure more smooth) and vice versa.

We accent this point as we show below that similar self-compensation takes place in premixed flames at the flamelet combustion mechanism where the turbulent flame velocity can be presented as $U_t = S_L(\overline{A}/A_0)$. Increasing of S_L results in smoothing of the instantaneous flame ("flamelet sheet") wrinkles and hence decreasing of the dimensionless flamelet sheet area $(\overline{A}/A_0)$. We explain below relatively weak dependence of U_t on the chemical kinetics (in comparison with S_L) in actual flames by this partial self-compensation.

Our approach to overcoming of the turbulent combustion challenge and the principles of modeling conceptually in fact only slightly modified Kolmogorov principles of turbulence modeling: we first express relevant effects of the small-scale coupling of turbulence and chemistry in terms of large-scale turbulent parameters u', L and the characteristic chemical time of combustion τ_{ch} assuming statistically equilibrium small-scale structure non only of the turbulence eddies, but also of the reaction zones, second we used these theoretical expressions for designing of the combustion model equations which together with the turbulence model equations in fact describe characteristics of large-scale both turbulence and reaction zone.

Small-scale equilibrium structures of the reaction zone are (i) the flamelet structure that in our case is not the laminar flame but thickened by the microturbulence one and (ii) small-scale flamelet sheet wrinkles giving the main contribution in $(\overline{A}/A_0)$ and hence U_t. At the same time in general case nonequilibrium in general case large-scale structure of the reaction zone is presented by the large wrinkles of the flamelet sheet controlling the turbulent flame ("flame brush") width δ_t.

Interpretation of the turbulent premixed flame in terms of equilibrium-nonequilibrium and small-scale and large-scale flamelet sheet structures opens the way to original and seems theoretically justified classification of the turbulent flame stages at strong turbulence

$u' >> S_L$. But before we remind some general properties of the passive surface in uniform turbulent field. Assume that in the initial time this surface is plain, $(\overline{A}/A_0)=1$. In time takes place stretch and bending of the surface, take place turbulent mixing of divided by the surface gases with increasing dispersion σ^2 and the surface area, $(\overline{A}/A_0)>1$ and forming a turbulent mixing. Though the mixing layer width increases relatively slow $\delta_m \sim (\sigma^2)^{1/2} = (2D_t t)^{1/2}$ in accordance with the turbulent diffusion theory [14][1] at the same time the area of the passive surface in accordance with the Batchelor estimation [15] increases exponentially in time $(\overline{A}/A_0) \sim e^{\alpha t}$, i.e. much faster. The reason of it is connected with the fact that the dispersion is controlled mainly by the large-scale wrinkles while the area is controlled mainly by the small-scale wrinkles, which are generated very fast in the case of developed turbulence.

In the case of combustion we have during some times practically the same (as for the nonreacting mixing layer) increasing of the flame width $\delta_t = (\sigma^2)^{1/2} = (2D_t t)^{1/2}$ (we define the flame brush width as $\delta_t = (\sigma^2)^{1/2}$). The reason is that sheet movement with slow speed $S_L << u'$ cannot strongly influence on the large-scale wrinkles which are controlled by the large-scale eddies with the characteristic velocity u'. At the same time the exponential increase of the speed $U_t = S_L(\overline{A}/A_0)$ impossible as moving flamelet sheet strongly influence on the small-scale wrinkles, "consumes" them in spite of the relatively small speed S_L.

We distinguish three stages of the turbulent premixed flame:

1. *Short initial stage*: Both small-scale and large-scale wrinkles are not equilibrium, i.e. takes place linear (in accordance with the Taylor theory [14], see the note 1) increasing of the brush width $\delta_t = (\sigma^2)^2 = u't$ and relatively fast increases of U_t.
2. *Intermediate stage (transient flames):* The small-scale winkles are already quasi-equilibrium (quasi complete balance between the sheet area increasing due to turbulence and the area decreasing due to consumption of the small-scale wrinkles by the moving flamelet sheet) while the large-scale wrinkles still practically nonequilibrium, in fact it is an intermediate asymptotic state. These transient flames are characterized by nearly constant flame speed and at the same time by increasing brush width. We called them in [16] the intermediate steady propagation (ISP) flames.

---

[1] The Taylor theory [14] states that for small times ($t << T_L$) the dispersion is $\sigma^2 = (u't)^2$ and for large times ($t >> T_L$) it is $\sigma^2 = 2D_t^{eq}t$, $D_t^{eq} = u'^2 \mathrm{T}_L$. The consequence is that D_t during initial stage is transient $D_t(t) = u'^2 \int_0^t R_L(t)dt$. We put below this fact for modeling of the initial stage of the premixed flame so used here different notation for the equilibrium and transient turbulent diffusion coefficient D_t^{eq} and D_t will be convenient.

3. *Final steady state stage*: Both the small-scale and the large-scale wrinkles are equilibrium; it is the final asymptotical state. It is steady state (SS) flames analyzed in many publications (usually in form of the classical 1D stationary turbulent flame).

The first stage is practically significant mainly in the spark ignition (SI) engines where after electrical charge takes place forming of the developed turbulent flame. It is special subject and we touch it only slightly. Analysis of the experimental data show that namely ISP flames (the second stage) dominate at developed turbulence while steady state SS flames (the third stage) practically unattainable in real condition, see, for example, the recent review [17]. ISP flames are observed not only in simple hydrodynamics situations such as combustion in uniform turbulent flows. But even the impingent flame reminding the 1D SS one due to contraction of the broadening flame brush by a specific velocity field is conceptually closer (as we show below) to the ISP than to the SS flame. The reason of this predominance is connected with the fact that the completion of combustion (due to the flame intersection or touching the wall) usually takes place long before forming completely equilibrium steady state flames corresponding to the third stage of combustion.

Our statement is that the transient ISP flames (corresponding to the second stage in our classification) constitute a special class of the practically significant flames that must be modeled by developing a special combustion model. The simplified assumption at designing of such combustion model was that we put complete equilibrium small-scale structures and at the same time the large-scale flamelet sheet wrinkles are assumed completely nonequilibrium. It permits to deduce theoretically some expressions for the turbulent flame speed U_t and the flame brush width δ_t for the ISP flame that were directly used for closure of the basic equations. The model name "the turbulent flame closure (TFC) model" reflects this fact. All numerical RANS and LES illustrations in the context of the TFC model refer to this case.

Nevertheless significant attention is devoted also to analysis of the initial and the final stages of combustion and deducing a set of equations describing in the context of the TFC model all three stages of combustion. We associate forming of the turbulent flame during the initial stage with developing of the turbulent diffusion coefficient and work out additional expressions for the speed U_t^{st} and width δ_t^{st} of the SS flame assuming complete equilibrium both the small-scale and the large scale flamelet sheet wrinkles. We use them directly in the set of relaxation expression describing the change of the flame stages.

Now we briefly explain listed above apparent paradoxes of the turbulent premixed combustion in the context of the TFC model which are thoroughly analyzed in the main text:

The first paradox: Observed chemistry dependence of the turbulent flame speed U_t is connected with the fact that actual premixed flames are transient. The flamelet speed and the sheet area control the flame velocity U_t. Relatively weak chemistry dependence of U_t is connected with the self-compensation mechanism (increasing of the flamelet speed results in reducing of the sheet area and vice versa). Due to growing flame brush width, the front edge speed $U_{f.e.} > U_t$ and this peed does not control the volume consumption rate U_t. Strictly speaking the parabolic TFC model equation results in infinite edge speed $U_{f.e.}$ and here there is no contradiction with finite U_t. In contrast the SS flame speed is controlled by the front edge speed $U_t^{st} = U_{f.e.}$. This speed for strong turbulence ($u' >> S_L$) was estimated in [3], [4] by the order of magnitude as $U_t^{st} \sim u'$. More accurate estimation of the speed U_t^{st} as the speed of the turbulent diffusion front that practically equal to $U_{f.e.}$ when $u' >> S_L$ results in that $U_t^{st} = U_{f.e.} = u'$, i.e. the flame speed not only has the order but practically equal to the root-mean-square turbulent pulsation velocity. We notice also that the growing brush width of the ISP flame δ_t does not depend directly on the chemistry while in contrast δ_t^{st} directly depends on it. The latter adapts to the speed $U_t^{st} = u'$ to yield the flamelet sheet area corresponding to this speed.

The second paradox: We explain the existence of the premixed flame with increasing brush width δ_t and nearly constant U_t in the context of the RANS version of the TFC model by existence of quasi-equilibrium small-scale and at the same time nonequilibrium large-scale structures of the strongly wrinkles flamelet sheet. In accordance with observations these flames dominate in premixed burners so, we believe, they constitute a special and practically significant class of premixed flame. We notice that in context of the RANS approach it needs special modeling while based on the TFC model our LES simulations give growing brush width and nearly constant flame inclination without special modeling.

The third paradox: We explain growing of the brush width in spite of the counter-gradient mean scalar transport by assumption that the counter-gradient transport phenomenon is not turbulent but gasdynamical one and connected with different acceleration of cold reactants and hot product by the transverse pressure gradient generated due to combustion. So the dispersion of the flamelet sheet is controlled by the random pulsation of the velocity, i. e. by the turbulent component of the transport that in the context of the TFC model is described by normal gradient diffusion in terms of the turbulent diffusion coefficient D_t. The counter-gradient gasdynamical component of the transport is in fact controlled by the pressure gradient difference due to the pressure gradient of the convective fluxes of reactants and products conceptually it is not some kind of a “turbulent counter-gradient diffusion” that can neutralize the actual turbulent gradient one. In the TFC model only the latter component controls directly the flame brush width so we always have growing brush width: when in the transport prevails gasdynamical counter-gradient, or turbulent gradient component, or they are in balance, i.e. when the general transport is counter-gradient, gradient or neutral. We stress that continuous growing the brush width at transition of the transport along the flame from gradient to counter-gradient one takes place without any special modeling.

The forth paradox: We show that successive growing of the thickened flamelet width due to falling more and more large eddies in more and more thickened flamelet has a limit when the convective flux of reactants into the flamelet, microturbulent transport across it and the reactants consumption due to combustion are of the same order of magnitude quite similar to the normal laminar flame. Assumption of the Kolmogorov inertial turbulence controlling both microturbulent transport inside the flamelet and the thickened flamelet stretch permits to deduce expressions for the flamelet speed U_f, width δ_f and parameters of the sub-flamelet turbulence u'_f, L_f, D_f used in the TFC model. For typical situations our estimations give $\delta_f << L$, i.e. thickened flamelet remains thin and combustion takes place in strongly wrinkles flamelet sheet. In accordance with these our formulas the situation when $\delta_f \sim L$ in principle at very strong in fact hypothetical turbulence, but our estimations below demonstrate that well before the combustion extinction would takes place due to the flamelet stretch.

In conclusion of the introduction we formulate the problem of the turbulent premixed combustion modeling as it follows:

1. Designing of the basic turbulent combustion model on the theoretically justify background that does not contain empirical approximations and is applicable to both RANS and LES approaches. It must be consistent with the basic turbulence models that now are the "$k-\varepsilon$" model in the RANS and the Smagorinsky model in the LES approaches.
2. This model must describe the turbulent flame including mentioned above paradoxical features of the premixed flame. It is obvious that some of them in the context of the RANS approach must be modeled specially (such as the counter-gradient transport phenomenon, specific transient flame with increasing brush width and nearly constant flame speed) while in the context of the LES approach they do not need modeling and appear as a result of averaging of the instantaneous LES fields.
3. Though the LES approach is justly considered to have more perspective in comparison with the RANS approach because the nonequilibrium structure of turbulence and the flamelet sheet do not need modeling, complete replacement of the RANS approach by the LES one is unlikely justified in practical simulations not only because of more economical calculations but also because of the fact that expression of the effects of unresolved small-scale coupling of turbulence and chemistry in terms of resolved turbulent parameters are theoretically more justified in terms of the RANS integral than in terms of LES subgrid turbulent parameters. So one of the problems is to develop a joint RANS/LES method combining strong sides of both approaches.

The chapter states the equations of the TFC combustion model which in the context of the RANS formulation describes the premixed flame at thickened flamelet combustion mechanism including the initial stage; the transient ISP flame with nearly constant and reasonably dependant on chemistry speed U_t and controlled by turbulence increasing brush width δ_t in spite of predicted by the model actual predominantly counter-gradient scalar transport, and finally the steady state stage of the flame with independent on chemistry speed.

But the main attention was devoted to RANS and LES modeling and numerical simulations of the transient flames in the context of the TFC model which found application including industrial use and were implemented in the commercial codes Fluent, CFX and StarCD.

2 Basic Unclosed Moment Equations and Principles of Modeling

Combustion equations in general cases are formulated in terms of the actual chemical species and the temperature containing the source terms controlling by the actual chemical kinetics. Our analysis is based on the statistical moment description of the base equations so the combustion is described by the fields of the Favre average chemical species concentrations and the temperature.

All average equations in these terms contain the unknown mean scalar transport and mean chemical source term make the modeling problem besides too cumbersome. This is the reason why the premixed turbulent flame now usually is interpreted simplistically in term of only one parameter the progress variable c that in fact is a generalized reacting species. So the combustion problem reduces to one differential equation in terms of $\tilde{c}$ containing one unknown mean scalar flux and one unknown mean source. But even in this case theoretically justified modeling of these unknown functions are very difficult problem keeping in mind the paradoxical properties of the premixed flame listed in the introduction.

2.1 General Combustion Equations

The basic first order Favre average moment equations describing a turbulent flame at large Reynolds (when the molecular component of the transport can be ignored) are the following:

$$\partial\bar{\rho}\tilde{Y}_i / \partial t + \nabla \cdot \bar{\rho}\tilde{\vec{u}}\tilde{Y}_i = -\nabla \cdot \bar{\rho} < \vec{u}''Y_i'' >_F + \bar{\rho}\tilde{W}_{Y_i} \tag{1}$$

$$\partial\bar{\rho}\tilde{T} / \partial t + \nabla \cdot \bar{\rho}\tilde{\vec{u}}\tilde{T} = -\nabla \cdot \bar{\rho} < \vec{u}''T'' >_F + \bar{\rho}\tilde{W}_T \tag{2}$$

$$\bar{p} = \bar{\rho}R\tilde{T} \tag{3}$$

$$\partial\bar{\rho}\tilde{\vec{u}} / \partial t + \nabla \cdot \bar{\rho}\tilde{\vec{u}}\tilde{\vec{u}} = -\nabla \cdot \bar{\rho} < \vec{u}''\vec{u}'' >_F - \nabla\bar{p} \tag{4}$$

$$\partial\bar{\rho}/\partial t + \nabla \cdot \bar{\rho}\tilde{\vec{u}} = 0 \tag{5}$$

These equations are Favre averaged, i.e. weighted by the density ensemble averaged (we used notations $\tilde{a} = \overline{\rho a}/\bar{\rho}$ and $< ab >_F = \overline{\rho ab}/\bar{\rho}$). Eqs. (1)-(3) relate to the combustion part of the problem while Eqs. (4) and (5) relate to the hydrodynamics one. Eqs. (1) describe the average fields of actual chemical species concentrations Y_i, Eq. (2) is the energy and describe the temperature field, Eq. (3) is the thermodynamic state equations. The last two hydrodynamics impulse Eq. (4) and continuous Eq. (5) equations are connected with the combustion part of the problem through common pressure $\bar{p}$, density $\bar{\rho}$ and velocity $\tilde{\vec{u}}$. Eqs (1) - (5) are unclosed as they contain the unknown average scalar fluxes of the species $\overline{\rho\vec{u}''Y_i''} = \bar{\rho} < \vec{u}''Y_i'' >_F$ and the temperature $\overline{\rho\vec{u}''T''} = \bar{\rho} < \vec{u}''T'' >_F$ of the combustion sub-problem and the unknown average stresses of the hydrodynamic part of the problem as well as the unknown average sources $\tilde{W}_{Y_i}$ and $\tilde{W}_T$.

The aim of turbulent combustion modeling is in fact to express these unknown fluxes and sources in terms of describes by the set known variables. As it is well known this modeling can have a form of some algebraic expressions or differential equations. In both cases following sets of closed model equations give an opportunity to describe the fields of the mean variables $\bar{\rho}, \bar{p}, \tilde{\vec{u}}, \tilde{Y}_i, \tilde{T}$, the fluxes $< \vec{u}''Y_i'' >_F, < \vec{u}''T'' >_F$ and the sources $\tilde{W}_{Y_i}, \tilde{W}_T$. At the same time for completeness of description we need missing Reynolds average variables $\bar{\vec{u}}, \bar{Y}_i, \bar{T}$, components of the fluxes $\overline{\rho' Y_i'}, \overline{\rho' T'}, \overline{\vec{u}' Y_i'}, \overline{\vec{u}' T'}, \overline{\rho' \vec{u}'}, \overline{\rho' \vec{u}' Y_i'}, \overline{\rho' \vec{u}' T'}, \overline{\rho' \vec{u}' \vec{u}'}$ and the sources $\bar{W}_{y_i}, \bar{W}_T, \overline{\rho' W_{Y_i}'}, \overline{\rho' W_T'}$. Estimation of the Reynolds average characteristics in parallel with the Favre average parameters are useful as experimental data frequently correspond namely to the Reynolds averaging.

2.2 Equations in Terms of the Progress Variable

In the case of premixed combustion generally accepted simplification is using only a single scalar variable, so called the progress variable c, instead of the actual chemical species and the temperature. This variable is in fact a generalized reacting species describing the combustion completeness ($c = 0$ in reactants and $c = 1$ in products) and in terms of the density it is defined as

$$c = (\rho_u / \rho - 1)/(\Theta - 1), \quad (6)$$

where $\Theta = \rho_u / \rho_b$, ρ_u and ρ_b are the densities of reactants (unburned gas) and products (burned gas). So in the context of the progress variable c the combustion part of the problem is described instead of several Eqs. (1) – (3) only by two equations that are the following:

$$\partial \bar{\rho}\tilde{c} / \partial t + \nabla \cdot \bar{\rho}\tilde{\vec{u}}\tilde{c} = -\nabla \cdot \bar{\rho} < \vec{u}''c'' >_F + \bar{\rho}\tilde{W} \text{ (a)}, \ \bar{\rho} = \rho_u / (1 + \tilde{c}(\Theta - 1)) \text{ (b)}. \quad (7)$$

Eq. (7b) directly follows from the definitions of c (Eq. (6)) and Favre averaging $\tilde{c} = \overline{\rho c} / \bar{\rho}$.

This combustion subset contains only two unknown parameters: the scalar flux $\overline{\rho \vec{u}''c''} = \bar{\rho} < \vec{u}''c'' >_F$ and the chemical source $\overline{\rho W} = \bar{\rho}\tilde{W}$. The hydrodynamics part of the problem is described by the subset (Eqs. (4,5)) and it contains unknown stress tensor $\overline{\rho \vec{u}''\vec{u}''} = \bar{\rho} < \vec{u}''\vec{u}'' >_F$.

We notice that Eqs. (7) are valid irrespective of the specific combustion mechanism which in extreme cases is the following:

- the surface mechanism (combustion takes place in strongly wrinkled and extremely thin sheets) so the progress variable PDF is bimodal, i.e.

$p(c) = P_u\delta(c) + P_b\delta(1-c)$, where P_u and P_b are the probabilities of the unburned (reactants) and burned (products) gases ($P_u = (1-\overline{c})$ and $P_b = \overline{c}$ as $\overline{c} = 0 \cdot P_u + 1 \cdot P_b = P_b$ and $P_u = 1 - P_b$).

- the volume or quasi-laminar combustion mechanism when $p(c) = \delta(c - \overline{c})$, i.e. it takes place stirred-reactor type process with negligible scalar pulsations.

2.3 The Bimodal Approximation

In the case of the flamelet combustion mechanism (combustion takes place in strongly wrinkled laminar flame) the assumption of the bimodal PDF is a good approximation and it is used, for example, in the BML combustion model. We analyze the cases of the laminar as well as the thickened by microturbulence flamelet, but as this thickening is relatively small and the thickened flamelet width remains much less (as it will be shown below) than the integral scale of turbulence L, we also use the bimodal approximation for estimation of the Favre and Reynolds average parameters. These expressions are the following:

$$\widetilde{Q} = Q_u(1-\widetilde{c}) + Q_b\widetilde{c} \text{ (a)}, \ \overline{Q} = [Q_u(1-\widetilde{c}) + Q_b\widetilde{c}(\rho_u/\rho_b)][1+\widetilde{c}(\rho_u/\rho_b - 1)]^{-1} \text{ (b)}, \quad (8)$$

where Q denotes ρ, T, Y_i, Q_u and Q_b are parameters of reactants ρ_u, T_u, Y_i^u and products ρ_b, T_b, Y_i^b assumed constant (low-Mach-number approximation). Eqs. (8) reasonably predict concentration of the main species and obviously cannot be used for accurate estimation of small concentrations of species presented mainly inside the flamelet such as nonequilibrium radicals and other active particles or significant in applications atmospheric contaminants such as NO_x and CO, which are connected with actual chemical nonequilibrium. Practical resolution of this difficulty at numerical simulation usually takes place on the post-process stage using, for example, the known Zel'dovich thermal mechanism for NO_x and so on. Eqs. (8) also can be used for estimation of the Favre average pressure $\widetilde{p}$ and Reynolds average speed $\overline{\vec{u}}$. Necessary for this conditional average $\overline{p}_u, \overline{p}_b$ and $\overline{\vec{u}}_u, \overline{\vec{u}}_b$ are strongly connected with the gasdynamics effects controlling so-called counter-gradient transport phenomenon in turbulent premixed flames (different acceleration of the more heavy reactants and more light products by the pressure gradient generated by the density reduction due to heat release). We touch this problem below.

2.4 Our Approach to the Turbulent Premixed Combustion Modeling

There are two main in fact classical now approach to designing of the combustion models: the first is based on some algebraic approximations of the scalar flux and the chemical source; second is based on producing of differential model equations in terms of the flux and the source.

1. The main difficulty of the algebraic modeling is connected with the fact that actual premixed combustion takes place in thin and strongly wrinkled sheets (a flamelet sheet). In this case the scalar transport is predominantly counter-gradient due to different gasdynamical acceleration of the reactants and product by the generated by combustion pressure gradient. So usual gradient presentation of the flux in terms of the turbulent diffusion coefficient $\overline{\rho \vec{u}''c''} = -\overline{\rho} D_t \nabla \widetilde{c}$ is not theoretically justified. Nevertheless this approximation is widely used in practical simulations. Simplified algebraic approximations of the chemical source also are not theoretically justified (and also used in practical simulations) as they do not follow from an analysis of actual small-scale coupling of turbulence and chemistry. Obvious attraction of this approach is based on it simplicity and small number of appearing empirical constants.
2. There is more sophisticated way namely designing of the differential model equations in term of the scalar transport and the chemical source term using as a base the unclosed differential equations for them. Usual in the literature turbulent-type approximations of unknown terms in the scalar transport equation, i.e. in fact interpretation of the counter-gradient transport phenomenon as a counter-gradient turbulent diffusion, bring several empirical constants and at the same time does not permit describe it as a gasdynamical one. In the previous chapter we analyzed some questionable moments in using of the flame surface density (FSD) concept and unclosed surface average $\Sigma -$equation for modeling of the Favre average source terms. And though we deduced appropriate for the flamelet combustion mechanism unclosed Favre average equations in terms of the chemical source and scalar flux (Eqs. (14) and (16) in the chapter ?, we do not touch them as the analyzed here TFC model does not use them.

Our approach differs from both of them.

- We interpret the scalar transport in the premixed $\overline{\rho \vec{u}''c''}$ not as a pure turbulent diffusion phenomenon but as one that is controlled by two different mechanisms:

a) By the turbulent pulsations of the velocity and this turbulent diffusion component of the transport is approximated using the standard gradient approximation in terms of D_t, i.e. it can be model in the context of usual turbulence model (in our simulations below we used the standard "$k-\varepsilon$" turbulence model).
b) By the gasdynanical mechanism namely by different pressure-driven acceleration of cold reactants and hot products due to the pressure falling across the turbulent flame due to reduction of the mean density. This component is oriented in direction opposite to the normal turbulent diffusion and is described in the context of our original gasdynamical model that at least in first approximation does not bring additional empirical constant.

The ratio between these components defines the type of the transport (gradient or counter-gradient) and controls actual transition of the transport from gradient to counter-gradient along the flame. The consequence of this interpretation is that we can use the standard turbulence models in spite of the counter-gradient transport phenomenon.

- We interpret the flamelet sheet as a random surface and explain successive stages of the flame with different statistical states (equilibrium or nonequilibrium) large-scale and small-scale sheet wrinkles:

a) Initial stage (forming the developed turbulent flame structure) both small-scale and large-scale wrinkles of the flamelet sheet are nonequilibrium.
b) Transient flame with increasing brush width and nearly constant flame speed (ISP stage) takes place when small-scale wrinkles are already quasi-equilibrium while large-scale wrinkles are nonequilibrium yet.
c) The final stage corresponds to equilibrium of both small-scale and large-scale wrinkles.

In the real flame initial stage practically is not significant. And the reason is not only that it is short but due to effects of the boundary layer, the flow nonuniformity and so on this stage usually is not clearly defined and could be ignored (with the exception of combustion in SI engines). The final stage is practically unattainable and it also can be ignored in modeling many real situations. So the main attention below will be paid to modeling and numerical simulations of the transient intermediate steady propagation (ISP) flames.

We express relevant effects of unresolved by model equations small-scale coupling of turbulence and chemistry ("the challenge of turbulent combustion") in terms of resolved large-scale turbulent parameters and the chemical combustion time assuming a statistical equilibrium in developed turbulent flames of the small-scale reaction zone, i.e. our theoretical analysis and designing of the combustion model fulfilled in the context of developed for turbulent combustion Kolmogorov ideas.

It should be stressed that modeling of the model chemical source terms for RANS and LES formulation of the TFC model is based on the same basic idea of the small-scale equilibrium, in fact in both cases the combustion model is identical, different only resolved by RANS or LES model equations large-scale turbulent parameters.

Below we analyze these questions in greater detail.

3 The Theoretical Concepts of the TFC Combustion Model

The theoretical concepts embodied in our approach to the premixed combustion modeling at strong turbulence and fast chemistry in fact follow from the necessity to describe several well established apparent paradoxical experimental facts and theoretical statements that we discussed in the introduction. So our requirement imposed on the combustion model is the following:

1. The model must predict thickening of the instantaneous reaction zone but the flamelet sheet must remain thin in comparison with the integral scale L and hence strongly wrinkled, i.e. the model must predict existence of thin flamelet combustion

mechanism in spite existence of the developed turbulence with the continuous spectrum of turbulence.

2. The model must predict existence of well-defined and in fact prevailing in experiments specific transient flame with growing brush width δ_t and constant flame speed U_t (the intermediate steady propagation (ISP) flame), and it must predict this growing also in the case of the counter-gradient transport that usually dominate in experiments.
3. The model must predict existence in the premixed flame of the gradient (the scalar flux direction corresponds to the normal gradient turbulent diffusion), the counter-gradient one (the scalar flux direction is opposite to the normal gradient turbulent diffusion), and the neutral one (when the flux is close to zero).
4. The model must yield the final steady state flame with independent of chemistry flame speed U_t^{st} which follows the ISP flame, and explain why in accordance with many observations this stage is practically unattainable, i.e. explain why real flames are mainly transient.
5. The model must describe at least in outline preceding the ISP flame the initial stage of the flame when it takes place forming of the developed wrinkled flamelet sheet structure accompanied by increasing flame speed and more fast (in comparison with the ISP flame) growing of the brush width.

Three main original basic concepts have been embodied in the TFC model and they are connected with the mechanism of combustion, the combustion regime and the counter-gradient scalar transport phenomenon. They are the following:

1. A concept of the thickened and strongly wrinkled flamelet sheet;
2. A concept of the intermediate steady propagation (ISP) transient regime;
3. A concept of non-turbulent but gasdynamical nature of the counter-gradient scalar transport phenomenon.

3.1 A Concept of the Thickened and Strongly Wrinkled Flamelet Sheet

The laminar premixed combustion takes place in a thin flame with the speed S_L and δ_L. These parameters are controlled by the molecular transport coefficients and chemical kinetics. Assuming that all molecular transport coefficients are equal $D = \chi$ and characterizing the chemistry by the chemical time τ_{ch} we can write from the dimension reasoning as it follows:

$$S_L \approx (\chi / \tau_{ch}) \text{ (a)}, \ \delta_L \approx (\chi \tau_{ch}) \text{ (b)}, \tag{9}$$

where χ is the molecular coefficient of thermal diffusivity. The term the flamelet combustion mechanism of turbulent combustion usually imply that instantaneous combustion takes place in a strongly wrinkled laminar flame, so-called flamelet sheet, Fig. 1a. We will use this term also in the case when combustion takes place in thin and strongly wrinkled microturbulent or thickened flamelet sheet, Fig. 1b. We notice that the width of the instantaneous microturbulent flamelet is not a constant as at some places it is smaller and at some places it is larger, alternatively and randomly. So used below the width of thickening flamelet δ_f is a result of "partial averaging". The second remark is that microturbulent flamelet always is stretched by outer eddies so δ_f can be smaller (as we see below) than the width of the normal laminar flame, but larger that a hypothetical laminar flame stretch by these outer eddies, i.e. the microturbulent flamelet is thickened in comparison with this stretch laminar flamelet.

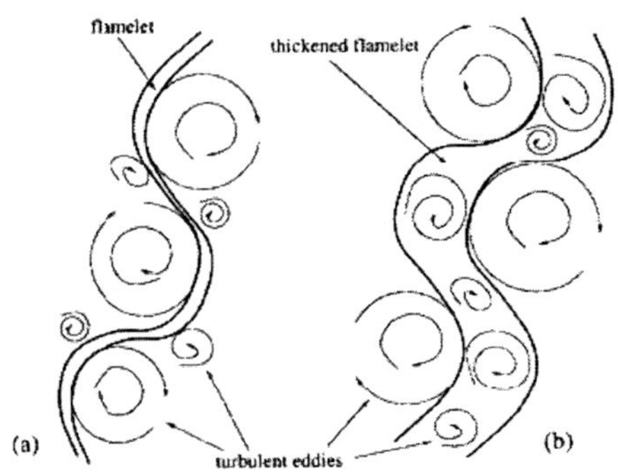

Fig.1. Laminar flamelet (a) and microturbulent (thickened) flamelet (b) combustion mechanisms

Model representations of the mechanics of combustion are used extensively in analyzing turbulent combustion. Two limit cases of combustion corresponding to the characteristic scales of turbulence L, which is much larger or much smaller in comparison with the thickness of the normal laminar combustion front in the fuel mixture under consideration δ_L, are distinguished [3]. In the first case ($L >> \delta_L$) combustion occurs in thin fronts, strongly curved because of the turbulent motion of the medium (surface mechanism of combustion). The combustion surface is hence often represented as an ordinary laminar flame front being propagated at the velocity S_L relative to the fuel mixture. In the second case ($L << \delta_L$), the action of turbulence reduces to increase of the transfer coefficients in the combustion zone (volume mechanism of combustion). A qualitative investigation of combustion in this case reduces to the analysis of the laminar front with the replacement of the molecular by turbulent transfer coefficients.

In the case of large turbulence Reynolds numbers $\mathrm{Re}_t = u'L/\nu$ (u' is the root-mean-square value of the velocity pulsation characterizing the turbulence energy, L is the characteristic size of the energy containing vortexes, usually taken equal to the integral scale of turbulence, and ν is the kinematical coefficient of molecular viscosity), a continuous turbulent spectrum is characterized, consequently, by a developed turbulence small-scale structure. So in the flow with $L >> \delta_L$ can exist turbulent vortexes less than δ_L. They will exert substantial influence on the structure of the instantaneous combustion front and intensify the transport processes in it. This results in a need to examine an intermediate model conserving the features of both limit cases, in which a strongly curved flamelet exists in the combustion zone, which is thin in comparison to the average width of the heat liberation zone (this front is considered as a random surface below). However, the thickness of such an instantaneous combustion front and its propagation velocity relative to the fresh mixture (reactants) are greater than the normal laminar front. In Russia in the late 50th there were several discussions about possible premixed combustion mechanism at strong turbulence. Between this mechanisms advocated were K. I. Shchelkin and A. G. Prudnikov while many believed in the volume mechanism. Shchelkin speaks in [18] about their arguments in the following expressions: "By returning to fine-scale turbulence, we recall the discussion found in the literature which is transformed into a good example of sophistry when it is not accompanied by quantitative estimates. The discussion is the following. Fine-scale turbulence, which is always present in the spectrum of scales, extends the normal flame combustion zone. Coarser scales become small relative to the expanded zone. By falling into the combustion zone, they extend it still more, etc. All the scales gradually are drawn into the combustion zone, which is not and cannot be in a turbulent flame". The author performed such quantities estimations in 1979 [19] and they show that such successive expanding of the reaction zone has a limit, i.e. in typical situations stirred reactor type combustion is impossible.[2] We analyze below the case of constant density and yielded resulted will by used for designing of the combustion model intended for using for real combustion.

3.1.1 Parameters of the Normal Thickened Instantaneous Flame

At developed turbulence with continuous spectrum the minimal size of eddies η (the Kolmogorov microscale of turbulence) is defined as

$$\eta = L\,\mathrm{Re}_t^{-3/4} \tag{10}$$

When $\eta < \delta_L$ small eddies fall in the flamelet, intensify transport process and thicken it, i. e. in this case we will have thickened one ($\delta_f > \delta_L$). At the same time at $\eta > \delta_L$ we can expect that the flamelet is laminar and the condition $\eta = \delta_L$ is the boundary between these combustion mechanisms. In our asymptotical analysis we assume that $\eta << \delta_f$ and ignore the laminar molecular diffusion so only small size eddies from the Kolmogorov inertial spectrum $E(k) = C\varepsilon^{2/3}k^{-5/3}$ (expressed in terms of large-scale turbulent parameters as $\varepsilon = Au'^3 / L$, $C \sim 1$, $A \sim 1$) control this transport. Hence we see that the r.m.s. pulsation u'_f, scale L_f and micro-turbulent transport coefficient $D_f \approx \chi_f$ inside the 1D normal thickened flamelet (we denote them below as $u'_{nf}, L_{nf}, D_{nf} \approx \chi_{nf}$) are as it follows:

[2] We notice that that nevertheless stirred reactor type combustion models are used in applications and they are presented in known commercial codes (strong appeal of these model is connected with opportunity of simple installation of the chemical kinetics equations and possibility to simulate directly different types of combustion (the cases of premixed, non-premixed, partially premixed combustion). Below we will touch this point and analyze physical meaning of this approach and general model equation describing it.

$$u_{nf}^{'2} \approx \int_{1/\delta_{nf}}^{\infty} E(k)dk \approx \varepsilon^{2/3}\delta_{nf}^{2/3},\ L_{nf} \approx \int_{1/\delta_{nf}}^{\infty} k^{-1}E(k)dk\left(\int_{1/\delta_{nf}}^{\infty} E(k)dk\right)^{-1} \approx \delta_{nf},\ \chi_{nf} \approx u'_{nf}L_{nf} \approx \varepsilon^{1/3}\delta_{nf}^{4/3} \quad (11)$$

We notice that the last expression is similar to the well-known in the atmospheric physics Richardson's four/third-law describing dependence of the turbulent diffusion coefficient inside a relatively small cloud on the characteristic size of it, $D_d \sim \varepsilon^{1/3} d^{4/3}$ [20 p.551] and this coincidence is connected with the fact, that in both cases the transport is controlled by the Kolmogorov inertial turbulence. The speed U_f and width δ_f of the 1D normal thickened flamelet are functions of χ_f (as at large Re_t $\chi_f >> \chi$ so we can ignore contribution of the molecular diffusion in the transport inside the flamelet) and the characteristic chemical time τ_{ch}. From reasoning of the dimension theory we can write

$$U_{nf} \approx (\chi_{nf} / \tau_{ch})^{1/2},\ \delta_{nf} \approx (\chi_{nf}\tau_{ch})^{1/2} \quad (12)$$

and hence using Eqs. (11) we can write the parameters of thickened flamelet as it follows:

$$\chi_{nf} \approx \varepsilon\ \tau_{ch}^2,\ U_{nf} \approx u'_{nf} \approx (\varepsilon\ \tau_{ch})^{1/2},\ \delta_{nf} \approx L_{nf} \approx \varepsilon^{1/2}\ \tau_{ch}^{3/2} \quad (13)$$

Talking into account, that $\varepsilon \approx u'^3 / L$, $D_t \approx u'L$, $Da = \tau_t / \tau_{ch}$, $\tau_t = L / u'$ we can rewrite Eqs. (11) as it follows for the 1D normal thickened flamelet characteristics

$$U_{nf} = a_U^n\ u' Da^{-1/2},\ \delta_{nf} = a_\delta^n L\ Da^{-3/2} \quad (14)$$

and for the sub-flamelet turbulence parameters

$$u'_{nf} = a_{u'}^n\ u' Da^{-1/2},\ L_{nf} = a_L^n L\ Da^{-3/2},\ \chi_{nf} = a_\chi^n\ D_t\ Da^{-2}, \quad (15)$$

where $a_U^n \approx 1$, $a_\delta^n \approx 1$, $a_{u'}^n \approx 1$, $a_L^n \approx 1$ and $a_\chi^n \approx 1$ are empirical constants.

We present Reynolds number scaling for the parameters of the normal thickened flamelet and turbulent characteristics inside it

$$U_{nf} \approx u'_{nf} \approx u'(u'/S_L)\mathrm{Re}_t^{-1/2},\ \delta_{nf} \approx L_{nf} \approx L(u'/S_L)^3\ \mathrm{Re}_t^{-3/2}, \quad (16)$$

$$\chi_{nf} \approx u'_{nf}L_{nf} \approx D_t(u'/S_L)^4\ \mathrm{Re}_t^{-2},\ \mathrm{Re}_t^{nf} = u'_{nf}L_{nf}/\nu \approx \mathrm{Re}_t/Da^2.$$

For the sake of definiteness we assume $a_U^n = a_\delta^n = a_{u'}^n = a_L^n = a_\chi^n = 1$ in our definition of the normal thickened flamelet parameters. We notice that $U_{nf} = u'_{nf}$, i. e. the flamelet speed is equal to the turbulent velocity inside the flamelet. It also should be particular emphasized that inside the normal flamelet the turbulent time equal to the chemical one $\tau_{nf} = L_{nf} / u'_{nf} = \tau_{ch}$, i.e. the normal flamelet Damköhler number is unit $Da_{nf} = \tau_{nf} / \tau_{ch} = 1$. Here we see analogy with the Kolmogorov theory where it is assumed for definition of the size of the minimal eddies η that corresponding Reynolds number is equal to unit $\mathrm{Re}_\eta = u'_\eta \eta / \nu = 1$, where $u'_\eta = u' \mathrm{Re}_t^{-1/4}$ is the turbulence inside this eddies. Hence we see that significant for the turbulent combustion theory Kolmogorov miscoscale η equal to $\eta = L \mathrm{Re}_t^{-3/4}$ where $\mathrm{Re}_t = u'L/\nu$.

A relation between parameters of normal thickened and laminar flamelets follows from Eqs. (9) and (14):

$$U_f / S_L \approx \delta_f / \delta_L \approx Da^{-1} \mathrm{Re}_t^{1/2} \tag{17}$$

The condition $U_f / S_L \approx \delta_f / \delta_L \approx 1$ becomes after Eqs. (17), (9) and (10) have been used $\delta_L \approx \eta$. In means that in the model there is no a break in the values of the flamelet speed and width on the boundary between thickened and laminar flamelet combustion mechanisms $\delta_L = \eta$ at transition from one mechanism to another.

Obviously, existence of two different mechanisms on both sides of the boundary is a model schematization. In reality in the vicinity of the boundary $\delta_L = \eta$ inside the flamelet the transport is controlled both by microturbulence and molecular viscosity, and the microturbulent diffusion coefficient does not corresponds to the 5/3 inertial spectrum. It is less than Eq. (15) gives due to direct influence of the molecular viscosity in the range of dissipation wave numbers $k \sim 1/\eta$.

As an example for possible for industrial large-scale burners parameters: $u' \approx 4m/s,\ L \approx 0.5cm,\ \nu \approx 0.2cm^2/s$, $\mathrm{Re}_t \approx 10^3$ the thickened flamelet combustion mechanism takes place at $Da < \mathrm{Re}_t^{1/2} \approx 32$. Estimated with Eqs. (14) the flamlelet width is much smaller in comparison with the turbulence integral scale L and the flamelet speed is significantly smaller than the r.m.s. velocity pulsation u':

$$Da = 5:\ U_{nf} \approx 0.5u',\ \delta_{nf} \approx 0.09L;\ Da = 10:\ U_{nf} \approx 0.3u',\ \delta_{nf} \approx 0.03L;$$

$$Da = 20:\ U_{nf} \approx 0.2u',\ \delta_{nf} \approx 0.01L;\ Da = 30:\ U_{nf} \approx 0.18u',\ \delta_{nf} \approx 0.006L.$$

It means that in spite of thickening the microturbulent flamelet remains thin and hence strongly wrinkled by the large-scale turbulent structure. At the same time it is an upper estimate of the flamelet speed and width as actual wrinkled thickened flamelet is not the normal thickened one but stretched by turbulent eddies with the sizes larger than the flamelet width. In accordance with a result of the next paragraph this stretch reduces the flamelet speed and width by one half.

In comparison with the laminar flamelet parameters the estimate gives the following:

$$Da = 5: \; U_{nf}/S_L \approx \delta_{nf}/\delta_L \approx 6.3; \; Da = 10: \; U_{nf}/S_L \approx \delta_{nf}/\delta_L \approx 3.1;$$

$$Da = 20: \; U_{nf}/S_L \approx \delta_{nf}/\delta_L \approx 1.6; \; Da = 30: \; U_{nf}/S_L \approx \delta_{nf}/\delta_L \approx 1.05.$$

Assuming $S_L = 20cm/s$ (lean mixture) we have an estimate $\delta_L \approx \chi/S_L \approx 0.1mm$ while the Kolmogorov's microscale η associated with the minimal size of the turbulent eddies is less and in our case is equal to $\eta = L\,\mathrm{Re}_t^{-3/4} \approx 0.03mm$, i.e. it is smaller than the normal laminar flame width.

3.1.2 Stretched Thickened Flamelet

In our analysis in the previous paragraph we in fact assumed that in the turbulent medium there are only eddies with the size $l_e \le \delta_f^n$ while eddies with the size $l_e > \delta_f^n$ does not exist at all. Only in this case we can speak about normal thickened flamelet. In reality, obviously, there are eddies with the size $l_e > \delta_f^n$ and they stretch the thickened flamelet. It inevitably results in reduction of the speed U_f and width δ_t of an actual thickened flamelet, i.e. $U_f < U_{nf}$, $\delta_f < \delta_{nf}$. We assume that eddies with the size $l_e > \delta_f^n$ controlling the flamelet stretch fall in the range of the Kolmogorov inertial 5/3 spectrum as well as the eddies controlling the microturbulent transport inside the thickened fall in this range. This spectrum is controlled by the dissipation rate ε. It means that the speed, width and turbulent parameters inside the actual thickened and stretched flamelet are controlled by ε and τ_{ch}, i.e. $U_f = (\varepsilon, \tau_{ch})$, $\delta_f = (\varepsilon, \tau_{ch})$, $u'_f = (\varepsilon, \tau_{ch})$, $L_f = (\varepsilon, \tau_{ch})$ and $\chi_f = (\varepsilon, \tau_{ch})$. The reasoning of the dimension theory results in that

$$U_f \approx u'_f \approx (\varepsilon\,\tau_{ch})^{1/2}, \; \delta_f \approx L_f \approx \varepsilon^{1/2}\,\tau_{ch}^{3/2}, \; \chi_f \approx \varepsilon\,\tau_{ch}^2. \tag{18}$$

These equations are the same so we can write expressions (similar to Eqs. (14)) as it follows for the thickened stretched flamelet characteristics

$$U_f = a\;u' Da^{-1/2}\text{ (a)}, \; \delta_f = a\;L\;Da^{-3/2}\text{ (b)}, \; \chi_{nf} = a^{4/3} D_t\;Da^{-2}\text{ (c)}, \; a = 1/2 \tag{19}$$

where new coefficients for the stretch flame a is less than for the normal thickened flamelet, i.e. $a < 1$. In this connection the following question arises; is this decreasing of the coefficient significant in the context of our analysis or it can be ignored. A theoretical estimate presented below demonstrates that $a = 1/2$, i.e. reduction of the flamelet speed and width by outer eddies from inertial interval spectrum $E(k) = C\varepsilon^{2/3}k^{-5/3}$ is significant. We notice here that this result will be used below for explanation in the context of the TFC model of the fact that at real Damköhler numbers the speed the transient flame $Da \sim 10$ is less than equal to u' the speed of the steady state flame $U_t \approx (1/2)u'Da^{1/4} < U_t^{ss} \approx u'$.

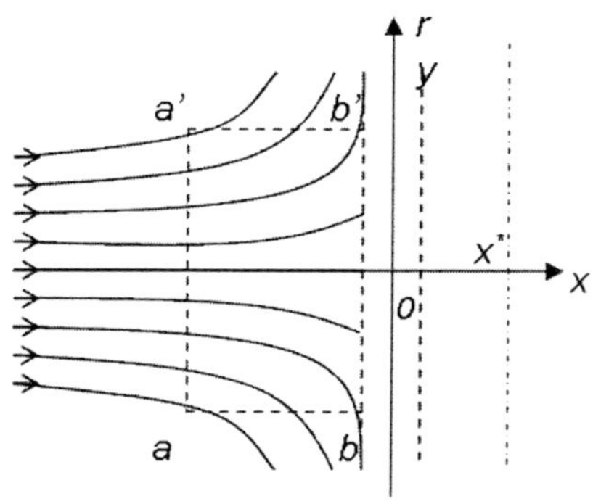

Fig. 2 Flamelet in a diverging flow of fuel mixture

Our estimation of this effect is performed in the context of the theoretical analysis for the stretch of the laminar flame described in the book [21, p. 332]. We will analyze the problem in the moving coordinate system where the flame is the stationary one. In the flow just before the flamelet is nonuniform over a scale length comparable to the width of the flame front δ_f due to existence of the turbulent eddies with sizes larger than δ_f, the thickened flamelet structure is different than the normal thickened flame. We analyze a hydrodynamic field that corresponds to the axisymmetric flow near a critical point. Fig. 2 shows mean streamlines and a coordinate system in which the critical point lies at $x = x^*$, $r = 0$. We analyze the irrotational flow of an incompressible flow obeying to the continuity equation $\partial \bar{u}_x / \partial x + (1/r)\partial(r\bar{u}_r)/\partial r = 0$.

The flow velocity increases linearly with distance from the critical point, i.e.

$$\bar{u}_x = -(x - x_*)/\tau_h = \bar{u}_* - (x/\tau_h) \;\; (x_* = \bar{u}_*\tau_h),\; \bar{u}_r = 0.5r/\tau_h, \tag{20}$$

where $\bar{u}_x$ and $\bar{u}_r$ are the orthogonal and radial components of the flow velocity, τ_h is the characteristic time controlling the flow divergence. Taking into account, that in our in fact quasi-laminar approach to the thickened flamelet modeling the reaction takes place in a narrow zone in the back part of flamelet due to strong dependence of the rate on the temperature (this zone is denoted by the dash lines near the plane $x = 0$), u_* is equal to the volume consumption speed by this reaction zone, i.e. $\bar{u}_* = U_f = a\ u'Da^{-1/2}$. We assume that mean flow divergence is controlled by the turbulence from the inertial spectrum so it is controlled by the dissipation rate ε and the flamelet speed U_f and δ_f, which in turn are controlled by ε and the chemical time τ_{ch}:

$$\tau_h = g[\varepsilon,\ U_f(\varepsilon,\ \tau_{ch}),\ \delta_f(\varepsilon,\ \tau_{ch})] = g_1(\varepsilon,\ \tau_{ch}) \approx \tau_{ch}, \qquad (21)$$

where the result $\tau_h \approx \tau_{ch}$ follows from the function $g_1(\varepsilon,\ \tau_{ch})$ using the Π – theorem of the dimension theory.

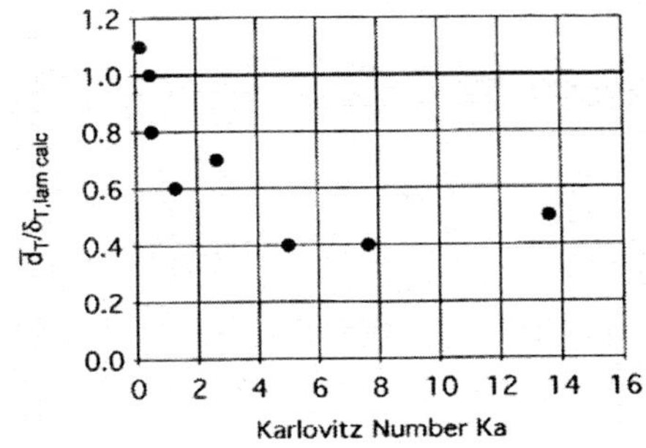

Fig. 3. The flamelet width in the turbulent flame from [22]

We shall now examine the stationary thickened flamelet in the divergent field described by Eq. (20). Estimation of the stretch-effect we begin by isolating a volume element $aa'b'b$ adjacent to the flame front on the side of the unburned mixture, Fig. (2). We denote the radius of this element as r_0 and its thickness is equal to the width of the thickened flamelet $\delta_f = a\ L\ Da^{-3/2}$. We now set up the balance of reactants in this volume of gas. Reactants flux entering in this volume through the surface aa' is $Q_1 = (\delta_f / \tau_{ch} + U_t)\pi r_0^2$. It is equal to the sum of the consumed by the reaction zone flux $Q_2 = U_t \pi r_0^2$ and the flux $Q_3 = (0.5 r_0 / \tau_{ch}) 2\pi r_0 \delta_t$, which is carried out through $a'b'$, ab and is burnt at $r > r_0$. U_t and δ_t is described by Eqs.(19) and it results that $Q_1 / Q_2 = 2$. It means the stretch effect reduces the flamelet speed and the width two times in comparison with the normal thickened flamelet, i.e. in Eqs. (19) the coefficient $a = 1/2$. Quite the same result we have analyzing the plane stretch. It is interesting to note that at the front edge of the stretch flamelet ($x = -\delta_f$) the reactants speed is equal to the velocity of the normal thickened flamelet U_{nf} and the Damköhler of the stretch flamelet remains the same $Da_f = 1$.

In the case when $\eta >> \delta_L$ the flamelet can be treated as the laminar flame as the stretch effect due large (in comparison with the width δ_L) turbulent eddies is not significant. At the same time on the boundary $\eta = \delta_L$ there are eddies with sizes comparable with δ_L. For estimate of the stretch-effect we assume that at $\eta = \delta_L$ the laminar flamelet is stretch by eddies belonging to the inertial interval of turbulence. In this case the estimate $\tau_h = \tau_{ch}$ remains valid in Eqs. (20) also for this case and the flame stretch reduces the width and speed also by half, $\delta_L / \delta_L^n = 0.5$ and $S_L / S_L^n = 0.5$. Increasing of the Kolmogorov microscale

($\eta > \delta_L$) reduces the stretch and at $\eta >> \delta_L$ this effect disappears. Optical experiments with the Bunsen methane-air premixed flames ($U = (4.7 - 16.5)m/s, \Phi = 0.56 - 0.8$) in [22] and with two dimensional V-shaped methane-air premixed flames ($U = 3.5 m/s, \Phi = 0.5 - 0.8$) seems to confirm this prediction. Visualization show that in Fig. 3 from [22] the maximal Karlovitz number for the laminar flamelet is $Ka = 5$ ($\delta_L / \eta \approx 2$) with the flamelet width equal to $0.5\delta_L^n$ (for comparison measurement in [23] gives for the V-flame gives $\approx 0.6\delta_L^n$ at $Ka = 5$). At decreasing of the Karlovitz number effect of the flamelet stretch reduces and experiment demonstrate practically normal laminar flamelet at small Ka. The point $Ka = 5$ is called in [22] a limit of the "flameler regime". But these experiments demonstrate that at larger Karlovitz number ($Ka = 13.6$) the sheet between reactants and product becomes microturbulent with slightly larger width. We interpret it as the thickened flamelet. Some different interpretation of these data is contained in [24].

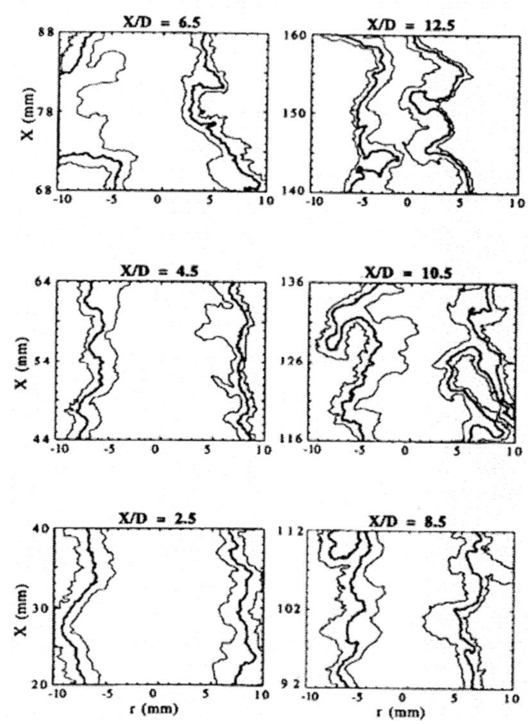

Fig. 4. Instantaneous flamelet sheet for the flame F1 from [24]

Instantaneous picture from [25] presented in Fig. 4 demonstrates the thickencd flamelet sheet in the Bunsen high velocity ($U = 65 m/s$) stoichiometric methane-air flame. Thin lines in the figure are isotherms corresponding to the minimal and maximal T_{max} temperatures, (front and back edges of the flamelet) the thick lines correspond to the intermediate isotherm with $T = (T_{\min} + T_{\max})/2$. Fig. 5 from [25] demonstrates dependence of the average width of the thickened sheet (here denoted as $\overline{\delta}_F$) on turbulence and at least qualitatively this result corresponds to the theoretical prediction by Eq. (19): increasing u' results in the width growing. The points denoted in Figs. 5 as F1 correspond to the flame shown in Fig. 4.

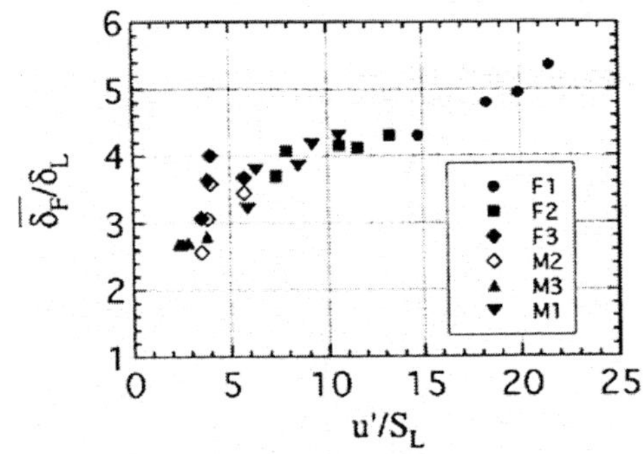

Fig. 5. Mean flamelet width for different flames from [25]

The paper [26] also contains experimental data for lean propane/air flames at high turbulence significant broadening of the flamelet. Here we do not correlate quantitatively our estimate of the flamelet width with existing experimental data as the magnitude of the width strongly depends on used definitions of the thickened flamelet width, which in fact were different in the experimental works and in our analysis. In particular we in fact estimated the Zel'dovich thickness of the flamelet while the experimental papers usually contain data about the thermal flamelet width and in addition estimated with different methods, i.e. not always comparable.

3.1.3 Remarks about the Thickened Flamelet Concept

1. This remark is devoted to explanation of the fact that combustion takes place in relatively thin and strongly wrinkled flamelet sheets in spite of existence of developed turbulence with continuous energy spectrum. And though both theoretical estimations and experimental data demonstrate thickening of the flamelet it remains relatively thin and strongly wrinkled[3]. This raises the question why there is no successive involving more and more large eddies in a more and more thickened flamelet till forming the distribute combustion with the speed and width corresponding as a first approximation to the quasi-laminar combustion mechanism with

$$U_t \sim (u'L/\tau_{ch})^{1/2} \sim u'Da^{1/2} \text{(a)}, \ \delta_t \sim (u'L\tau_{ch})^{1/2} \sim L\ Da^{-1/2} \text{(b)}. \qquad (22)$$

It is easy to show from Eqs. (14) for U_{nf} and δ_{nf} that the flamelet coordinate system the microturbulent scalar flux in the flamelet, convection, and the effect of the chemical reaction have the same order of magnitude. It is so as from the Eqs. (14, 15) it follows that

[3] We put the theoretical estimate before results of the experimental observations as this estimate as well as the theoretical explanation of this fact were published in [19] long before experimental results.

$$\chi_{nf}(1/\delta_f) \sim U_{nf} \sim W\delta_{nf}, \tag{23}$$

where $W \approx 1/\tau_{ch}$ is the mean reaction rate inside the flamelet (we remind that in our analysis we assumed constant density). Therefore, the expansion of the flamelet has a limit, involvement of more and more coarse vortices occurs until establishing equilibrium between the convection, heat conduction, and chemical reaction processes. This situation is similar to the normal laminar flame where namely equilibrium between these processes control S_L and δ_L, in fact the only difference is that in the thickened flamelet the microturbulent transport coefficient χ_f depends on the flamelet width δ_f while in the laminar flame the diffusion flux is controlled by independent on δ_L molecular transport coefficient χ. This explanation was published first in [19].

2. We point out that in our model we estimate the chemical time as $\tau_{ch} = \chi / S_L^2$ so the normal laminar velocity S_L and the molecular heat transfer coefficient χ appear in the combustion model with non-laminar flamelet sheet as the physico-chemical characteristics of the initial unburned mixture. The consequence of it is that the chemistry dependence for the laminar and thickened flamelet is the same. Identity of the chemical time in fact means that we assume invariability of the reaction time during laminar and turbulent combustion. This invariability holds, for instance, for an identical function form of the temperature and chemical species distributions across the laminar flame and thickened flamelet and in the absence of noticeable temperature pulsations in the latter $\overline{T'^2}/\Theta^2 << 1$ (Θ is the characteristic temperature range of the chemical reaction). The fact is that in conformity with Eq. (L_{nf}), the characteristic scale of the vortices in the instantaneous flame or flamelet is commensurate with its width, which can result in a high pulsation level. An estimate of the temperature pulsation intensity requires the reliance on balance equations for the temperature pulsations $\overline{T'^2}$ in the coordinate system in which the flamelet is fixed. Equating in terms of c (instead of T): the convective transfer (i), the generation associated with the mean gradient of the progress variable (ii), the micro-turbulent diffusion (iii) and dissipation of $\overline{c'^2}$ (iv), which can be estimated respectively as

$$U_{nf}\frac{d\overline{c'^2}}{dx} \sim U_{nf}\frac{\overline{c'^2}}{\delta_{nf}},\ \chi_{nf}\left(\frac{d\overline{c}}{dx}\right)^2 \sim \chi_{nf}\left(\frac{1}{\delta_{nf}}\right),\ \frac{d}{dx}\left(\chi_{nf}\frac{d\overline{c'^2}}{dx}\right) \sim \chi_{nf}\frac{\overline{c'^2}}{\delta_{nf}},\ \frac{\overline{c'^2}u'_{nf}}{L_{nf}} \tag{24}$$

in order of magnitude, and using Eqs. (11 and 13) we obtain $\overline{c'^2} \sim 1$. This estimate shows that the thickened flamelet most probably differs in structure from in fact the quasi-laminar model used here. So we must keep in mind that our assumption about identity of the chemical time in turbulent and laminar combustion is sufficiently strong and it correspond to quasi-laminar model of the thickened flamelet. More sophisticated models of the thickened flamelet could yield more accurate chemistry dependence for U_f and δ_f, and explain strong influence of the molecular transport effects (for example, the preferential diffusion effect) observed in the experiments also at strong turbulence (see, for example a review paper [27]) when we intuitively expect that influence of the molecular transport on the global properties of the premixed flame is negligible.

3. In our estimate of the flamelet parameters we following to the Kolmogorov theory also ignore the fact of the dissipation rate ε pulsations. At the same time influence of actual strong pulsation of ε resulting in appearing of large local instantaneous gradients of the velocity and hence strong flamelet stretch and even local extinction is well known. For example, the statistical characteristics of the dissipation pulsations was used by K. Bray [28] for analysis of the local flamelet extinction: when the instantaneous dissipation rate is more than some critical value $\varepsilon > \varepsilon_{cr}$ takes place a local extinction of combustion. So the probability of the local extinction is equal to the probability $P(\varepsilon > \varepsilon_{cr})$. We used in our simulations this Bray approach in our model for explanation of "bending effect" (reduction of the turbulent flame speed U_t at large u') [29] and global extinction of combustion. We touch this question in more details below.

4. This remark is devoted to more general question namely to a problem of the theoretically justified definition of the thickened flamelet parameters: the flamelet speed U_f, width δ_f, microturbulent parameters u'_f, L_f, $,\chi_f$ and flamelet sheet area $(\overline{A}/A_0)$ controlling the turbulent flame speed U_t. Concept of the thickened flamelet is based on some averaging of the instantaneous picture. This natural question arises as a definition of the flamelet parameters strongly depends on the scale of averaging so theoretical estimations must depend on this scale. In principle there are several sets of mentioned parameters in accordance with chosen scales of averaging.

We will show the physical meaning of our definition and point out some alternative definitions:

- Our estimate of δ_f in fact corresponds in fact to the flamelet width definition such that convection, microturbulent transport and chemical consumption inside the flamelet are of the same order of magnitude. This width includes not only thin sheet of intensive combustion (visualized in experiments, for example, by the location, where observed instantaneous OH radical shows strong gradients [22], [23]), but also more thickly preheating zone. Our estimate of the flamelet speed U_t, Eq. (19) and performed below estimate of the thickened flamelet sheet area $(\overline{A}/A_0)$, Eq. (43), correspond to namely this definition of the flamelet width. Namely this correspondence gives an opportunity to present the mass consumption rate of the flame as $\rho_u U_t = \rho_u U_f (\overline{A}/A_0)$ and deduce below the expression for the turbulent flame velocity, Eq. (44). We note that using of the same reactants density ρ_u in the both in right and left sides of the presented expression is caused by the fact that the width δ_t is large enough so we have reactants with $\rho = \rho_u$ on the front edge of the flamelet.

- Hypothetically we can define the flamelet width equal to the width of the thin reaction zone δ_f^r, i.e. some small vicinity of the instantaneous surface with the maximal reaction rate W^r, the density ρ_r and speed U_f^r. For simplicity assume very strong temperature dependence of the chemical reaction (the energy activation energy $E \to \infty$) that is a reasonable approximation of real situation. In this case instantaneous infinite reaction rate takes place on a strongly wrinkled surface (the flamelet of zero width) moving with some velocity U_r. The density on this surface ρ_r practically equal to the density of the products so the turbulent flame speed can be expressed as $\rho_u U_t = \rho_b U_f^r (\overline{A}/A_0)_r$ where $(\overline{A}/A_0)_r$ is the mean dimensionless area of this surface. Correct estimate of U_f^r and $(\overline{A}/A_0)_r$ results in correct U_t. We do not know (in contrast to our approach) how to estimate these speed and area. Nevertheless potentialities of this way could be promising for modeling of well-documented molecular effects on the turbulent combustion [27].
- In the case of LES we assume that effective flamelet width is equal to the mesh size Δ, the corresponding speed of this effective flamelet U_t^Δ we estimate by averaging of our flamelet sheet inside the mesh (below we deduce an expression for U_t^Δ which we used in our simulations) so we can write $U_t = U_t^\Delta (\overline{A}/A_0)_\Delta$ where the area of the effective flamelet follows from averaging of the LES.

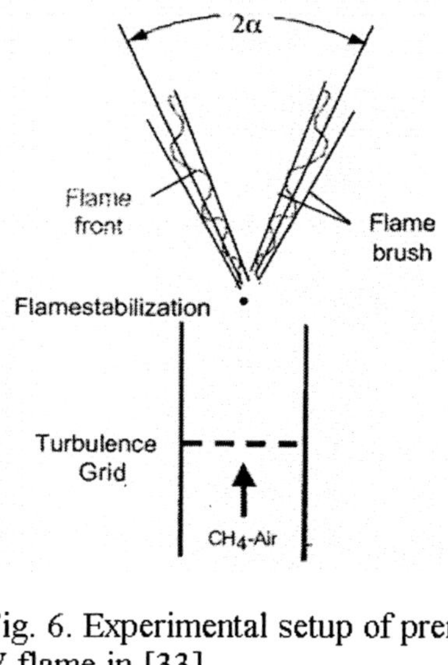

Fig. 6. Experimental setup of premixed V-flame in [33]

The flamelet width as we define it is on one hand is much smaller than the integral scale of turbulence $\delta_f << L$ so we describe actual mechanism when combustion takes place in thin and strongly wrinkled sheet. From the other hand the width is much larger in comparison with the Kolmogorov microscale $\delta_f >> \eta$ so using of Kolmogorov idea of the small-scale equilibrium for description of the unresolved by model equations seems theoretically justified.

We notice a strange at first glance peculiarity our thickened flamelet namely the chemistry dependence of the flamlet speed $U_f \sim \tau_{ch}^{1/2}$, Eq. (19). It means that in contrast to the laminar flamelet ($S_L \sim \tau_{ch}^{-1/2}$) the speed of the thickened flamelet is larger for rich and lean mixtures in comparison with the stoichiometrical one. It is caused by strong chemistry dependence for the thickened flamelet width $\delta_f \sim \tau_{ch}^{3/2}$ (for comparison $\delta_L \sim \tau_{ch}^{1/2}$) as it results in strong chemistry dependence for the flametet microturbulent transport coefficient $\chi_f \sim \delta_f^{4/3} \sim \tau_{ch}^2$, Eq. (18). So the expression $U_f \approx (\chi_f / \tau_{ch})^{1/2}$ the effect of the τ_{ch} – dependence for χ_f prevails. We have no now experimental data for U_f but corresponding estimate below for the flamelet sheet area, Eq. (43), gives $(\overline{A}/A_0) \sim \tau_{ch}^{-3/4}$ and results in $U_t \sim \tau_{ch}^{-1/4}$, i.e. the chemistry dependence for U_t is less by half in comparison with S_L and it in general corresponds to the experimental observations.

3.2 A Concept of the Transient Intermediate Steady Propagation (ISP) Flame

The transient ISP flames, i. e. the flames with increasing brush width δ_t and nearly constant turbulent flame speed U_t observed in many experiments. This experimental fact was probably first documented by A. G. Prudnikov [30], [31], who experimentally showed that in actual premixed flames in turbulent flow increasing δ_t took place in accordance with the turbulent diffusion law and the flame width is controlled mainly by the turbulent diffusion coefficient in cold reactants flow. In [32] he showed using original optic-diffusion method that in the premixed flames turbulent transport of nonuniform passive addition is gradient in contrast with counter-gradient transport of reacting spices founded in similar flames at a later date.

Resent optical investigations [33] of the V-shape premixed flame in uniform turbulent flow behind the rod stabilizer, Fig. 6, confirm, Fig. 7, that the flame have practically constant inclination of the flame to the flow and at the same time looks like turbulent mixing layer with increasing brush width. Fig. 8 presents results of numerical simulation of this flame performed in [33], we touch this result in the section 7 below. Visualization of the instantaneous flamelet sheet in [34] in the Bunsen premixed flames at the high pressure, Fig. 9, demonstrate that instantaneous flamelt sheet is wrinkled (a) and an ensemble of instantaneous sheets constitute the flame brush with increasing width (b). At the same time the mathematical expectation of the flamelet sheet is nearly straight line (c, d, f), i.e. here also the turbulent flame has practically constant inclination of the flame to the flow. In Fig. 9 (a), (b) and (c) refer to the same flame while (d) and (f) refer to another flames with different pressure and turbulence intensity. The flame F1 from Fig. 4 also has increasing brush width, Fig. 10. The figure shows curved-fitted mean flame-front contour of the flame and the PDFs of the displacement relative to mean flame-front positions, and horizontal scaling of the PDF diagrams is identical to the X-axis scaling for flame contour [34].

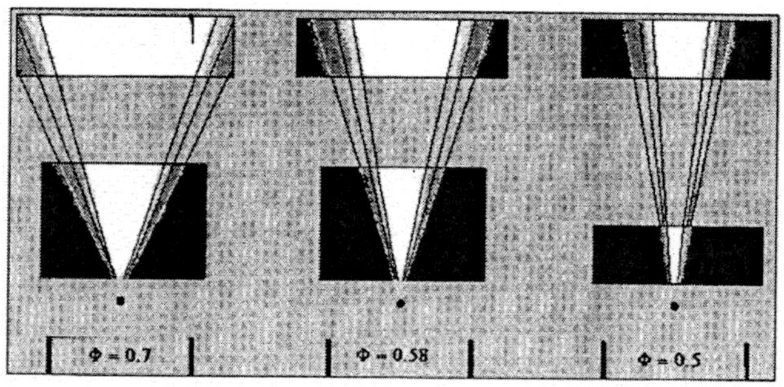

Fig.7 Experimental mean reaction progress variable $\bar{c}(x,y)$ in the plane perpendicular to the flame holder for different stoichiometric ratios [33]. Fitted contour lines $\bar{c}$=0.1, 0.5, 0.9.

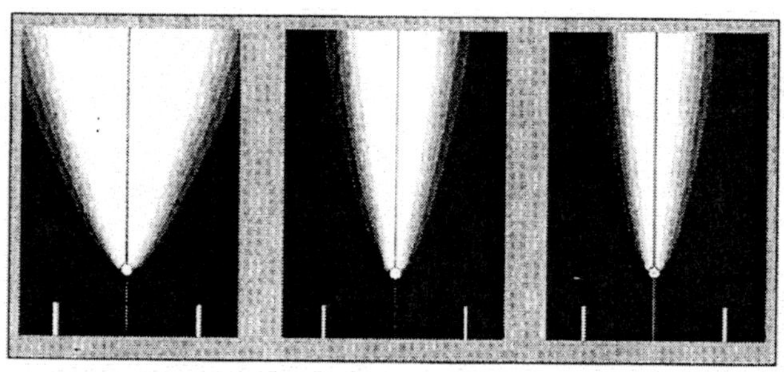

Fig.8 Mean reaction progress variable $\tilde{c}(x,y)$ for stoichiometric ratios as in Fig.7 from numerical simulation in [33] with our TFC model. Fitted contour lines $\tilde{c}$=0.1, 0.3, 0.5, 0.7, 0.9.

Analysis of many others experimental data in the review paper [14] also confirms that real flame is transient with increasing brush width.

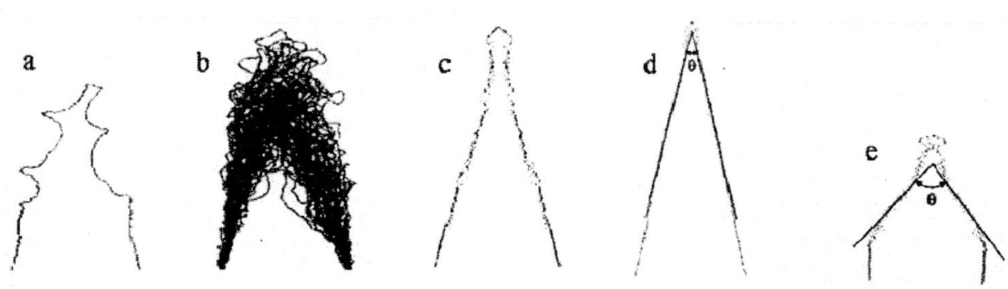

Fig.9 Instantaneous images of the flamelet sheet (a,b) and its mathematical expectation (c)-(e)[34].

We will see from a theoretical analysis below that the small-scale sheet wrinkles give the main contribution in the sheet area while the large-scale sheet wrinkles give the main contribution in the flame width. Simple estimation below shows that in real flames the former are already quasi-equilibrium and while the latter are non-equilibrium yet and their size increases and is controlled by turbulent diffusion.

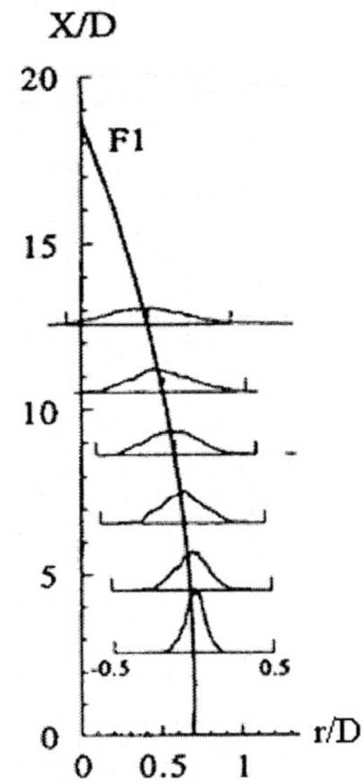

Fig.10 Increasing brush width of the flame F1 from [24]

So there are strong experimental and theoretical grounds to believe that these flames (in fact running with nearly constant speed turbulent mixing layers) constitute a well-defined class observed flames, we called them in [16] the intermediate steady propagation (ISP) flames. In fact the ISP flame is a steady moving turbulent mixing layer while the steady state (SS) flame (1D stationary flame) is in fact a running stationary wave.

For explanation of such paradoxical behavior and modeling of the ISP flame we need a theoretical concept. As we mentioned above our explanation is based on different statistical states of the small-scale and large-scale wrinkles. There is good reason to believe that at strong turbulence unresolved small-scale and resolved large-scale processes control different parameters of the turbulent flame, namely, the turbulent flame speed U_t is controlled mainly by unreserved flamelets and small-scale sheet wrinkle structures while $|\nabla \tilde{c}|$ (i.e. $\tilde{c}$ distribution) is controlled mainly by large-scale wrinkles statistically described by turbulent diffusion. It follows from our simple estimation of the sheet area: at strongly wrinkled flamelet sheet when $(\overline{A}/A_0) >> 1$ the area of 1-D flame can be estimated as it follows (below $F(k)$ is the spectrum of random flamelet sheet $x = h(y,z,t)$)

$$(\overline{A}/A_0) = \overline{(1+|\nabla h|^2)^{1/2}} \approx \overline{|\nabla h|} \approx \left(\overline{|\nabla h|^2}\right)^{1/2} = (\int_0^\infty k^2 F(k)dk)^{1/2}, \qquad (25)$$

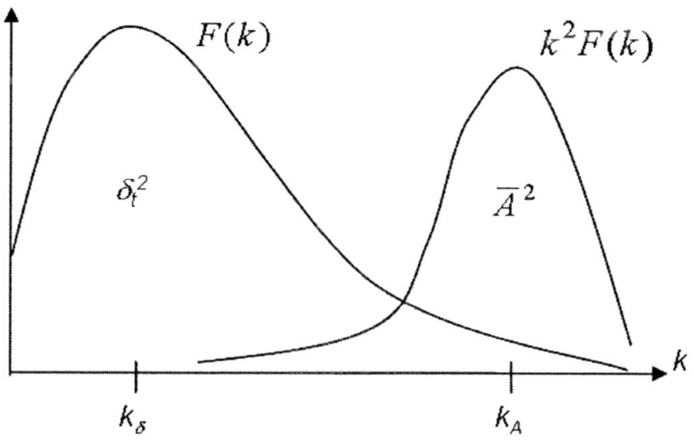

Fig.11 Spectral contributions in δ_t and $\overline{A}$

Eq. 25 demonstrates that the main contribution in $(\overline{A}/A_0)$ (i.e. in $U_t = U_f(\overline{A}/A_0)$) gives the part of the spectrum with large wave numbers $k \sim k_A$ (small wrinkles) while the main contribution in δ_t gives the part of spectrum with small wave numbers $k \sim k_\delta$ (large wrinkles) since as $\delta_t = (\sigma_f^2)^{1/2}$, where $\sigma_f^2 = \overline{(x-\overline{x})^2} = \int_0^\infty F(k)dk$ is the dispersion of the flamelet sheet, Fig. 11. An obvious consequence is that $\overline{\rho W}$ connected with U_t by the expression $U_t = 1/\rho_u \int \overline{\rho W} dx$) depends on the overall spectrum or on both the small- and large-scale processes.

In our estimations presented above flamlet speed U_f was controlled by the equilibrium inertial part of the turbulent energy spectrum $E(k)$. This inertial 5/3-th spectrum is controlled by the dissipation rate ε that in its turn is controlled by u' and L (and hence $\varepsilon \approx u'^3/L$). Here we additionally assume that the spectrum $F(k)$ of the random reacting surface is also equilibrium at large k and is controlled by u', L and τ_{ch}. It results in that at stationary and uniform turbulence $U_t = const$ in spite of increasing brush width as $F(k)$ at small k is non-equilibrium.

At $u' >> U_f$ for some period time $t << \tau^*$ influence of small flamelet speed on the brush width is small so the dispersion of the flamlet sheet is controlled by the turbulent diffusion coefficient $\sigma_f^2(t) \cong 2D_t t \approx u'Lt$. We estimate the characteristic time τ^* from the condition $(u'L\tau^*)^{1/2} \approx U_f \tau^*$ (the transport due to turbulent velocity pulsation and one due to the flamelet movement are of the same order). So at $t << \tau^* \approx \tau_t(u'/U_f)^2$ the flame is transient with growing brush width grows and at $t >> \tau^*$ we have asymptotically the steady state flame. In our estimations and we assume that at $t < \tau^*$ the flame is the ISP one with constant speed U_t^{isp} and increasing $\delta_t^{isp} \approx (u'Lt)^{1/2}$ and at $t > \tau^*$ the flame is steady state with constant speed U_t^{ss} and width δ_t^{ss}. Obviously, τ^* is different for the laminar flamelet τ_{lf}^* and for thickened one τ_{tf}^* so using appropriate expressions for the flamelet speed we will have the following conditions for the ISP flame:

$$t < \tau_{lf}^* \approx \tau_t (u'/S_L)^2 \text{ (a)}, \; t < \tau_{tf}^* \approx 4\tau_t Da \text{ (b)}. \qquad (26)$$

Using $\tau^\bullet$ from Eqs.(26a) and (26b) we can directly estimate the flame brush width at the end of the ISP stage, $\delta_t^* \approx (u'L\tau^*)^{1/2}$, which is in fact the width of the steady state flame $\delta_t^{ss} \approx \delta_t^*$. For the laminar and thickened flamelet mechanisms these widths are as it follows:

$$\delta_{t,lf}^{ss} \approx \delta_{t,lf}^* \approx L\, u'/S_L \text{ (a)}, \; \delta_{t,tf}^{ss} \approx \delta_{t,tf}^* \approx 1/2 \; L\, Da^{1/2} \text{ (b)}. \qquad (27)$$

Below we show that this estimate of the steady state width coincide with an independent direct estimate of it that is not connected with the flame prehistory of the flame but based on assumption of the small-scale equilibrium inside the steady state turbulent flame.

Comparing of the characteristic residence time in the premixed flame τ_r with τ^* gives plausible explanation why observes premixed flames are predominantly transient. At $t > \tau_{isp}$ the flame could be steady state, but in real experiment we usually do not observe this stage. Very simple estimation explains the reason of it.

Assuming the flame speed $U_t \approx 0.5u'$ and parameters of the flow turbulence $u' \approx 10^{-1}U$ and $L \approx 10^{-1}H$ (U and H are the speed and width of the flow) we estimate from simple geometrical consideration of the inclined flame the length an open flame or a flame in the channel equal to $X \approx 10H$ (at this length the flame crosses or reaches the wall and combustion finishes). It corresponds to the residence time inside the flame $\tau_r \approx 10H/U$ while the turbulent time $\tau_t = L/u' \approx H/U$, i.e. the residence time by an order of magnitude greater that the turbulent time. The time for the SS flames, for example, in the case $u'/S_L = 10$ or $Da = 10$ is $\tau^* \approx 10^2\tau_t$, i.e. it is greater by an order of magnitude than the residence time. So at intensive turbulence and large Damköhler numbers the combustion is complete long before formation of the steady state flame so the flame in this condition inevitably is transient.

The statement that sums up this consideration and provides the necessary background to our approach to the premixed combustion modeling is the following: in accordance with our basic concept the turbulent flame predominantly is not the running turbulent combustion front with constant speed and width (the classical steady state turbulent flame), but the running turbulent combustion mixing layer with increasing width nearly constant speed that we call ISP flame (observed in many experiments specific transient flame). In the premixed laminar combustion (as well as in many turbulent combustion model) such flames impossible: initial stage with increasing flame speed and width go over into the final stage with constant speed and width of the flame. Our model equations describe all three stages. But before producing them we deduce expressions, which describe the turbulent flame speed and width for these three stages in the cases of the thickened and laminar flamelet combustion mechanisms.

3.3 A Concept of a Gasdynamic Nature of the Counter-Gradient Scalar Transport

The counter-gradient scalar transport phenomenon is a paradoxical properties of the turbulent premixed flame. Really, numerous direct measurements demonstrate that in contrast to our expectations the flux of the progress variable and temperature $\overline{\rho \vec{u}''c''}$ and $\overline{\rho \vec{u}''T''}$ take place not from the products with $c = 1$ and high temperature to reactants with $c = 0$ and low temperature but in the opposite direction. The fact that the flux $\overline{\rho \vec{u}''c''}$ is described by the turbulent type unclosed equations provokes interpret this phenomenon as a turbulent one, it is called "the counter-gradient turbulent diffusion" and known from the literature scalar flux models are based on turbulent type approximations of unknown terms of the unclosed scalar transport equation. The phenomenon of the counter-gradient or negative viscosity turbulent viscosity in known, for example, in meteorology and the cosmic physics, see, for example, a monograph [35] but our concept of the counter-gradient transport phenomenon is based not on a turbulent, but on a gasdynamic nature of it. Strictly speaking this idea is not new. In fact, this phenomenon, as has been know for long time, is connected with a dominant contribution in the total transport of the non-viscous gasdynamic effect of different acceleration of heavy reactants and light products under the pressure drop across the flame generated by heat release [36]. The originality of our concept is connected with gasdynamic (but not turbulent) modeling of this phenomenon and on analysis of role of it on the flame parameters.

Developed in [37] gasdynamic model for the pressure-driven counter-gradient contribution to the total transport in turbulent premixed flames based on direct estimation from the gasdynamic equations of this factor on the conditional averaged velocities of burned (products) $\overline{\vec{u}}_b$ and unburned (reactants) $\overline{\vec{u}}_u$ gases directly related to the scalar transport $\overline{\rho \vec{u}''c''}$. In the context of this concept the total scalar flux is presented as the sum of two terms that are controlled by different physical mechanisms and could be estimated independently: the gasdynamical pressure driven component caused by different acceleration of the reactants and products and the turbulent diffusion component caused by the random pulsations of the velocity. In the context of the bimodal approximation the scalar flux $\overline{\rho \vec{u}''c''}$ is expressed in terms of the difference of the conditional average velocities of burned (products) and unburned (reactants) gas $\overline{\vec{u}}_b - \overline{\vec{u}}_u = \Delta\overline{\vec{u}}$ [38]:

$$\overline{\rho \vec{u}''c''} = \tilde{c}(1-\tilde{c})(\overline{\vec{u}}_b - \overline{\vec{u}}_u) = \tilde{c}(1-\tilde{c})\Delta\overline{\vec{u}} . \qquad (28)$$

In our approach we present the scalar flux the sum of the gasdynamic pressure-driven and the turbulent diffusion contributions

$$\overline{\rho\vec{u}''c''} = \overline{\rho\vec{u}''c''}_{p-d} + \overline{\rho\vec{u}''c''}_{td} = \tilde{c}(1-\tilde{c})(\Delta\vec{\overline{u}}_{p-d} + \Delta\vec{\overline{u}}_{td}), \tag{29}$$

where $\Delta\vec{\overline{u}}_{p-d} = ((\vec{\overline{u}}_b - \vec{\overline{u}}_u)_{p-d}$ and $\Delta\vec{\overline{u}}_{td} = (\vec{\overline{u}}_b - \vec{\overline{u}}_u)_{td}$ are contribution of these mechanisms in the velocity difference

$$\Delta\vec{\overline{u}}_{p-d} + \Delta\vec{\overline{u}}_{td} = \vec{\overline{u}}_b - \vec{\overline{u}}_u. \tag{30}$$

In the hypothetical case of the constant ($\rho_u = \rho_b$) we have only the gradient turbulent diffusion mechanism, i.e. $\overline{\rho\vec{u}''c''}_{td} = -\overline{\rho}D_t\nabla\tilde{c}$ and hence

$$\Delta\vec{\overline{u}}_{td} = -[\overline{\rho}D_t\nabla\tilde{c}]/[\tilde{c}(1-\tilde{c})]. \tag{31}$$

In the case of 1D stationary turbulent flame with $d\tilde{c}/dx > 0$ (flow from left to right) this mechanism yields $\Delta\overline{u}_{td} = (\overline{u}_b - \overline{u}_u)_{td} < 0$, i.e. the conditional average reactants velocity $\overline{u}_u$ is larger than the products one $\overline{u}_b$ (though the unconditional average speed $\overline{u}$ in this case is constant).

In the case of $\rho_u > \rho_b$ takes place falling of the mean pressure across the flame $(d\overline{p}/dx < 0)$ that yields opposite tendency due to more intensive acceleration of the light products in comparison with heavy reactants. So we have at the same time $\Delta\overline{u}_{td} < 0$ and $\Delta\overline{u}_{p-d} > 0$. It is clear the direction of the total scalar flux $\overline{\rho u''c''}$ (from products to reactants or vice versa) is defined by prevailing turbulent or gasdynamic mechanisms. In contrast to the turbulent component the gasdynamic component of the flux is not proportional to the mean progress variable gradient but in accordance with our model it is controlled by U_t^{ss} and $\Theta = \rho_u/\rho_b$, $\overline{\rho u''c''}_{p-d} = \rho_u U_t^{ss} f(\Theta,\tilde{c})$, and for case of 1D stationary combustion front the problem has an analytical solution [37].

The gasdynamic conservation equations for mass and moment, and the condition of the total pressure conservation inside the flame are as it follows:

$$\rho_u\overline{u}_u P_u + \rho_b\overline{u}_b P_b = \rho_u U_t^{ss}, \tag{32}$$

$$(\overline{p}_u + \rho_u\overline{u}_u^2)P_u + (\overline{p}_b + \rho_b\overline{u}_b^2)P_b = \overline{p}_{-\infty} + \rho_u(U_t^{ss})^2, \tag{33}$$

$$\overline{p}_u + \rho_u\overline{u}_u^2/2 = \overline{p}_{-\infty} + \rho_u(U_t^{ss})^2/2, \tag{34}$$

where the probabilities of reactants and products are equal to $P_u = (1-\overline{c})$ and $P_b = \overline{c}$. These equations are written in the coordinate system traveling with the flame front and contribution of turbulence was ignored. Under assumption that the conditional averages pressures are equal $\overline{p}_u(x) = \overline{p}_b(x) = \overline{p}(x)$ the system Eqs. (32)-(34) the system can be solved with simple algebraic manipulations. It yields the following expressions for the conditional averaged velocities $\overline{u}_u$ and $\overline{u}_b$:

$$\overline{u}_b / U_t^{ss} = \frac{-\beta + (-4\alpha\gamma + \beta^2)^{1/2}}{2\alpha}, \quad \overline{u}_u / U_t^{ss} = \frac{1 - (\overline{u}_b / U_t^{ss})\overline{c} / \Theta}{1 - \overline{c}}, \tag{35}$$

$$\alpha = \frac{\overline{c}}{\Theta}\left(\frac{0.5 - \overline{c}}{(1 - \overline{c})^2} \cdot \frac{\overline{c}}{\Theta} + 1\right), \quad \beta = -2\frac{\overline{c}}{\Theta} \cdot \frac{0.5 - \overline{c}}{(1 - \overline{c})^2}, \quad \gamma = \frac{0.5 - \overline{c}}{(1 - \overline{c})^2} - 0.5 . \tag{36}$$

Taking into account Eq. (28) and the expression $\overline{c} = \tilde{c}\Theta / [(\Theta - 1) + 1]$ (following from the definition $\tilde{c} = \overline{\rho c} / \overline{\rho}$) we have dependences for the gasdynamic mechanism that do not contain empirical parameters: $\overline{\rho u'' c''} / \rho_u U_t^{ss} = f(\tilde{c}, \Theta)$, $\overline{u}_u / U_t^{ss} = f_1(\tilde{c}, \Theta)$, $\overline{u}_b / U_t^{ss} = f_2(\tilde{c}, \Theta)$.

There is an upper estimate of the gasdynamic effect corresponding to the hypothetical case $\overline{u}_u(x) \equiv U_t^{ss}$ and $\overline{u}_b(x) \equiv \Theta U_t^{ss}$. Using Eq. (28) this estimate is as it follows:

$$\frac{\overline{\rho u'' c''}}{\rho_u U_t^{ss}} = \frac{\tilde{c}(1 - \tilde{c})(\Theta - 1)}{1 + \tilde{c}(\Theta - 1)} . \tag{37}$$

We present numerical results for conditional velocities and the scalar transport together with the classical Moss data for the open flame, Fig. 12a, [39] in Fig. 13. Before discussing them we make three remarks:

We see that (i) so called counter-gradient flux (more exactly the counter-gradient component of it) does not depend on the progress variable gradient (in contrast to the turbulent component proportional to $\nabla\tilde{c}$), (ii) though in the 1D flame this and turbulent components are collinear with opposite directions this is not like this in general 3D case when the angle between the pressure and progress variable gradients is not zero. So a term "the counter-gradient scalar flax" is in fact a misnomer. Nevertheless we will use it below as it is universally accepted term.

Used above assumption $\overline{p}_u = \overline{p}_b = \overline{p}$ is reasonable in the cases when the turbulent flame speed is significantly larger than the flamelet one. Really, this difference cannot exceed the pressure drop across the flamelet that is equal to $\Delta p_f = \rho_u S_L^2(\Theta - 1)$ (for definiteness the flamelet is the laminar flame), so $\overline{p}_u - \overline{p}_b < \rho_u S_L^2(\Theta - 1)$. At the same time the pressure falling in the turbulent flame $\Delta\overline{p}_t = \overline{p}_{-\infty} - \overline{p}_{+\infty}$ is equal to $\Delta\overline{p}_t = \rho_u (U_t^{ss})^2(\Theta - 1)$. As $U_t^{ss} \approx u'$ we have an estimation $(\overline{p}_u - \overline{p}_b) / \Delta\overline{p}_t < (S_L / u')^2$, i. e.

at strong turbulence $u' >> S_L$ the effect of the difference of $\overline{p}_u$ and $\overline{p}_b$ can be ignored at estimations of the scalar flux in comparison with influence of the mean pressure falling $\overline{p}(x)$. At the same time there are situations where it is not like this. For example, in impingent flame with not sufficiently large U_t the flame is situated close to the obstacle so the pressure falling across the flame can be significantly suppressed by it. This case was analyzed specially in [40] and it was shown that for quantitative modeling of known from the literature experimental data it is necessary to take into account difference between conditional averaged $\overline{p}_u$ and $\overline{p}_b$.

Formulas Eqs. (35)-(36) were deduced for the 1D stationary flame. Real turbulent premixed flames are transient. Our statement is that the effect of the increasing brush width due to the turbulent diffusion on the pressure distribution across the flame $\overline{p}(x)$ can be ignored in comparison with the pressure falling due to combustion, i.e. the pressure distribution across the flame corresponds to the quasi-stationary situation. The physical reason of such incompressible formulation of the problem is connected with the fact that mean and pulsation velocities are very small in comparison with the sound speed. The pressure falling is connected with the reduction of the average density due to heat release so without combustion the corresponding turbulent mixing layer is practically isobaric. Support this statement simple consideration is the following. As we mentioned above the transient flame with increasing width has practically constant flame speed U_t. Such 1D flame in the coordinate system traveling with this speed is the turbulent mixing layer. At $x = -\infty$ we have the pressure $p_{-\infty}$ and the flow speed U_t, and at $x = +\infty$ correspondingly $p_{+\infty}$ and ΘU_t. The difference $\Delta p = p_{-\infty} - p_{+\infty}$ for the transient flame follows from the impulse equation $\Delta p = \rho_u U_t^2(\Theta - 1)$. In incompressible formulation the flow before the front and behind the back edges of the flame have constant parameters corresponding $x = -\infty$ and $x = +\infty$, so the pressure falling across the transient flame is practically the same as across the steady state flame with the same U_t and $\Theta = \rho_u / \rho_b$.

Finally we used present above estimation of the gasdynamic pressure-driving contribution $(\overline{\rho \vec{u} c''})_{p-d}$ (assuming that it is pointed in direction of the pressure gradient) to the transient ISP flame while the turbulent diffusion component is expressed in terms of the turbulent diffusion coefficient $(\overline{\rho u'' c''})_{td} = -\overline{\rho} D_t \nabla \tilde{c}$ and general scalar flux is presented as the sum of them

$$\overline{\rho \vec{u} c''} = (\overline{\rho \vec{u} c''})_{p-d} + (\overline{\rho u'' c''})_{td} \tag{38}$$

The experiments in the open flame demonstrate as the counter-gradient transport phenomenon in the orthogonal to flame direction [39] as well as the transitions from the gradient transport to counter-gradient one [41] and Eq. (38) gives an opportunity to explain these observations.

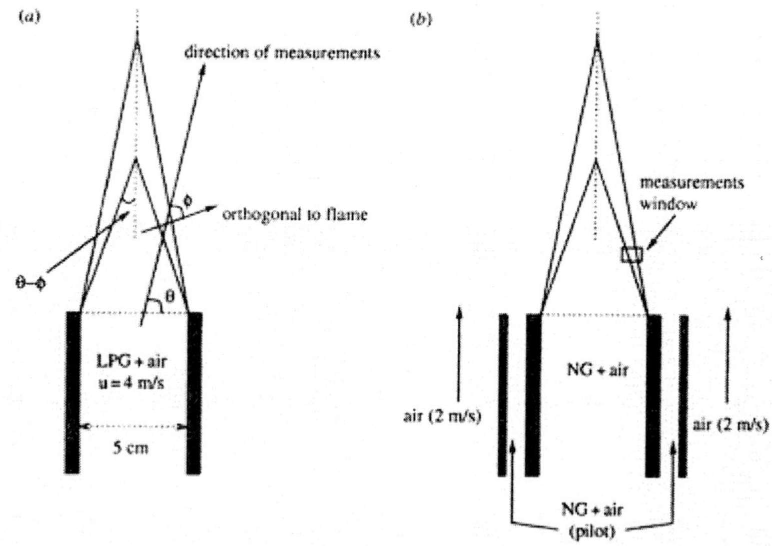

Fig. 12 (a) Schematics of combustor in Moss experiments [39] and (b) Frank et al [41]

The schematic of the Moss experiment is shown in Fig. (12a). A burner pipe of 60cm in length was fed with air and LPG (approximately 70% propane and 30% butane), which become fully premixed at the pipe end. The velocity at the burner exit was approximately ten times the laminar flame speed. The r.m.s. velocity at the center of the pipe exit was $u' \approx 0.5m/s$. Measurements have been taken across the flame brush along a line inclined at 23° with the pipe axis. The angle observed between the flame and the burner axis was approximately $\theta - \phi = 16^{\circ}$. The author used a light scatter technique and the laser Doppler method for measurements of the concentration (the progress variable) and the velocity. The Moss experimental results were subsequently postponed in [42] and this data was used for comparison with the theoretical prediction. Presented in [37] analysis shows that in the Moss experiments the order of magnitude of the turbulent diffusion component within the total progress variable flux is small in comparison with the gasdynamic component generated by the pressure drop across the turbulent flame brush so it is not strange that Fig. 13 demonstrates in general quite reasonable agreement of experimental nondimensional flux $\overline{\rho u''c''}/\rho_u U_t$, and nondimensional conditional average velocities $\overline{u}_u/U_t$ and $\overline{u}_b/U_t$ in orthogonal to flame direction with calculations using the gasdynamic model. Nevertheless we notice that the gasdynamic model give small overestimation of the experimental scalar flux probably related to having neglected here small opposite effect of gradient diffusion as well as predicted $\overline{u}_u/U_t$ and $\overline{u}_b/U_t$ are on average slightly smaller and larger, respectively, that the corresponding measured values which might be again the consequence of having neglect this factor which slightly increases the former and decreases the latter. At the same time the upper estimation gives two to four times larger values of the flux compared to the experiments.

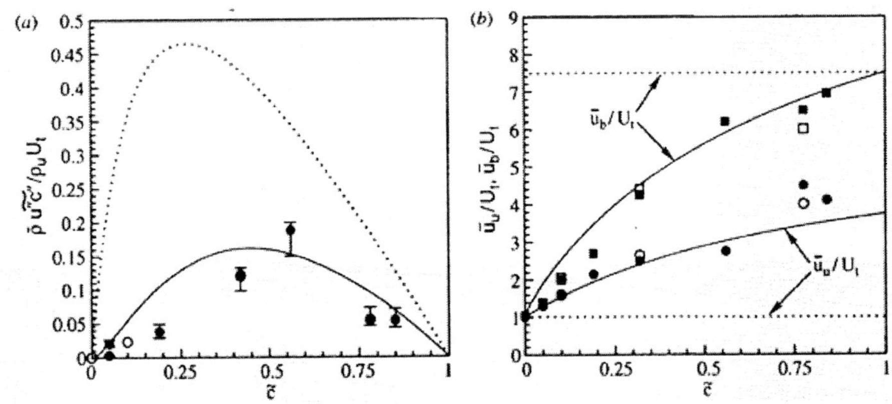

Fig.13 Comparison of (a) progress variable flux and (b) conditional velocities from Moss experiment. Simulations: (····) results from upper estimation of $\overline{\rho u''c''}$, (——) results from the gasdynamic model. Experiments: (a) (•) calculated in [42] from Moss measured conditional mean velocities, (o) calculated in [42] from Moss measured $\overline{\rho u''c''}$, (b) (•) measured experimental u_u, (o) u_u from measured $\overline{\rho u''c''}$, (■) measured experimental u_b, (□) u_b from measured $\overline{\rho u''c''}$.

Frank *et al* in [41] present experimental evidence of the transition from counter-gradient to gradient transport in the turbulent premixed flame. Measurements where performed in a fixed area (measurement window) of axisymmetric turbulent premixed natural gas/air open flame stabilized in a co-flowing air stream by a pilot jet of hot gases. The conditional average velocities $\overline{u}_u$ and $\overline{u}_b$ were measured for different values of the ratio u'/S_L (due to mainly by changing the fuel-air mixture composition) and a transition from the gradient transport to the counter-gradient one (i.e. changing of the scale flux direction from products to reactants to the opposite one) was observed for decreasing u'/S_L. Qualitatively this behavior corresponds to our concept: as small S_L (weak combustion) the gasdynamic effect is not significant and we have transport as in the turbulent mixing layer; as large S_L (strong combustion) the gasdynamic effect is so strong and it dominate over the turbulent one. So these experimental data are therefore of particular interest for the quantitative validation of the concept. The schematic of these experiments is shown in Fig. 12b. The fuel used is natural gas and the fuel/air mixture flows out fully premixed from the 36 mm diameter piloted axisymmetric burner. The pilot gases have a very small flow rate (about 1.5% by volume of the main flow). A perforate plate with 4 mm diameter holes spaced 7 mm apart is placed inside the burner to generate turbulence. The radial and axial components of the conditionally averaged velocities and the Reynolds averaged progress variable $\overline{c}$ were measured within a 2.8×4.2 mm region of the flame located approximately 2.7 cm downstream of the burner mouth (see Fig. 12 b)). In used for analysis four experimental samples the flames were practically parallel to the burner axis so 1D theory for modeling the pressure-driven transport component of the flux was applied in the radial practically orthogonal to the flame direction. Comparison of the experimental data with theoretical prediction is shown on the Fig. 14.

In should be stressed that different pressure-driven acceleration of reactants and products results in not only the counter-gradient transport phenomenon, but also significantly deforms the chemical source term $\widetilde{W}$ in Eq. 7a. The paper [37] contains some result on validation of this deforming presented in Fig. 4 in [37]. It makes questionable to model directly the chemical source controlled not only by chemistry and turbulence, but also specific for flamelet combustion mechanism gasdynamics.

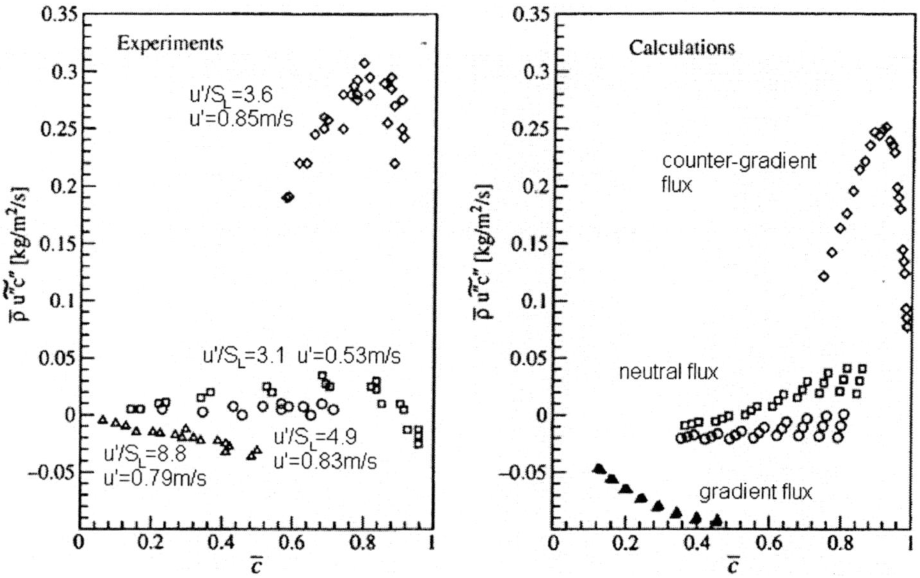

Fig.14 Comparison of calculated progress variable radial flux with experimental data from [39]

4 The Flame Speed and Width for Different Stages of the Flame

In this section we examine 1D nonstationary turbulent premixed flame that begins in the initial moments $t = 0$ with the stepwise distribution of the progress variable and finishes with the steady state flame. We analyze both thickened and laminar flamelet mechanisms. In this connection we notice that for thickened flamelet combustion mechanism we practically have the situation when $u'/U_f >> 1$. It follows from Eq. (19) $u'/U_f = 2Da^{1/2}$ and estimated below of possible minimal values of the Damköhler numbers following from the bow off boundary, Eq. (71a), which result in that $Da > Da_{\min} \approx 0{,}1\,\mathrm{Re}_t^{1/2}$. For example, for $\mathrm{Re}_t = 10^3$ $Da_{\min} \approx 3$ so $(u'/U_f)_{\min} \approx 3.5$; for more realistic situation $Da = 10-20$ we have $u'/U_f = 6-9$. So for thickened flamelet we have only the situation $u'/U_f >> 1$ while for the laminar two could be both situation with $u'/S_L >> 1$ and $u'/S_L \le 1$. But here we analyze cases of strong turbulence with $u'/S_L >> 1$.

We produce expressions for the flame speed U_t and width δ_t for the an initial stage U_t^i, δ_t^i (forming of the developing flamelet sheet), for the well documented in the experiments transient ISP stage when the speed U_t^{isp} is nearly constant and the width δ_t^{isp} increases, and for the final stage when the steady state flame has constant speed U_t^{ss} and width δ_t^{ss}. We will do it both for the thickened and laminar flamelet combustion mechanisms.

Successive description of all combustion stages reduces to designing of model equations that describe these mentioned stages of combustion and in fact interpolate intermediate states. We demonstrate this equation in the next section.

4.1 The Flame Speed and Brush Width of the Transient ISP Flame

Results of this paragraph refer only to the transient ISP flames with increasing brush width δ_t and constant flame speed $U_t = U_f(\overline{A}/A_0)$. In accordance with Eq. (25) the area can be presented as

$$(\overline{A}/A_0) \approx \sqrt{\sigma_t^2}/\lambda, \tag{39}$$

where the dispersion of the random flamelet sheet σ_t^2 and its Taylor microscale λ in terms of the spectrum of this sheet $F(k)$ are as it follows:

$$\sigma_t^2 = \int_0^\infty F(k)dk \text{ (a)}, \; \lambda^2 \approx \int_0^\infty F(k)dk / \int_0^\infty k^2 F(k)dk) \text{ (b)}. \tag{40}$$

We remind that Eq. (40b) is a general definition (with presison of a multiplyer) of the Taylor mocroscale λ [20, p. 55].

We see that the dispersion σ_t^2 is controlled directly by large-scale turbulence $\sigma_t^2 = 2D_t t$ while the micro-scale λ we present in general form as a function of all controlling parameters $\lambda = f(L, u', U_f, \delta_f, t)$. We will analyze the latter expression in the context of the dimension theory using well-known $\Pi-$theorem, as number of variables is too large for usual intuitive similarity considerations. Direct applying this theorem result dimensionless for of the expression

$$\lambda/\delta_f = f_1(u't/\delta_t, u'/U_f, L/\delta_f). \tag{41}$$

We take into account Eq. 19 for U_f and δ_f, and use a condition $(\overline{A}/A_0) \approx const$ due to assumption of equilibration of the spectrum $F(k)$ in Eq. 25 for large k controlled by u', L and τ_{ch}. Hence we obtain that

$$\lambda / \delta_f = (u't / \delta_f)^{1/2} f_2(Da). \tag{42}$$

In our case of large Damköhler numbers $Da >> 1$ we assume that $f_2(Da) \approx f_2(\infty) = const \sim 1$[4]. Hence $\lambda / \delta_f \approx (u't / \delta_f)^{1/2}$ and the expressions for the flamelet sheet are $(\overline{A}/A_0)$ is as it follows:

$$(\overline{A}/A_0) \approx \sigma_t / \lambda \approx (u'/U_f)^{3/2} \approx (Da)^{3/4}. \tag{43}$$

Eq. (43) demonstrates that increasing the flamelet speed decreases the sheet area and vice versa. This self-compensation mechanism is a significant hydrodynamic mechanism controlling the chemistry dependence of U_t.

Using Eq. (19a) and the estimation for the chemical time $\tau_{ch} = \chi / S_L^2$ we have desired expressions for the speed of the transient ISP flame (we denote it here U_t^{isp}) that are the following:

$$U_t^{isp} = U_f(\overline{A}/A_0) = \mathrm{A}(1/2)u'Da^{1/4} = \mathrm{A}(1/2)u'^{3/4}S_L^{1/2}\chi^{-1/4}L^{1/4} \tag{44}$$

where a priori $\mathrm{A} \sim 1$ and its magnitude cannot be estimated theoretically in the context of our approach. Analysis a large body of Karpov's data in spherical bomb with artificial turbulization of the mixture shows that this empirical coefficient is practically the same for methane, ethane, propane and even for hydrogen (if to exclude for it data for very lean and very rich $H_2 - air$ mixtures) [43], [44] (see the section 7 below). In our simulations we used for different fuels the same value $\mathrm{A} = 1.02$ (in the mentioned papers and results below we used the value $A = 0.5\mathrm{A} \approx 0.51$) following from the standard experiments in the stirred bomb analyzed in the subsection 7.1. So practically the empirical coefficient can be assumed to be equal to $\mathrm{A} \cong 1$.

In connection with this result we make three particular remarks:

Appearing in Eq. (44) of the laminar field of the normal laminar speed S_L and the molecular heat transfer coefficient must not confuse: in the analyzed combustion regime the

---

[4] We remind that similar reasoning is used in the classical aerodynamics for producing the expression for aerodynamic drag of the sphere. The aerodynamic force F depends on the sphere diameter d, the flow velocity U, the density ρ and the molecular viscosity coefficient μ, i. e. $F = f(d, U, \rho, \mu)$. Using the $\Pi -$theorem results in the expression $F\rho^{-1}U^{-2}d^{-2} = f_1(Ud\rho\mu^{-1})$ that usually is presented in the standard form $F = C_x(\mathrm{Re})(\rho U^2/2)S$, where $\mathrm{Re} = \rho UD/\mu$ and $S = \pi d^2/4$. $C_x(\mathrm{Re})$ is the resistance coefficient that can be presented also as $C_x(\mathrm{Re}/\mathrm{Re}_{cr})$, where Re_{cr} is the critical Reynolds number. It is assumed at large $\mathrm{Re}/\mathrm{Re}_{cr} >> 1$ that $C_x(\mathrm{Re}/\mathrm{Re}_{cr}) \approx C_x(\infty) = const \sim 1$ in the same manner as we assume that $f_2(Da) \approx f_2(\infty) = const \sim 1$.

flamelet is not laminar and at assumed assumption of the strong turbulence we ignore in the thickened flamelet the molecular diffusion contribution in comparison with the microturbulent transport. S_L and χ are the chemico-physical characteristics of the reactants as well as u' and L are the large-scale characteristic controlling in our model turbulent premixed combustion, i.e. appearing in Eq. (44) the speed S_L does not mean that the flamelet is laminar.

We keep the multiplier factor of the order of magnitude 1/2 describing decreasing of the thickened flamelet speed due to the stretch effect of the outside inertial turbulence to stress that for the Damköhler numbers when the model is valid ($Da \sim 10$ for real turbulence as we show below) $U_t^{isp} \approx 1/2\ u'Da^{1/4} < u'$ that is in agreement with the experimental data. As the speed of the stationary state flame is $U_t^{ss} \approx u'$ keeping this coefficient stresses the speed of the transient ISP flame is lower than of the steady state flame, i.e. developing of the ISP flame into the SS flame is accompanied by increasing of the turbulent flame speed.

Strictly speaking used for the flamelet sheet area estimation Π – theorem yields also the second solution. Really, installing $\delta_f \approx L\ Da^{-3/2}$ from Eq. (19b) into Eq. (42) we will have $\lambda \approx (u'Lt)^{1/2} f_3(Da)$ where $f_3(Da) = Da^{-3/4} f_2(Da)$. Assuming $f_3(Da) = const \approx 1$ at $Da >> 1$ and keeping in mind that $\sigma_t^2 \approx u'Lt$ we have the second result $(\overline{A}/A_0) \approx 1$ and $U_t \approx U_f$. Obviously, this estimate is not appropriate for the case of large Damköhler numbers so we reject it. We notice that in analyzed in the footnote 4 aerodynamic example also there is the second solution and it also is not reasonable for the case of large Re.

4.2 The Flame Speed and Width of the Steady State Flame

The reasoning of the pioneers of the turbulent combustion theory Damköhler and Shchelkin, [3], [4], predicted that at strong turbulence $u' >> S_L$ the turbulent flame velocity does not depend on chemistry $U_t \sim u'$. They meant 1D steady state flame and connected the flame speed with the dynamitic of large-scale turbulent eddies containing the main part of turbulent energy. Zel'dovich in his presentations spoke about "leading points", the most advanced by large-scale turbulence elements of the flamelet sheet in reactants. These leading points constitute the front edge of the turbulent flame and the velocity of this front edge by the reactants is the velocity of the steady state flame U_t^{ss}. The flame width δ_t^{ss} must yield such the flamelet sheet area $(\overline{A}_{ss} / A_0)$ that results in $U_f(\overline{A}_{ss} / A_0) \sim u'$. Obviously, δ_t^{ss} depends both on turbulence and chemistry.

4.2.1 The Steady State Flame Velocity and the Speed of the Turbulent Diffusion Front

Strong support of the classical result $U_t^{ss} \sim u'$ follows from analysis of the speed of the turbulent diffusion front U_{df}. This speed and the flame speed in the case at $u' >> S_L$ are practically identical. More accurate approximation of the turbulent diffusion by the hyperbolic differential equation, which describes the turbulent diffusion front, is used in the turbulence theory [45, p. 676]. Quantitative theoretical analysis of this problem performed in [46] show, that the coefficients of the hyperbolic equation yield the diffusion front speed equal to u' not only by the order of magnitude but practically exactly $U_{df} \cong u'$ at least in the context of the premixed combustion problem.

Below we shortly reproduce that part of our analysis in [46], which is relevant to the combustion problem. The unclosed 1D equations in terms of the passive concentration $\overline{z}(x,t)$ and the scalar flux $\overline{u'z'}(x,t)$ are the following:

$$\partial \overline{z} / \partial t + \partial \overline{u'z'} / \partial x = 0\,, \tag{45}$$

$$\partial \overline{u'z'} / \partial t + \partial \overline{u'^2 z'} / \partial x + (1/\rho)\overline{z'\partial p' / \partial x} + \overline{u'^2}\partial \overline{z} / \partial x + 2\nu \overline{(\partial z' / \partial x_\alpha \cdot \partial u' / \partial x_\alpha)} = 0\,. \tag{46}$$

In these equations $\rho = const$, the mean velocity $\overline{u} = 0$, the molecular diffusion is ignored as at large Re_t it is small in comparison with the turbulent one. At the same time in the latter equation (that follows from the instantaneous and N-S equations) we keep the last dissipation term as even at small molecular transport coefficients $\nu \approx D$ this term cannot be ignored. We use the simplest hypothesis to close the system Eqs. (45) and (46): we assume that the sum of the nonlinear and dissipative terms of Eq. (46) is a function of $\overline{u'z'}$, $u' = (\overline{u'^2})^{1/2}$ and the turbulence integral scale L (the legitimacy of this assumption for the term containing the pressure results from the possibility of expressing the pressure fluctuation p', from dimensional considerations, in terms of the velocity fluctuation $p' \sim \rho u'^2$. Then the functional dependence of unknown terms on the listed parameters can be expressed accurate to a multiplier on the basis of the Π – theorem of dimensional analysis

$$\partial \overline{u'^2 z'} / \partial x + (1/\rho)\overline{z'\partial p' / \partial x} + \overline{u'^2}\partial \overline{z} / \partial x + 2\nu \overline{(\partial z' / \partial x_\alpha \cdot \partial u' / \partial x_\alpha)} = C(u'/L)\overline{u'z'}\,, \tag{47}$$

where C is a multiplier of the order of unity that generally depends on time.

Substituting Eq.(47) into Eq. (46) and eliminating $\overline{u'z'}$ from the equation system with the aid of Eq. (45), we obtain

$$C \cdot \partial \overline{z} / \partial t + (L/u')\partial^2 \overline{z} / \partial t^2 = u'L\, \partial^2 \overline{z} / \partial x^2\,. \tag{48}$$

It is the telegraph equation that describes turbulent diffusion with a finite velocity of the diffusion front $U_{d.f.}$. Analysis of the characteristics of this equation using the standard methods [47] results that $U_{d.f.} = u'$.

As $u' \to \infty$, Eq. (48) changes asymptotically to the parabolic equation usually used to approximate turbulent diffusion. If the Lagrangian integral turbulence scale $l_L = u'T_L$ is taken instead of the Eulerian integral scale L, the turbulent diffusion coefficient at large times $t >> T_L$ equal to $D_t = u'^2 T_L = u' l_L$ and $C = 1$ in the parabolic equation. At such diffusion times, this value of C is also valid for finite u', since the ratio of the distance traveled by the concentration front to the rms displacement of a particle under the action of the turbulent fluctuations during the diffusion time equals $u't/(2D_t t)^{1/2} = (t/2T_L)^{1/2} >> 1$, i.e. a concentration front situated far out on the periphery of the mixing zone has little influence on the concentration profile in the bulk of this zone. But in the vicinity of the concentration front, the influence of the term proportional to $\partial^2 \overline{z} / \partial t^2$ is decisive. So finally more accurate hyperbolic equation describing turbulent diffusion with finite velocity of the concentration front at $t \geq T_L$ is as it follows:

$$\partial \overline{z} / \partial t + (D_t^{eq} / \overline{u'^2}) \partial^2 \overline{z} / \partial t^2 = D_t^{eq} \partial^2 \overline{z} / \partial x^2 . \tag{49}$$

It should be noted that the elementary closing hypothesis Eq. (47) is evidently inadequate for description of diffusion at small times $t < T_L$. This is because use of the telegraph equation means that the coefficients of the equation must depend on time and as the solution of the equation must obey the Taylor exact kinematical equation $\sigma^2 = 2u'^2 \int_0^t (t-s) R_L(s) ds$ [14] , it means that they are not independent. It appears physically reasonable to assume that the speed of the concentration front at initial moments when diffusion starts ($t << T_L$) equal to the maximum instantaneous velocity fluctuation $u'_{\max}$ possible in the medium (usually it is assumed that $u'_{\max} \approx 3(\overline{u'^2})^{1/2}$), and then decreases with time, tending to constant value that is equal, given validity of Eq. (48) and (49) to $U_{d.f.} = u'$.

As we already mentioned above this analysis yields more accurate estimation of the velocity of the steady state flame U_t^{ss} namely that at $u' >> S_L$ (or $u' >> U_f$ in the case of the thickened flamelet) it is not only equal to u' by the order of magnitude $U_t^{ss} \sim u'$ in accordance with the reasoning in [3] and [4], but $U_t^{ss} \cong U_{d.f.} \cong u'$ (the velocity is close to u') as the flame velocity is practically equal to the speed of the turbulent diffusion front and this controlled by the large-scale turbulence speed can be estimated theoretically with reasonable correctness. As $U_t^{ss} \cong u'$ we have $(\overline{A}/A_0) \sim u'/U_f$ (or $(\overline{A}/A_0) \sim u'/S_L$ for the laminar flamelet), i.e. this hydrodynamics mechanism yields complete self-compensation. It makes clear why U_t^{ss} does not depends on chemistry in spite of strong chemistry dependence of the flamelet speed.

In the case $u' \sim S_L$ (or $u' \sim U_f$) we have $U_t^{ss} \approx u' + S_L$ (or $U_t^{ss} \approx u' + U_f$). There are DNS results [48, 49] that confirm these conclusions. In these papers 1D stationary premixed flames were simulated. During these simulations the reactants flow speed was tuned (increased) to have finally stationary picture of the flame in the simulated domain. The paper contains simulations for two cases: $u'/S_L = 4.8$ in [48] and $u'/S_L = 0.88$ in [49]. While in the former case this speed was close to u', in the letter case the final speed of the flow was about $1.2m/s$ for $S_L \approx 0.6m/s$ and hence for $u' \approx 0.5.m/s$. It results in the steady state speed $U_t^{ss} \approx 1.1m/s$ that was close to the flow velocity.

4.2.2 The Width of the Steady State Flame

In the subsection 3.2 we estimated the width of the ISP flame at the end of this stage δ_t^* that is in fact is an estimation of the width of the SS flame. This result ($\delta_t^{ss} \approx \delta_t^* \approx 1/2\ L\,Da^{1/2}$, Eq. (27b) corresponds to an intuitive expectation: the steady state flame width by order of magnitude could be several times larger than the integral scale of turbulence.

Now we produce an independent estimate of δ_t^{ss} based on assumption of the equilibrium small-scale structure of the reaction zone so δ_t^{ss} can be expresses in terms of large-scale parameters of turbulence u' and L, and the chemical combustion time τ_{ch}. For steady state flame in the expression $(\overline{A}/A_0) \approx \sigma/\lambda$, Eq. (39), $\sigma = \sigma_{ss} = const$ and $\lambda = \lambda_{ss} = const$ where $\sigma_{ss} = (\sigma_{ss}^2)^{1/2}$ (σ_{ss}^2 is the dispersion of the random flamelet sheet of the SS flame so $\delta_t^{ss} \approx \sigma_{ss}$) and λ_{ss} is the Taylor micro-scale of the random flamelet sheet. In general form $\lambda_{ss} = f(u', L, U_f, \delta_f)$ where f is some function. The Π – theorem of the dimension theory yields that $\lambda_{ss}/L = f_1(u'/U_f, L/\delta_f)$. As both arguments u'/U_f and L/δ_f are controlled by the Damköhler number we can write that $\lambda_{ss}/L = f_2(Da)$. Taking into account that $(\overline{A}_{ss}/A_0) \approx u'/U_f$ we can produce an expression for the flame brush width as it follows:

$$\delta_t^{ss} \approx \sigma_{ss} \approx \lambda_{ss}(\overline{A}_{ss}/A_0) \approx \lambda_{ss}(u'/U_f) \approx L(u'/U_f) f_2(Da). \quad (50)$$

Assuming at $Da >> 1$ that $f_2(Da) \approx f_2(\infty) \sim 1$ and using Eq. (19a) we will have the desired result

$$\delta_t^{ss} \approx 1/2\ L \cdot Da^{1/2} \text{ or } \delta_t^{ss} = 1/2\ \mathrm{B} \cdot L \cdot Da^{1/2}, \quad (51)$$

where B is an empirical coefficient of the order of unit.

4.3 The Flame Speed and Brush Width During the Initial Stage of Combustion

The initial stage is a stage when the plane in initial moment flamelet sheet is transformed by turbulence into a developed turbulent flame with strongly wrinkled flamelet sheet. There is sub-initial stage: forming the plane thickened flamelet sheet from plane step-wise distribution of the parameter. If estimate this time as δ_f / U_f using Eqs. (18) results in this time is $\delta_f / U_f = \tau_{ch}$, i.e. this time at $Da >> 1$ is small so this sub-stage can be we ignored.

4.3.1 Duration of the Initial Stage

We associate acceleration of the turbulent flame during initial stage when takes place forming of the developed wrinkled flamelet sheet with growing of the turbulent diffusion during initial stage of diffusion in accordance with the Taylor theory of the turbulent diffusion. So duration of the initial stage τ_i is equal to the Lagrange integral turbulent time T_L. This time is often assumed in accordance with observations equal to half the turbulent time $T_L \approx (1/2)\tau_t$ with physical interpretation that large eddies break up after half-turn. It should be noticed that that in our estimation the duration of the initial stage does not depend on chemistry. The author proposed this idea in ones time to A. Lipatnikov and analyzed by him in [50] existing experimental data seems indicate in favor of this assumption at least as the first approximation. We realize here this idea in a little different way that do not need an empirical parameters at least as a first approximation. But we do not pay special attention to this stage as it is not important in many applications and is significant only for combustion in the SI engines.

During initial stage at $t << T_L \approx (1/2)L/u'$ $\sigma^2 = u'^2 t^2$ in accordance with the Taylor theory of the turbulent diffusion [14] the dispersion is equal to $\sigma^2 = u'^2 t^2$. It means that the turbulent diffusion coefficient increases during the initial stage: $D_t^i = 1/2\ d\sigma^2 / dt = u'^2 t$. We remind that at large time we have asymptotically equilibrium constant turbulent diffusion coefficient that in accordance with the Taylor theory is exactly equal to $D_t = u'^2 T_L = u' l_L$ where T_L and l_L are the Lagrange time and space integral scales. For description of the ISP and SS flame we use approximate expression for D_t in terms of the Eulerian integral scale of turbulence $D_t \approx u'L$.

4.3.2 The Flame Speed and Brush Width

We connect increase during initial stage of the turbulent flame speed U_t^i with increase of the turbulent diffusion coefficient D_t during initial period of mixing. In accordance with the Taylor theory of turbulent diffusion this increase takes place during $t \le T_L$ and $D_t(t)$ becomes equilibrium $D_t(\infty) = D_t^{eq} = u'^2 T_L = const$ at $t >> T_L$. So we present the flame speed in the exponential form $U_t \approx u' Da^{1/4} \overline{D}_t^n$ where $\overline{D}_t(t) = D_t(t) / D_t^{eq}$. Produced below relaxation differential equations for $\overline{D}_t$ and U_t, Eqs. (g, k) describing initial stage give $\overline{D}_t = U_t = 1 - \exp(-t/T_L)$ that corresponds to the exponential Lagrange correlation function and the value of the exponent $n = 1$. So we have following schematic approximation for the speed and width during initial stage that takes place at this approximation at $0 \le t \le T_L$:

$$U_t^i \approx (1/2)\text{A}u' \; Da^{1/4}(t/T_L) \text{ (a)}, \; \delta_t^i = u't \text{ (b)}, \tag{52}$$

We will have accurate description of U_t^i and δ_t^i if to attract an expression for the turbulent diffusion coefficient in terms of the Lagrange autocorrelation function $R_L(t)$ following from the Taylor theory of turbulent diffusion[14] (See also [45, p.614]:

$$U_t^i(t) \approx u' Da^{1/4} T_L^{-1} \int_0^t R_L(s)ds \text{ (a)}, \; \delta_t^i(t) \approx \left(2u'^2 \int_0^t (t-s)R_L(s)ds\right)^{1/2} \text{ (b)}, \tag{53}$$

where $T_L = \int_0^\infty R_L(s)ds$ is the Lagrange integral time scale. Solutions of the Eqs (52) correspond to these expressions if to use the exponential function $R_L(t) = 1 - \exp(-t/T_L)$.

We point out that the flame in the initial stage can have significant curvature (in SI engines, for example) and it influences on the flame speed [17], but in our first approximation approach we ignore this factor. In conclusion we notice that strictly speaking in the initial moment $t = 0$ must be $U_t^i = U_f$ and $\delta_t^i = \delta_f$ instead of zero values but as at strong turbulence $U_f << U_t$ and $\delta_f << \delta_t$ we ignore this factor.

4.4 The Flame Speed and Brush Width for the Laminar Flamelet Combustion Mechanism

Here we produce expressions for three stages of flame in the case of when the flamelet is laminar. It takes place when the Kolmogorov microscale is larger than the laminar flame width $\eta = L\,\text{Re}_t^{-3/4} >> \delta_L = (\chi\tau_{ch})^{1/2}$. In theoretical considerations this flamelet assumed to be identical to the normal laminar flame with known speed S_L. In our analysis we assume that at $\eta > \delta_L$ and $\eta < \delta_L$ we have respectively the laminar and thickened flamelet, and impose the condition that estimated in the context of the thickened and laminar flamelet turbulent speed and width coincide at $\eta = \delta_L$ (or for $\nu = \chi$ at $Da = \text{Re}_t^{1/2}$). We remind that the controlling parameters in our formulation of the problem are u', L, ,$S_{L,}$ δ_L or u', L, ,$S_{L,}$ ν as the molecular kinematical viscosity coefficient ν is equal to

$\nu \approx \chi \approx S_L \delta_L$. We analyze the case $u' >> S_L$ and only for completeness make some remarks concerning probably more difficult for analysis case when $u' \le S_L$.

4.4.1 The Flame Speed and Brush Width for the Transient ISP Flame

The flame speed for this stage is $U_{t,lf}^{isp} = S_L (\overline{A} / A_0)$. Similar to our analysis in 4.1 of the thickened flamelet sheet area the area of the dimensionless flamelet sheet can be presented as $(\overline{A} / A_0) \approx \sigma / \lambda$ where the Taylor microscale of the sheet $\lambda = f(u', L, S_L, \nu, t)$. The Π – theorem yields $\lambda / L = f(S_L t / L,\ u' / S_L,\ u'L / \nu)$. Keeping in mind that $U_{t,lf}^{isp} \approx const$ (and hence $\lambda \sim t^{1/2}$), we rewrite the last expression in the form $\lambda / L = (S_L t / L)^{1/2} f_1(u' / S_L,\ u'L / \nu)$. It yields following intermediate expression for the speed $U_{t,lf}^{isp} = (u'S_L)^{1/2} f_1(u' / S_L,\ \mathrm{Re}_t)$. Now the key point is the chemistry dependence of $U_{t,lf}^{isp}$. We assume the simplest assumption namely that the chemistry dependence of the turbulent flame speed for laminar and thickened flamelet combustion mechanism is the same. From Eq. (44) follows that $U_t^{isp} \sim S_L^{1/2}$. It means that $f_1(u' / S_L,\ \mathrm{Re}_t) = f_2(\mathrm{Re}_t)$. Naturally on the boundary between two combustion mechanisms ($\eta = \delta_t$) there is no discontinuity, i.e. on this boundary the speeds $U_{t,lf}^{isp} = U_t^{isp}$, Equating $U_{t,lf}^{isp}$ and U_t^{isp}, Eq. (44), on this boundary we have an expression for the unknown function f_2 that is the following:

$$f_2(\mathrm{Re}_t) = \frac{1/2 \mathrm{A} u' Da^{1/4}}{(u'S_L)^{1/2}}. \tag{54}$$

Remembering that $S_L \approx (\chi / \tau_{ch})^{1/2}$ and $\nu \approx \chi$ we have finally

$$f_2(\mathrm{Re}_t) = 1/2 \mathrm{A}\, \mathrm{Re}_t^{1/4} \tag{55}$$

so the desired result for the ISP flame at laminar flamelet combustion mechanism is the following:

$$U_{t,lf}^{isp} \approx 1/2\ \mathrm{A}(u'S_L)^{1/2} \mathrm{Re}_t^{1/4} \text{ (a)}, \quad \delta_t^{isp} \approx (u'Lt)^{1/2} \text{ (b)}, \tag{56}$$

where A is the same empirical constant the same as in Eq. (44).

It is very significant that producing Eq. (56a) we in fact did not used the condition $Da = \mathrm{Re}_t^{1/2}$ equivalent to $\eta = \delta_t$. It means that deduced for the case of the thickened flamelet mechanism $U_t^{isp} = 1/2Au'Da^{1/4}$, Eq. (44), is identical to Eq. (56a)[5], i.e. it is valid also in the case of the laminar flamelet combustion mechanism. We see that assumption of equal chemistry dependence for the turbulent flame speed for both combustion mechanisms resulting in the identity of the expressions for the speed, i.e. we can use the same expression for U_t^{isp}, Eq. (44), equally for the thickened and the laminar flamelet combustion mechanisms.

In principle could be different opportunities but it seems that now there are no firm experimental evidences or theoretical base for it. On the other hand our experience in combustion simulations demonstrates (and show some results below) that using Eq, (44) in the combustion model equation gives good results both in the situations with the thickened flamelet (high velocity and large scale combustor) and with the laminar one (laboratory burners).

It is easy to check that in this case the hydrodynamical self-compensation mechanisms are described by the expression $(\overline{A}/A_0) \sim (u'/S_L)^{1/2}$. In fact it is the only mechanism responsible in Eq. (52a) for different chemistry dependence of $U_{t,lf}^{isp}$ in comparison with it for the laminar flame.

We notice that in the case $u' \leq S_L$ the ISP stage practically can practically disappear and the steady state stage with $U_t^{ss} \cong u' + S_L$ follows directly after the initial stage, i.e. after some period of increasing of U_t and δ_t they become constant simultaneously.

4.4.2 The Flame Speed and Brush Width for the Steady State Flame

As we discussed above at $u' >> S_L$ the steady state flame speed is equal to $U_t^{ss} \approx u'$, i.e. for both combustion mechanisms the steady state flame speed is the same. But the flame width in this case of the laminar flamelet is different than for the thickened flamelet combustion mechanism.

In the subsection 3.2 we estimated the width of the ISP flame at the end of this stage δ_t^* that is in fact is an estimation of the width of the SS flame. This result for the case of the laminar flamelet combustion mechanism $\delta_{t,lf}^{ss} \approx \delta_t^* \approx Lu'/S_L$, Eq. (27a) is confirmed by direct analysis of the steady state flame. Our analysis here is similar to it in the paragraph 4.2.2.

Using the expression for the flame speed $U_{t,lf}^{ss} = S_L(\overline{A}/A_0)_{lf} \approx S_L(\sigma_{lf}^{ss}/\lambda_{lf}^{ss}) \approx u'$, where the Taylor micro-scale of the steady state random flamelet sheet is $\lambda_{ss} = f(u', L, S_L, \nu)$. Using the $\Pi-$ theorem results in an expression in terms of dimensionless variables $\lambda_{lf}^{ss}/L = f_1(u'/S_L,\ u'L/\nu)$. Keeping in mind that $\delta_{t,lf}^{ss} \approx \sigma_{lf}^{ss}$

[5] More accurately we must say *practically identical* as here there is a subtle point connected with possible influence of the stretch on the laminar flamelet. Strictly speaking S_L in Eq. (56a) is the speed of the normal laminar flame only when $\eta >> \delta_L$ and the speed of the stretch laminar flame at $\eta \geq \delta_L$. In the paragraph 3.1.2 we gave the upper estimate of this effect that for practical modeling does not seem significant.

we can write that $\delta_{t,lf}^{ss} \approx (Lu'/S_L) f_1(u'/S_L, \mathrm{Re}_t)$ where f_1 is an unknown function. We find it using the imposed condition that δ_t^{ss} for laminar and thickened flamelet combustion mechanisms are equal at $\eta/\delta_L = Da^{1/2}/\mathrm{Re}_t^{1/4} = 1$. Using Eq. (51) and remembering that $S_L \approx (\chi/\tau_{ch})^{1/2}$ and $\nu \approx \chi$ we have

$$f_1(u'/S_L, \mathrm{Re}_t) \approx (L\, Da^{1/2})/(Lu'/S_L) \approx Da\ \mathrm{Re}_t^{-1/2} \approx 1. \tag{57}$$

So we write the expression for the steady state width as it follows

$$\delta_{t,lf}^{ss} = \mathrm{B} L(u'/S_L). \tag{58}$$

Assuming the empirical parameters B the same as in Eq. (51) we have the exact equivalence of Eqs. (51) and (58) when $\delta_L = \eta$. At the same time at $\delta_L \neq \eta$ these equations are not equivalent. (in contrast to the equations for the speed of the transient ISP flames Eqs. (44) and (56a).

4.4.3 The Flame Speed and Brush Width during the Initial Stage

Similar to the case of the thickened combustion mechanism, Eqs (53, 53) the expressions for the flame speed and width are (for $t \leq T_L \approx \tau_t$)

$$U_t^i \approx (1/2)\mathrm{A}(u'S_L)^{1/2}\,\mathrm{Re}_t^{1/4}(t/T_L)\ \text{(a)},\ \delta_t^i \approx u't\ \text{(b)}. \tag{59}$$

This expression can be produces more accurately if to take into account actual dimensionless Lagrange autocorrelation function $R_L(t)$. In this case expression for the flame speed and width can be written is it follows:

$$U_t(t) \approx (u'S_L)^{1/2} T_L^{-1} \int_0^t R_L(s)ds\ \text{(a)},\ \delta_t(t) \approx (2u'^2 \int_0^t (t-s)R_L(s)ds)^{1/2}\ \text{(b)}. \tag{60}$$

In the case $u' \leq S_L$ we cannot assume that $\delta_t(t)$ during initial stage is independent on S_L and in general this case is more difficult for description: used in our analysis assumption of statistical equilibrium could be questionable, an effect of the Landau thermal flame instability could be significant and so on.

4.5 A Summary of the Theoretical Method

We outline here a logical scheme of our analysis based on several physical assumptions and using as the method for deducing of the theoretical expression the $\Pi-$ theorem of the dimension theory. The analysis from technical viewpoint is simple. But from the conceptual viewpoint it is not obvious as it is based not only on the Kolmogorov idea of the small-scale equilibrium, but also contains several others physical assumptions.

1. We assume that in the turbulent flame at strong turbulence and fast chemistry not only large-scale but also small-scale processes in the turbulent premixed flame are controlled by two large-scale turbulent parameters u' and L, and two physico-chemical parameters of the combustible mixture η $(=\nu)$ and τ_{ch}. It is consequence of the assumption 1 that all small-scale hydrodynamical structures of the turbulence and reaction zones have a statistical equilibrium (it is in fact a generalized for premixed turbulent combustion the main Kolmogorov idea of small-scale equilibrium widely used in turbulence modeling).
2. The velocity U_t and width δ_t of the turbulent premixed flame are controlled by these parameters and in general case depend on time t (in contrast to the steady state flame with constant speed U_t^{ss} and width δ_t^{ss}). The Π – theorem yields in these cases the following:

$$U_t / u' = f_U(\mathrm{Re}_t,\ Da,\ t/\tau_t),\ \delta_t / L = f_\delta(\mathrm{Re}_t,\ Da,\ t/\tau_t) \tag{61}$$

$$U_t^{ss} / u' = f_u^{ss}(\mathrm{Re}_t,\ Da),\ \delta_t^{ss} / L = f_\delta^{ss}(\mathrm{Re}_t,\ Da), \tag{62}$$

If the molecular transport can be ignored we have the following:

$$U_t / u' = f_U(Da,\ t/\tau_t),\ \delta_t / L = f_\delta(Da,\ t/\tau_t) \tag{63}$$

$$U_t^{ss} / u' = f_u^{ss}(Da),\ \delta_t^{ss} / L = f_\delta^{ss}(Da), \tag{64}$$

3. Eqs. (61 - 64) are too general for application. So we present the speed as $U_t = S_L(\overline{A}/A_0)$ or $U_t = U_f(\overline{A}/A_0)$. We have in the case or strong turbulence and fast chemistry that $u' >> S_L \approx (\chi/\tau_{ch})^{1/2}$ and $L >> \delta_L \approx (\chi\ \tau_{ch})^{1/2}$, i.e. the laminar flamelet is relatively slow and thin, and hence strongly wrinkled $(\overline{A}/A_0) >> 1$. The fact that the thickened flamelet sheet reminds relatively slow $u' >> U_f$ and thin $L >> \delta_f$ follows from a theoretical estimation of U_f and δ_f. This estimation based on the fact that the microturbulence inside the thickened flamelet is controlled by the dissipation rate ε (consequence of the assumption 1) so ignoring an effect of the molecular diffusion in the microturbulent flamelet the dimension reasoning directly yields expressions for U_f and δ_f, Eqs. (18).
4. For strongly wrinkled flamelet sheet our estimation yields $(\overline{A}/A_0) \approx \sigma_t/\lambda$, Eq. (39). This is not an assumption but a general statistical result for the random surfaces

and in the case of the Gauss random surface the multiplayer in this expression is an exact number. This results means that the area of the flamelet sheet is controlled mainly by the small-scale wrinkles of it while the flame width is controlled mainly by large wrinkles. So we distinguish in the model three clearly defined successive stages of the turbulent flame:

(i) The initial stage when both small-scale wrinkles are statistically nonequilibrium: $U_t(t)$ and $\delta_t(t)$ increase in time;

(ii) The transient stage with an intermediate asymmetric when the small-scale wrinkles are already quasi-equilibrium while the large scale wrinkles are still practically nonequilibrium, the intermediate steady propagation (ISP) regime, $U_t^{isp} \approx const$, $\delta_t^{ist} \approx (2D_t t)^{1/2}$;

(iii) The final steady state stage asymmetric when both the small- and large-scale wrinkles are completely equilibrium, the classical 1D stationary flame with $U_t^{ss} = const$ and $\delta_t^{ss} = const$.

In our analysis we deduce expressions for the width and separately for ISP and SS stages. Formation in time of the ISP flame during initial stage we connected with the time formation of the turbulent diffusion coefficient in accordance with the Taylor theory of the turbulent diffusion, it is our assumption 2. We stress that the final stage is achieved only asymptotically so between the ISP and SS flame we have some intermediate state that in following equations in 5.1.1, in fact, was interpolated.

5. The flame speed of the steady state flame is controlled by the speed of the "leading points"(Zel'dovich term), i.e. the most advanced parts of the flamelet sheet. We connect this speed with the velocity of the turbulent diffusion front that more susceptible to a theoretical treatment. It yields more accurate estimate of the flamelet speed $U_t^{ss} \cong u'$ in comparison with correct estimate but only by the order of magnitude $U_t^{ss} \sim u'$ in [3, 4].
6. We use for deducing other parameters of the ISP and SS flames the Π – theorem and the condition of continuity of the flame parameters on the boundary between laminar and thickened flamelet mechanisms and assumed (assumption 3) that the chemistry dependence of the ISP flames at thickened and laminar flamelet combustion mechanisms are the same. We stress that our derivations contains common in such kind analysis assumption 4 namely that appearing on the intermediate stage of our analysis functions like $f(Da)$ can be estimated as $f(Da) \approx const \approx 1$ at $Da >> 1$. Assumption 4 from the mathematical viewpoint means that we expect the self-similarity of the first kind. But it is not always like this. For example, we know that $(\overline{A} / A_0) = g(u', \ L \ , \tau_{ch})$ so direct applying of the dimension reasoning result in $(\overline{A} / A_0) = h(Da)$. We cannot put $h(Da) \approx 1$ at $Da >> 1$, as this result would not be physically justified (it is justified at $Da << 1$). And really we see that our analysis, which is based on the intermediate presentation $(\overline{A} / A_0) \approx \ \sigma / \lambda$, gives

physically justified $Da >> 1$ result $(\overline{A} / A_0) \approx Da^{3/2}$, Eq. (43). Appearing of the power-law dependence at large Damköhler numbers is actually a manifestation of self-similarity of the second kind examined in [51, 52]. At this self-similarity we have at $Da >> 1$ instead of $h(Da) \approx 1$ different result namely $h(Da) \approx Da^n$ where n is a parameter that directly cannot be found in the context of the dimension theory. Using the intermediate presentation $(\overline{A} / A_0) \approx \sigma / \lambda$ we in fact reduced the second kind self-similarity problem to the first kind one and found in the context of the dimension theory the value of the exponent n.

7. Our theoretical analysis has an inherent limit connected (i) with inherent limit of used Kolmogorov ideas and (ii) with ignoring in our model of the molecular transport in the model though molecular effects (for example, the preferential diffusion one) are documented in experiments even at very strong turbulence.

   (i) In our analysis (as well as in the Kolmogorov one) pulsation of the dissipation rate ε are ignored. For combustion modeling this limitation is strong as controlled by instantaneous ε large instantaneous velocity gradient can cause local extinction and even further general combustion blow off.

   (ii) In our analysis we did not take into consideration molecular diffusion as turbulent diffusion including the microturbulent one in the preheat zone of the thickened flamelet prevails at strong turbulence. But molecular effects can be significant in the vicinity of the instantaneous reaction zone that due to strong dependence of the chemistry on the temperature can be much thinner then the flamelet width. One of them is co-called effect of the prudential diffusion when due to possible difference of the molecular diffusion coefficient of the chemical species the local compositions in instantaneous curved reaction zones are different than the mean one.

   Available experimental data shows significance in general case of these mechanisms. We resolved this problem in practical simulations by using existing in the literature approaches to these problems. We touch these questions in more details below when we will speak about the combustion model validation.

4.6 A Diagram of the Combustion Mechanisms and a Table Summarization of the Results

In this subsection present the diagram of the premixed combustion mechanisms and summarize all previous results in the form of a table. This information is used in the next section for designing model combustion equations.

4.6.1 The Diagram of the Combustion Mechanisms

R. Borghi proposed the diagram for combustion mechanisms in 1985 [53] (well known "Borghi diagram" and several modifications of it). This diagram demonstrate boundaries between the laminar flamelet, the thickened flamelet and the stirred-reactor type combustion mechanisms in terms of u'/S_L and L/δ_L. Borghi himself did not analyze the question why shown by his diagram intermediate mechanism between laminar flamelet and reactor type combustion is possible. In other words why at decreasing of the Kolmogorov microscale η after the laminar flamelet combustion mechanism at $\eta > \delta_L$ we have no directly distributed reactor type combustion due to involving at $\eta < \delta_L$ more and more coarse eddies in more and more widened instantaneous reaction zone. This question was resolved theoretically in [19] (see also [16])where it was shown why thin and strongly wrinkled microturbulent flamelet sheet exists at $\eta << \delta_L$ (this analysis is presented in the paragraph 3.1.3). The paper [19] (correctly cited by Borghi in [53]) contains also the analytical expressions for the boundaries between these three regimes in terms of the dimensionless time τ_t / τ_{ch} (notation Da was not used in [19]) and Re_t, and the boundaries presented in the Borghi diagram in terms of u'/S_L and L/δ_L are equivalent to them in [19]: expressions for the boundaries in terms of u'/S_L and L/δ_L directly follows from them in terms Da and Re_t assuming that $\tau_{ch} \approx \chi / S_L^2$ and $\nu \approx \chi$. We notice that Williams reformulated the Borghi boundaries in terms of Da and Re_t, and received our results.

Our analysis in [19] as well as the Borghi diagram [53] is not complete as it does not contain significant for analysis of actual combustion mechanisms flame extinction boundary. The diagram point out that for fixed L/δ_L at increasing of u'/S_L we have successively the laminar flamelet, the thickened one and the stirred reactor type combustion mechanism. Simple estimate added in the diagram in [16] shows that at increasing u'/S_L before forming the distributed reactor type combustion mechanism takes place the flame blow off, so in accordance with this estimation the distributed premixed combustion mechanism is unlikely. Experiments seems confirm this deduction, for example, results an experimental investigation of instantaneous reaction zones in the industrial gasturbine burners demonstrate that combustion takes place in thin and strongly wrinkled sheets.

Here we reproduce results for the boundaries from [19] and add the boundary for combustion extinction from [16]. The Borghi diagram does not deal with the different stages of combustion. It can reflect a 3D diagram containing additional axis t and we discuss this opportunity.

In our analysis of the thickened flamelet the vortices governing the turbulent transport processes in it are referred to the inertial interval of the turbulent spectrum

$$L >> \delta_f >> \eta \; (= L\,\mathrm{Re}_t^{-3/4} \approx \nu^{3/4}\varepsilon^{-1/4}). \tag{65}$$

The condition for the existence of a strongly thickened combustion front ($\delta_f >> \delta_L$), the right side of Eq. (57), becomes after Eqs. (9) and (18) have been used:

$$\mathrm{Re}_t^{3/4} / Da^{-3/2} (\approx (\varepsilon / \nu)^{3/4} (\chi / S_L^2)^{3/2}) >> 1 . \tag{66}$$

Using the expression for ε in terms of the turbulence Taylor miscroscale λ: $\varepsilon = 15\nu u'^2 / \lambda^2$ [20 p.56], we obtain $(u'/\lambda)^{3/2} >> (S_L^2/\chi)^{3/2}$ from Eq. (66), which is practically in agreement with the condition for the strong influence of turbulence on the laminar flame front $u'/\lambda > S_L^2/\chi$ proposed by Kovasznay [54]. By using the expression for η in terms of ε from (65) the inequality (66) is converted into the form $\eta^3 << (\chi/S_L)^3$, i.e. at least there should be $\eta < \delta_L$, which agrees physically with the initial hypotheses of the model. Let us also note that Eq. (66) yields an upper bound on Re_t, or $\mathrm{Re}_t << (u'/S_L)^3$, for fixed u' and S_L, which can be spoiled for too large L.

The left side of the inequality Eq. (65), the condition of strong wrinkling of the flamelet sheet, becomes after Eqs (9) and (18) have been used

$$Da^{3/2}(\approx (\varepsilon/u'^2 \cdot \chi/S_L^2)^{3/2}) << 1. \tag{67}$$

It follows from condition Eq. (67) that that the residence time of the substance in the thickened flamelet $\tau_r \approx \delta_f / U_f$ equal to $\tau_r \approx \tau_{ch}$, according to Eqs. (15 or 19), should also be much less than the dissipation time $\tau_d \approx u'^2/\varepsilon \approx L/u' \approx \tau_t$. This inequality as the condition for the existence of a surface combustion mechanism is mentioned in [18].

Remembering that from Eq. (9) $\chi \approx S_L\delta_L$ for $\nu \approx \chi$ the inequalities Eqs. (66) and (67) can be directly reformulated in terms of L/δ_L and u'/S_L, so the conditions that the flamelet is strongly wrinkled and that the flamelet sheet is strongly wrinkled are expressed by the following inequalities:

$$(u'/L)^{9/4} >> (L/\delta_L)^{3/4} \text{ (a)}, \; (u'/S_L)^{3/2} << (L/\delta_L)^{3/2} \text{ (b)}. \tag{68}$$

So the boundary between the thickened and laminar flamelet follows from Eqs. (66) and (68a) and respectively in terms of $(Da,\ \mathrm{Re}_t)$, $(u'/L,\ L/\delta_L)$ and $(Ka,\ \mathrm{Re}_t)$ are as it follows:

$$Da = \mathrm{Re}_t^{1/2} \text{ (a)}, \; (u'/L) = (L/\delta_L)^{1/3} \text{ (b)}, \; Ka = 1 \text{ (c)}, \tag{69}$$

that are equivalent to condition $\delta_L \approx \eta = L\,\mathrm{Re}_t^{-3/4}$. At the same time the boundary between the thickened flamelet and the stirred reactor combustion mechanisms following from Eqs. (67) and (68b) respectively in terms of $Da,\ \mathrm{Re}_t$ and are as it follows:

$$Da = 1 \text{ (a)}, \; (u'/L) = (L/\delta_L) \text{ (b)}, \; Ka = \mathrm{Re}_t^{-1/2} \text{ (c)} \tag{70}$$

that are equivalent to condition $\delta_f \approx L$ and $(\overline{A} / A_0) \approx 1$: the flamelet becomes thick and non-wrinkled. At $Da < 1$ takes place combustion described in first approximation by Eqs. (22).[6]

There is additional extinction boundary, which narrows the area of real combustion and makes unattainable not only the boundary between the thickened and stirred reactor type combustion (Eqs. (70) but combustion at significantly larger than unite Damköhler numbers. It is connected with flamelet quenching by the stretching and this mechanism is absent in the analyzed combustion model, i.e. the expressions gives an opportunity to reach this boundary formally.

We assume that the on the extinction takes place when the characteristic value of the maximal velocity gradient g (caused by the eddies of minimal sizes) is equal to some critical value g_{cr}. This critical gradient is controlled by the chemical time and can be estimated as $g_{cr} \approx 1/\tau_{ch}$ using the dimension theory reasoning. At the same time the value g is controlled by the spectrum of turbulence that in turn is controlled by the turbulent dissipation rate ε. So combustion extinction takes place when ε riches some value ε_{cr}[7]. The dissipation rate on the one hand is controlled by the large-scale turbulence $\varepsilon = \alpha\ u'^3 / L$, where an empirical coefficient $\alpha \approx 0.4 - 0.5$ at large Re_t. On the other hand ε is expressed exactly in terms of the small-scale characteristic of turbulence by the expression $\varepsilon = 15\nu\overline{(\partial u_1 / \partial x_1)^2}$ [20 p. 142] (we remind that it is valid in the case of the isotropic small-scale structure of turbulence). Estimating the gradient velocity as $g = \left(\overline{(\partial u_1 / \partial x_1)^2}\right)^{1/2}$ we have $0.5u'3/L \approx 15\nu g_{cr}$ and hence the equation of the on the extinction boundary (the curve C) in terms of $Da,\ \mathrm{Re}_t$ and $L/\delta_L,\ ,u'/S_L$ are as it follows:

$$Da = Q^{-3/2}\,\mathrm{Re}_t^{1/2}\ \text{(a)},\ u'/S_L = Q(L/\delta_L)^{1/3}\ \text{(b)},\ Ka = Q\ \text{(c)}, \tag{71}$$

where $Q = (30\chi^2 g_{cr}^2 / S_L^4)^{1/3}$ so taking into account that $g_{cr} \approx 1/\tau_{ch}$ we have $Q \approx 3$. It is easy to show that on the boundary C we have $U_f / S_L = \delta_f / \delta_L = Q^{2/3} \approx 5$. We remind that at the boundary B we had $U_f / S_L = \delta_f / \delta_L = 1$, i.e. at intensification of

[6] Using in practice stirred reactor type combustion models [55], [56] contains some sub-models of the micro-turbulent mixing which reduce mean chemical reaction rates. We demonstrate in the Chapter ??? that these models are described by quasi-laminar type differential equations with modified in comparison with the classical quasi-laminar equations chemical source terms. While in the classical quasi-laminar model in fact ignoring the scalar parameters pulsations the chemical source is described by the chemical kinetics expressions with average temperature and chemical species concentrations, the source terms in the reactor type models contains less than unite multipliers depending on turbulence. For example, in [55] multiplier is canalled by Re_t while in [56] it depends also on chemistry.

[7] Here there is a subtle point. Strictly speaking extinction impossible at constant molecular transport coefficients We in fact speak about the boundary of drastic reduction and practical ceasing of combustion that inevitably results in complete flame blow off due to heat losses and so on.

turbulence ($L = const$, u' increases and hence Da decreases) we have first transition from the laminar flamelet to thickened one combustion mechanism. Then takes place acceleration and broadening of the flamelet and in accordance with our estimation at achieving the width $\delta_f / \delta_L \approx 5$ combustion goes out. As en example in accordance with this estimation the minimal Damköhler numbers $Da_m \approx 6$ for $\mathrm{Re}_t = 10^3$ and $Da_m \approx 2$ for $\mathrm{Re}_t = 10^2$, i.e. the stirred reactor type combustion ($Da < 1$) practically unattainable. Fig. 15 presents these results graphically in different variables.

We must keep in mind that these boundaries are not something absolute, they follow from estimations of the order of magnitudes. For example, in Fig. 3 the boundary between the laminar and thickened flamelet corresponds to $Ka \approx 5$ or $\delta_L / \eta \approx 2$ so we can expect combustion extinction at three times larger Karlovitz number $Ka \approx 15$.

4.6.2 Table Summary of the Results

The Borghi diagram shows only the boundaries between combustion mechanisms. But for the same combustion mechanism there are also different stages of the combustion that were analyzed above. The table below summarized all previous results. We remind that they refer to 1D nonstationary flame for the case $u' >> U_f$ or $u' >> S_L$, i.e. for the case of sufficiently strong turbulence. This table presents the flame speed and flamelet brush width for the cases of the laminar and thickened flamelet mechanisms for initial, ISP and SS stage of combustion together with the

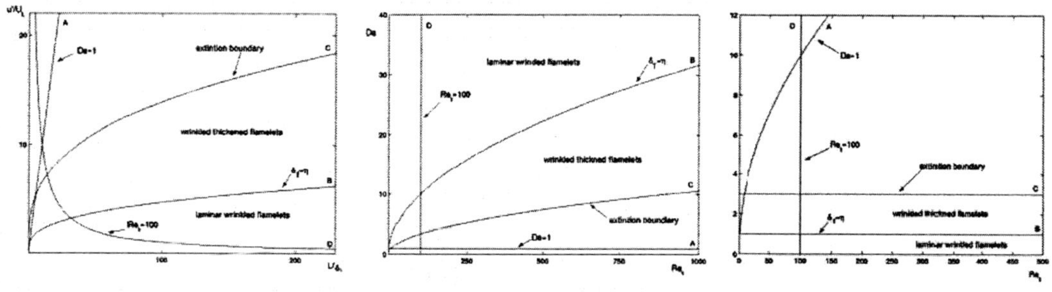

Fig.15 Diagrams of combustion mechanisms

ranges of the turbulent and physico-chemical parameters of the system and ranges of the time when this combustion regimes and flame studies take place. This table is as it follows:

Table 1. The summary plate of the turbulent flame parameters at laminar and microturbulent (thickened) flamelet combustion mechanisms for three stages of combustion.

Flame stages / Combustion mechanisms	Initial stage	Transient ISP stage	Final SS stage
Laminar flamelet combustion mechanism: $S_L \approx (\chi / \tau_{ch})^{1/2}$	$U_t^i \approx 1/2 A(u'S_L)^{1/2} \mathrm{Re}_t^{1/4} (t/\tau_t)$ $\approx 1/2\ u'Da^{1/4} (t/\tau_t)$	$U_{t\,\mathit{isp}}^{isp} \approx 1/2\ A(u'S_L)^{1/2} \mathrm{Re}_t^{1/4}$ $\approx 1/2\ u'Da^{1/4}$	$U_t^{ss} \approx u'$
$\delta_L \approx (\chi\ \tau_{ch})^{1/2}$ Conditions:	$\delta_t^i \approx u't$	$\delta_t^{isp} \approx (2D_t t)^{1/2} \approx (u'Lt)^{1/2}$	$\delta_t^{ss} \approx L(\mathrm{Re}_t / Da)^{1/2}$ $\approx L(u'/S_L)$
$u'/S_L < (L/\delta_L)^{1/3}$ $Da > \mathrm{Re}_t^{1/2}$	$0 < T_L \approx 0.5\tau_t$	$T_L \approx 0.5\tau_t < t < \tau_t (u'/S_L)^2$	$t >> \tau_t (\mathrm{Re}_t / Da)$ $>> \tau_t (u'/S_L)^2$
Thickened flamelet c. m.: combustion mechanism: $U_f \approx 1/2\ u'Da^{-1/2}$	$U_t^i \approx (1/2) A u'\ Da^{1/4} (t/\tau_t)$	$U_t^{isp} \approx 1/2\ u'Da^{1/4}$	$U_t^{ss} \approx u'$
$\delta_f \approx 1/2\ L\ Da^{-3/2}$ Conditions: $(L/\delta_f)^{1/3} < u'/S_L <$ $< 3(L/\delta_L)^{1/3}$ *or*	$\delta_t^i \approx u't$	$\delta_t^{isp} \approx (2D_t t)^{1/2} \approx (u'Lt)^{1/2}$	$\delta_t^{ss} \approx L\ Da^{1/2}$
$0.2\,\mathrm{Re}_t^{1/2} < Da < \mathrm{Re}_t^{1}$	$0 < T_L \approx 0.5\tau_t$	$T_L \approx 0.5\tau_t < t < 4\tau_t Da$	$t >> 4\tau_t Da$

5 Equations of the TFC Turbulent Premixed Combustion Model

In these sections we produce equations of the TFC model. The first part of the section is devoted to analysis of the hypothetical case of the constant density. It simplifies the problem, as the counter-gradient transport phenomenon does not take place here. In this part we propose equations that describes all the stages. The second part is devoted to the case of variable density. We concentrate our efforts on analysis and numerical simulations of the ISP flame in the context of the RANS and LES formulation of the problem.

5.1 RANS Equations for the Case of Constant Density Describing All Combustion Stages

The basic equations of the TFC combustion model are the unclosed moment one in terms of the mean progress variable and the kinematical equations describing initial, intermediate and steady state stages of the flame. In the previous section we produced the theoretical expressions describing turbulent flame speed and the flame brush width for this stages that directly used for designing of the model equation. Developing of the flame from one stage to the next takes place in the context of the relaxation process using theoretical estimations for the characteristic times of the stages.

5.1.1 Nonstationary 1D Turbulent Flame in Uniform and Stationary Turbulent Field

We assume that 1D turbulent flame propagates along the axis x in the turbulent medium with constant mean speed and uniform and stationary turbulent characteristics. The unclosed equation valid for all stages and the kinematical equations describing the transient ISP and steady state flames are the following:

$$\partial\bar{c}/\partial t + u\partial\bar{c}/\partial x = -\partial\overline{u'c'}/\partial x + \overline{W}\,, \tag{72}$$

$$\partial\bar{c}/\partial t + u\partial\bar{c}/\partial x = D_t^{eq}\partial^2\bar{c}/\partial x^2 + U_t^{isp}\,|\,\partial\bar{c}/\partial x\,|, \tag{73}$$

$$\partial\bar{c}/\partial t + u\partial\bar{c}/\partial x = U_t^{ss}\,|\,\partial\bar{c}/\partial x\,|, \tag{74}$$

where D_t^{eq} is the equilibrium (at $t > T_L \approx \tau_t$) turbulent diffusion coefficient $D_t^{eq} = u'^2 T_L \approx u'L = const$, $U_t^{isp} \approx 1/2\ u'Da^{1/4} = const$ and $U_t^{ss} \approx u' = const$ are the velocities of the ISP and SS flames. It is really so as in the coordinate system moving with the flame velocity U_t Eqs. (73) and (74) has the form

$$\partial\bar{c}/\partial t + u\partial\bar{c}/\partial x = D_t^{eq}\partial^2\bar{c}/\partial x^2 \text{ (a)},\ \partial\bar{c}/\partial t + u\partial\bar{c}/\partial x = d\bar{c}/dt = 0 \text{ (b)}. \tag{75}$$

Eqs. (75a) and (75b) describe the turbulent mixing layer and the steady state distribution.

Eq. (73) follows from Eq. (72) using following approximations for the scalar flux and the chemical source:

$$\overline{u'c'} = -D_t^{eq}\partial\bar{c}/\partial x \text{ (a), and } \overline{W} = U_t^{isp}\,|\,\partial\bar{c}/\partial x\,| \text{ (b)}, \tag{76}$$

so it means that here there is one-to-one correspondence between actual and model fluxes and the actual $\overline{W}$ and the "kinematical" $U_t^{isp}\,|\,\partial\bar{c}/\partial x\,|$ sources. At the same time a comparison of Eqs. (72) and (74) demonstrate that the "kinematical source" equal to

$$U_t^{ss}\,|\,\nabla\bar{c}\,| = \overline{W} - \partial\overline{u'c'}/\partial x \tag{77}$$

or using the gradient approximation for the scalar transport

$$U_t^{ss}\,|\,\nabla\bar{c}\,| = \overline{W} + D_t^{eq}\partial^2\bar{c}/\partial x^2\,. \tag{78}$$

So transformation in time of the ISP flame into SS one is accompanied by a progressive including of the transport term into the model source. A possible kinematical equation that describe transformation of the ISP flame into SS flame is the following:

$$\partial\overline{c}/\partial t + u\partial\overline{c}/\partial x = D_{ef}(t)\partial^2\overline{c}/\partial x^2 + U_t(t)\,|\,\partial\overline{c}/\partial x\,|, \qquad (79)$$

where an effective transport coefficient $D_{ef}(t)$ reduces in time and the kinematical source $U_t(t)\,|\,\partial\overline{c}/\partial x\,| = \overline{W} + (D_t^{eq} - D_{ef}(t))\partial^2\overline{c}/\partial x^2$ contains increasing in time part of the transport term. At $t \to \infty$ we have asymptotically $D_{ef}(\infty) = 0$ and $U_t(\infty) = U_t^{ss}$. Ignoring the short initial stage corresponds to assumption that at $t = 0$ we have beginning of the ISP flame, i.e. $\overline{c}(x \le 0) = 1$ and $\overline{c}(x > 0) = 0$ (for the flame moving from left to right), $D_{ef}(0) = D_t$ and $U_t(0) = U_t^{isp}$. We in our model we added relaxation equations describing transformation in time of the ISP flame into the SS one that are in our simple case the following:

$$dD_{ef}/dt = -D_{ef}/\tau^* \text{ (a)}, \quad dU_t/dt = -(U_t - U_t^{ss})/\tau^* \text{ (b)}, \qquad (80)$$

where τ^* and U_t^{isp} are expresses for the laminar or thickened flamelet mechanisms by the Eqs. (26, 44, 56). Eqs. (80) have simple solutions

$$D_{ef}(t) = D_t^{eq}\exp(-t/\tau^*) \text{ and } U_t(t) = U_t^{ss} + (U_t^{ist} - U_t^{ss})\exp(-t/\tau^*) \text{ (b)}, \qquad (81)$$

so the combustion is described by the linear equation with time dependant coefficients:

$$\partial\overline{c}/\partial t + \vec{u}\partial\overline{c}/\partial x = D_t^{isp}\exp(-t/\tau^*)\partial^2\overline{c}/\partial x^2 + (U_t^{ss} + (U_t^{ist} - U_t^{ss})\exp(-t/\tau^*))\,|\,\partial\overline{c}/\partial x\,|. \qquad (82)$$

We describe all three stages (together with transition from ISP flame to SS one) described by the kinematical Eq, (79) and the following relaxation equations in terms of the actual turbulent diffusion coefficient and the effective coefficient in the kinematical equation $D(t)$

$$dD_{ef}/dt = -(D_t - D_t^{eq})/T_L - D_{ef}/\tau^* \text{ (a)}, \quad dD_t/dt = -(D_t - D_t^{eq})/T_L \text{ (b)}, \qquad (83)$$

and the relaxation equations in terms the turbulent flame speed of the transient flame U_t^{tr} and describing all three stages of combustion speed $U_t(t)$

$$dU_t/dt = -(U_t^{tr} - U_t^{isp})/T_L - (U_t - U_t^{ss})/\tau^* \text{ (a)}, \quad dU_t^{tr}/dt = -(U_t^{tr} - U_t^{isp})/T_L \text{ (b)}. \qquad (84)$$

Solution of Eq. (83b) at the initial data $D_t(0) = 0$ and corresponding Lagrange autocorrelation function are the following

$$D_t(t) = D_t^{eq}(1 - \exp(-t/T_L)) \text{ (a)}, \quad R_L = \exp(-t/T_L) \text{ (b)}. \qquad (85)$$

The solution for $D_{ef}(t)$ with $D_{ef}(0)=0$ follows from the set Eqs. (83) and is the following

$$D_{ef}(t)=D_t^{eq}(\tau^*/(\tau^*-T_L)(\exp(-t/\tau^*)-\exp(-1/T_L)). \quad (86)$$

White similar Eq. (84 b) at the initial data $U_t^{tr}(0)=0$ results in

$$U_t^{tr}=U_t^{isp}(1-\exp(-t/T_L)) \quad (87)$$

and the solution of the set of Eqs. (84) with $U_t\ (0)=0$ is the following:

$$U_t(t)=D_t^{isp}(\tau^*/(\tau^*-T_L))(\exp(-t/\tau^*)-\exp(-t/T_L))+U_t^{ss}(1-\exp(-t/\tau^*)). \quad (88)$$

In the coordinate system moving with this velocity $U_t(t)$ the turbulent flame represents as a turbulent mixing layer with the width (dispersion) controlling by the effective transport coefficient D_{ef}. So the dispersion of the turbulent flame using Eq. (86) is the following:

$$\sigma^2(t)=2\int_0^t D_{ef}dt=D_t^{eq}[\tau^*/(\tau^*-T_L)][\tau^*(1-\exp(-t/\tau^*))+T_L(1-\exp(-t/T_L))]. \quad (89)$$

At $t\to\infty$ Eqs. (88) and (89) result in

$$U_t(\infty)=U_t^{ss}\approx u' \text{ (a)},\ \sigma^2(\infty)=D_t^{eq}\tau^*(1+T_L/\tau^*)/(1-T_L/\tau^*)\approx 4L^2Da \text{ (b)} \quad (90)$$

In Eq. (90b) we took into account that $D_t^{ef}\approx u'L$ and $\tau^*\approx 4\tau_t Da >> T_L$, Eq. (26b). So at $t\to\infty$ the width of the steady state width is equal to $\delta_t(\infty)=\delta_t^{ss}\approx 2L\cdot Da^{1/2}$ that corresponds to independent estimation in the previous section (Eq. (51)).

Fig. 16 demonstrates (solid lines) developing of the 1D flame including short initial stage, ISP stage and then asymptotically (at much larger times) the steady state stage (1D stationary flame). Pointed typical for combustors residence time explains why steady state stage practically unattainable at realistic turbulence and chemistry. All parameters in the figure are dimensionless.

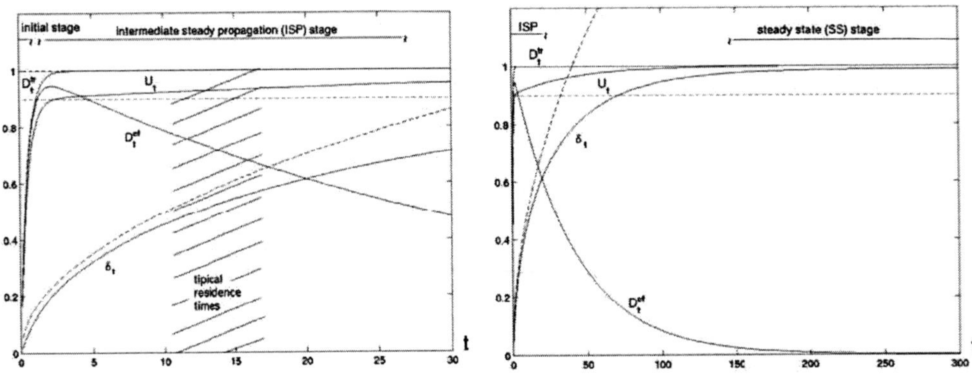

Fig. 16 Parameters of the 1D nonstationary turbulent premixed flame

Broken lines correspond to Eq. (73), which does not contain the initial and SS stages. Namely this simplified approach (instead of using full set of equations) we used in practical simulations as the initial stage in many applications are not significant and the SS stage as practically unattainable, i.e. in many cases does not need to be included in the combustion model.

5.1.2 Turbulent 3D Flame in Arbitrary (Nonuniform and Nonstationary) Turbulent Field

The kinematical equation of the ISP and SS flame in this case are as it follows:

$$\partial \bar{c} / \partial t + \vec{u} \nabla c = \nabla \cdot (D_t^{eq} \nabla \bar{c}) + U_t^{isp} \mid \nabla c \mid , \tag{91}$$

$$\partial \bar{c} / \partial t + \vec{u} \nabla \bar{c} = U_t^{ss} \mid \nabla \bar{c} \mid . \tag{92}$$

It should be particularly emphasized that in this case $U_t^{isp}(\vec{x},t)$ and $U_t^{ss}(\vec{x},t)$ are not the turbulent flame speed in strict sense. In fact $U_t^{isp}(\vec{x},t)$ and $U_t^{ss}(\vec{x},t)$ are notations for the expressions $u'Da^{1/4}$ and u' in the chemical source terms depending on the points $(\vec{x},t)$:

$$U_t^{isp}(\vec{x},t) \approx 1/2 \;\; u'(\vec{x},t)(Da(\vec{x},t))^{1/4} \text{ (a)}, \; U_t^{ss}(\vec{x},t) \cong u'(\vec{x},t) \text{ (b)}. \tag{93}$$

Namely in this sense the notations U_t^{isp} and U_t^{ss} are used below.

The general kinematical equation describing all three stages of combustion is the following

$$\partial \bar{c} / \partial t + \vec{u} \nabla c = \nabla \cdot (D_{ef} \nabla \bar{c}) + U_t \mid \nabla \bar{c} \mid , \tag{94}$$

where $D_{ef}(\vec{x},t)$ and $U_t(\vec{x},t)$ are described by a set of the relaxation equations:

$$\partial D_{ef} / \partial t + \vec{u}\nabla D_{ef} = -D_{ef} / \tau^* - (D_t - D_t^{eq}) / T_L \tag{95}$$

$$\partial D_t / \partial t + \vec{u}\nabla D_t = -(D_t - D_t^{eq}) / T_L \tag{96}$$

$$\partial U_t / \partial t + \vec{u}\nabla U_t = -(U_t - U_t^{ss}) / \tau^* - (U_t^{tr} - U_t^{isp}) / T_L \tag{97}$$

$$\partial U_t^{tr} / \partial t + \vec{u}\nabla U_t^{tr} = -(U_t^{tr} - U_t^{isp}) / T_L . \tag{98}$$

The general formulation of the 3D nonstationary problem is the following: in the initial moment $t = 0$ some part of space V_0 is filled with products $(c = 1)$ while all space out of this V_0 is filled with reactants $(c = 0)$, and $D_{ef}(\vec{x},0) = 0$, $D_t(\vec{x},0) = 0$, $U_t(\vec{x},0) = 0$, $U_t^{tr}(\vec{x},0) = 0$. The velocity $\vec{u}(\vec{x},t)$ as well as the flame speeds $U_t^{ss}(\vec{x},t) \cong u'(\vec{x},t)$, $U_t^{isp}(\vec{x},t) \approx 1/2 \; u'(\vec{x},t) \; (Da(\vec{x},t))^{1/4}$, the characteristic times $T_L(\vec{x},t) \approx 0.5\tau_t(\vec{x},t)$, $\tau^{\bullet}(\vec{x},t) \approx 4\tau_t(\vec{x},t)Da(\vec{x},t)$, the equilibrium turbulent diffusion coefficients $D_t^{eq} = (u'(\vec{x},t))^2 T_L(\vec{x},t) \approx u'(\vec{x},t)L(\vec{x},t)$ are follows from the hydrodynamical part of the problem. We see that the equations Eqs. (91-98) contains all three stages of combustion.

In connection with this formulation of the problem we make three comments:

1. In the particular case of the stationary situation ($\partial \bar{c} / \partial t = 0$) the equations, obviously, contain nonequilibrium turbulent effects, but in contrast to the particular nonstationary case (see Eqs. (82-84) where they take place in time) they are manifested along stream lines.
2. Description of the nonequilibrium effects by direct implementation in the model time-dependent functions (as it was made, for example, in [17]) is not theoretically justified: correct general model equations must be invariant so they cannot contain any time dependant coefficient.
3. We used as a first approximation the simplest relaxation equations that in fact yield Gauss transition functions. Obviously, the relaxation equations can be more sophisticated.

It should be particular emphasized that equations of this paragraph with all results presented in the Table 1 and expressions for the thickened flamelet speed, width and microturbulent transport Eqs. (19) constitute though a first approximation (as it is based on the simplest turbulence and combustion parameters u', L, $,\tau_{ch}$ and contains only the main controlling physico-chemical mechanisms of premixed combustion) but a complete theory (as it describes full pictures of the turbulent flame including all stages of combustion for both possible laminar and thickened flamelet combustion mechanisms). We stress that all exponents in Eqs. (19) and in expressions in the Table 1 are not result of some correlations of empirical data but follow from qualitative (i.e. not contain numbers) physical assumptions and further theoretical analysis based on the generalized to turbulent combustion Kolmogorov ideas of statistical small-scale equilibrium of the hydrodynamical structures, which in our case refer not only to turbulence but also to thin flamelet sheet that is treated as a random surface with developed spectrum of wrinkles. Above we attracted experimental results only to show that our qualitative assumptions correspond to observations. In the case of variable density the counter-gradient transport phenomenon appears in the equations (see subsection 5.2), but developed in [37] and presented in the subsection 3.3 approximate (containing qualitative physical assumptions) gasdynamics theory gives an opportunity to avoid appearing additional empirical constant. In fact the equations contain two empirical coefficients of the orders unite: a multiplier in the expression for the turbulent flame speed, Eq. (19a) (the coefficient A in Eq. (100)) and a multiplier in Eqs. (26) for the characteristic time boundary τ^* between ISP and SS stages). In calculations presented in Fig. 16 we put these two multipliers equal to unit. We see below that validation against experimental data in the stirred bomb gives $A \cong 1.0$. We do not know experimental data for estimation of the second multiplier in Eqs. (26) and the main reason of it that SS flames at strong turbulence and fast chemistry practically do not observed . As a consequence accurate estimation of the time τ^* practically is not necessary so assumption of the second multiplier equal to unit seems theoretically justified.

5.2 Equations for the ISP Stage in the Case of the Variable Density

For premixed combustion in turbulent flows with developed small-scale eddies combustion is completely controlled in general case by unresolved coupling of three factors: chemistry, micro-turbulent mixing and molecular transport. We assume that in our case of strong developed equilibrium small-scale turbulence the micro-turbulent transport inside thickened flamet dominates over the molecular diffusion so below we ignore the latter at estimation of the flamelet speed U_f and width δ_f. Instantaneous heat release takes place in strongly wrinkled flamelet sheet also with developed equilibrium small-scale structure of wrinkles and these small-scale wrinkles give (as we will see below) the main contribution in the flamelet sheet $\overline{S}$ and hence in the turbulent combustion velocity (the turbulent flame speed) $U_t = U_f(\overline{S}/S_0)$, while large-scale wrinkles controls the dispersion of the flamelet, i. e. the flame brush width $\delta_{t.}$.

5.2.1 RANS Equations for the ISP Stage

We analyze premixed flames at flamelet combustion mechanism in terms of the progress variable that at bimodal approximation is $\widetilde{c} = (\rho_b / \overline{\rho})P_b$, $\overline{\rho} = \rho_u P_u + \rho_b P_b$, and also $\widetilde{T} = T_u(1-\widetilde{c}) + T_b\widetilde{c}$), $\widetilde{C}_i = C_{i,u}(1-\widetilde{c}) + C_{i,b}\widetilde{c}$) where $\rho_u, T_u, C_{i,u}, P_u$ and $\rho_b, T_b, C_{i,b}, P_b$ are the density, temperature, species, probability of unburned (reactants, $\widetilde{c} = 0$) and burned (equilibrium products, $\widetilde{c} = 1$) gases. The bimodal approximation obviously does give an opportunity to predict species presented mainly in the flamelet (nonequilibrium radicals and so on).

Exact unclosed (Eq.99) and the TFC model (Eq.100) equations are as follows:

$$\frac{\partial \bar{\rho}\tilde{c}}{\partial t} + \frac{\partial \bar{\rho}\tilde{u}_i\tilde{c}}{\partial x_i} = -\frac{\partial \overline{\rho u_i'' c''}}{\partial x_i} + \bar{\rho}\tilde{W} \longleftarrow \boxed{\text{Actual source}} \quad (99)$$

GTD · CGT

$$\frac{\partial \bar{\rho}\tilde{c}}{\partial t} + \frac{\partial \bar{\rho}\tilde{u}_i\tilde{c}}{\partial x_i} = \frac{\partial}{\partial x_i}(\bar{\rho}D_t \frac{\partial \tilde{c}}{\partial x_i}) + \rho_u U_t |\nabla \tilde{c}| \longleftarrow \boxed{\text{Model source}} \quad (100)$$

where

$$U_t = \mathrm{A}(1/2)u'^{3/4} S_L \chi_u^{-1/4} L^{1/4}. \quad (101)$$

Validation against the Karpov's standard experimental data in the stirred bomb show (see Section 7 below) with good practical accuracy $\mathrm{A} \cong 1.0$ for all tested fuel including (with some limitation) even the hydrogen.

Namely Eqs. (101) and (102) where proposed in [57] and [43] for modeling of the premixed turbulent comnustion.

It is necessary to stress that conceptually Eq. (100) is a kinematical equation, which the turbulent combustion front moving with the speed U_t and having controlling by D_t increasing brush width. But strictly speaking it is so only in the 1D case, constant density and uniform and stationary turbulence. In general case variable $U_t(\vec{x},t)$ has no in strict sense physical meaning of the flame speed and it is in fact notation the expression in r.h.s. of Eq. (101) used for modeling the source.

In this approach the aim of turbulent combustion modeling is to simulate all variables of the unclosed equation (Eq. 1), i.e. fields of Favre or Reynolds averaged $\tilde{c}$, the flux $\overline{\rho u_i'' c''}$, $\bar{\rho}$, $\tilde{u}_i$, the reaction (source) term $\bar{\rho}\tilde{W} = \overline{\rho W}$ (and also $\tilde{T}$ and the main species $\tilde{C}_i$ presented in cold reactants and equilibrium products that could be expressed in terms of $\tilde{c}$).

Two main problems encountered in at physically correct modeling of the unknown terms:

- The source $\bar{\rho}\tilde{W}$ depends on both large-scale and small-scale processes as combustion takes place in thin flamelet, but their global distribution in space is controlled by large-scale eddies. Kolmogorov method is used for prediction of the model source in Eq. 2 but at transition from the model $\rho_u U_t |\nabla \tilde{c}|$ to actual $\bar{\rho}\tilde{W}$ source we meet the fundamental problem as in the general case $\overline{\rho u_i'' c''} \neq -\bar{\rho} D_t \nabla \tilde{c}$.
- The transport $\overline{\rho u_i'' c''}$ in premixed flames as a rule is counter-gradient, i.e. using the usual approximation in turbulence models of the transport term though the turbulent diffusion coefficient would be untrue. The physical reason for this that we will discuss below is that there are two opposing tendencies: the gradient turbulent diffusion (GTD) connected with random velocity pulsation and the pressure-driven

counter-gradient transport (CGT) connected with different acceleration of hot and cold volumes in the flame.

In the model kinematical equation Eq. 2 the transport term is the GTD part of $\overline{\rho u_i'' c''}$; another CGT part is included in the model source term. It means that at closure of Eq. 1 we did not approximate separately the transport and the source terms but in fact approximate the sum of them. A. Lipatnikov aptly called such approach a "joint closure", see Author's Reply in [43].

5.2.2 Joint RANS/LES Formulation of the Problem

In large eddy simulations (LES) of the premixed flamelet combustion we describe obviously not the actual instantaneous thickened flamelet sheet but an instantaneous smoothed (averaged on the subgrid level) flame sheet. The turbulence, which controls these flame characteristics, is the subgrid one and its parameters depend also on used mesh size Δ. LES have obvious advantage over the RANS modelling as the large-scale turbulence (described in our RANS simulations below by the "$k-\varepsilon$" model) as well as the counter-gradient phenomenon (describe in them by the original gasdynamics model) do not need modelling. But complete simulations of modern industrial combustors usually are expensive as usually they have complicated geometry, fuel-air mixing pre-chamber, air jets cooling the walls and so on. So in the industries the numerical modelling usually is limited to RANS simulations. At the same time some information about instantaneous picture and unsteady characteristics are necessary in many applications: unsteady combustion in gas turbine combustion, cycle variations in SI engines, connected with combustion vibrations and so on. For these cases we propose our joint RANS/LES approach.

The main idea of this approach: not LES instead of RANS simulation but using them together. Namely the first step is RANS simulation that gives the averaged picture, the second step is LES using RANS fields of the turbulent energy $\tilde{k}(\vec{x},t)$ and the dissipation $\tilde{\varepsilon}(\vec{x},t)$ for estimation of the subgrid turbulence and subgrid flame speed, the latter gives a nonstationary image corresponding to the former stationary picture.

Assuming Kolmogorov equilibrium inertial interval $E(k) = C\varepsilon^{2/3}k^{-5/3}$ we can directly estimate the subgrid turbulent velocity and scale

$$u'_\Delta \approx \int_{/\Delta}^{\infty} E(k)dk \approx \varepsilon^{1/3}\Delta^{1/3}\,,\; L_\Delta \approx \int_{/\Delta}^{\infty} k^{-1}E(k)dk \,/ \int_{/\Delta}^{\infty} E(k)dk \approx \Delta \qquad (102)$$

and hence the subgrid turbulent transport coefficients are equal to

$$D_\Delta \approx \nu_\Delta \approx \chi_\Delta \approx u'_\Delta L_\Delta \approx \varepsilon^{1/3}\Delta^{4/3} \qquad (103)$$

Using these turbulent parameters for estimation of the subgrid flame speed and applying them in Eq. (2) we have the expression for the flame speed in the model LES equation

$$U_t^\Delta = U_f(\overline{F}_\Delta / F_0) = A^* u'_\Delta S_L^{1/2} \chi_u^{-1/4} \Delta^{1/4} [\approx U_t(\Delta / L)^{1/2}] \qquad (104)$$

The model LES equation (ignoring the counter-gradient phenomenon on the subgrid level) is as it follows:

$$\partial(\overline{\rho}_\Delta \widetilde{c}_\Delta)/\partial t + \nabla \cdot (\overline{\rho}_\Delta \widetilde{u}_\Delta \widetilde{c}_\Delta) = \nabla \cdot (\overline{\rho} D_\Delta \nabla \widetilde{c}_\Delta) + \rho_u U_t^\Delta \left|\nabla \widetilde{c}\right| \qquad (105)$$

Obviously we can use expression Eq.104 of the TFC model for the subgrid flame speed in the context of traditional LES approach (LES instead of RANS), but as in this case we do not know the fields of ε and k to use in our "pure" LES simulations, the Smagorinsky model [58] that connected this turbulence with the strain-rate tensor $\widetilde{S}_{ij}$ resolved in LES and the mesh size Δ

$$u'_\Delta \approx \Delta \,|\, \widetilde{S} \,|, \; D_\Delta \approx \nu_\Delta = (c_s^\Delta \Delta)^2 \,|\, \widetilde{S} \,|, \; |\, \widetilde{S} \,| \equiv (2\widetilde{S}_{ij}\widetilde{S}_{ij})^{1/2}. \qquad (106)$$

In parallel with Eq. (104) we estimated the subgrid flame speed in LES using the same expression, as for RANS simulation, Eq. (101), with the same coefficient $\mathrm{A} = 1.0$ and had good results as effect of LES flame wrinkling was not very significant and it was compensated in fact by tuning the constant in the expression for the subgrid viscosity in Eq. (103) in joint RANS/LES and the coefficient c_s^Δ in LES, which were independent on RANS simulations. Numerical examples below correspond to namely this case.

6 Two Physico-chemical Effects Omited in the TFC Model

Our theoretical analysis does not predict two significant experimental effects:

- so called "bending effect", reduction of the flame speed U_t before combustion blow off at increasing u', Fig. 18 below;
- difference of the values of the air excess coefficients α_0 ($\alpha_0 = 1/\Phi$) of the mixtures corresponding to maximum of the turbulent U_t and laminar S_L speeds in the case of heavy and light fuels, see Fig. 17 reproduced from [59].

We connect this limitation of the model with two factors. First is that our theoretical analysis is based on the Kolmogorov ideas so it inherently does not contains pulsations of the dissipation rate and hence large instantaneous velocity gradients corresponding to large instantaneous dissipation rates. It causes large instantaneous stretch of the flamelet, which reduces of the flamelet speed and produces local extinction resulting in the bending effect and then final combustion blow off. Second we assume that transport in the thickened flamelet is controlled only by misroturbulence and in the laminar one all molecular transport coefficients are equal while it could be not like this: (i) instantaneous reaction zone in thickened flamelet is so thin ($<< \eta$ due to strong temperature dependence of chemistry) that so molecular transport influence also on the structure of instantaneous reaction zone of the thickened flamelet; (ii) the molecular transport coefficients could be different and connected with it so-called the preferential diffusion effect can be significant both for laminar and thickened flamelet. The experimental results, Fig. (17), clear demonstrate a shift of the composition corresponding to the maximum turbulent combustion rate to leaner composition hydrogen-air mixtures (b), and towards richer compositions for vaporized gasoline-air mixtures (a) in comparison with the laminar flame. In particular, under laminar conditions the burner rate of hydrogen-air mixture is greatest in a rich mixture when the excess air coefficient is $\alpha_0 \approx 0.6$, but in the turbulent flame at $u' \approx 3m/s$ it has been experimentally observed that this maximum shifts to $\alpha_0 \approx 1$. At the same time for heavy fuel the tendency was opposite.

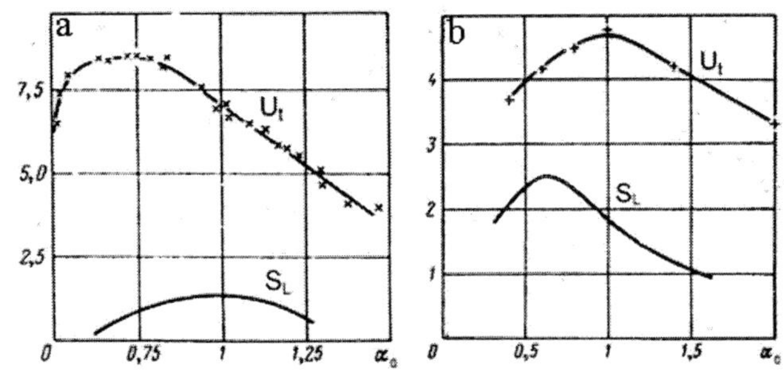

Fig.17 Turbulent and laminar flame speed in benzine-air (a) and hydrogen-air (b) mixture [59]

At validation of the model in [43] we meet this difficulty and A. Lipatnikov propose to evoke two submodels (strictly speaking these submodel referring to the laminar flamelet combustion mechanism) for description of these effects namely the Bray submodel for the stretch-effect in the turbulent field based on the log-normal PDF approximation for the dissipation pulsations [58] and the Kuznetzov-Sabelnikov theory [59] for description of the preferential diffusion effects on U_t based on the Zel'dovich concept of "leading points".

Using these submodels result in the following:

1. Additional multiplier ("stretch-factor") G in the expression for U_t, Eq. (101), which is $G \cong 1$ for moderate turbulence and decreases $G < 1$ at increasing u';
2. Variation of the mixture composition in the reaction zone from initial due to the preferential diffusion that results in variation of S_L in Eq. (101).

6.1 A Submodel of the Stretching Effect

In the submodel is assumed [28] that differently strained flamelets are reduced to either unquenched or unstrained ones, for which the absolute value of the velocity gradient g is less than some critical value g_{cr}, or highly strained flamelets, for which $g > g_{cr}$ and reactions rates are negligible, i.e. quenched flamelets. If G is the stretch-factor (the probability of unquenched flamelets), obtained by integrating the log-normal distribution of the dissipation rate ε of the turbulent energy (which is valid at high turbulence Reynolds number), then

$$G = 0.5\ erfc\{-(1/2\sigma)^{0.5}[\ln(\varepsilon_{cr}/\tilde{\varepsilon}) + \sigma/2]\}, \tag{107}$$

where $erfc$ is the complementary error function, $\sigma \cong \mu \ln(L/\eta)$ is the standard deviation of the log-normal distribution, μ is a constant, which was measured for inert flows ($\mu \approx 0.26$), the critical dissipation rate is equal to $\varepsilon_{cr} = 15\nu g_{cr}^2$ and the Kolmogorov microscale $\eta = L\,\mathrm{Re}_t^{-3/4}$. To account for the flame stretch, the model source term in Eq. (100) should be multiplied by the stretch factor $G \le 1$. For laminar and steady conditions the critical velocity gradient g_{cr} may be obtained numerically: see, for example, [60]. But difficulties emerge in using such kind results for applications. The fact is that, firstly, different steady-state problems, corresponding for instance to different geometries (single-flame or twin-flame configurations and so on) produce too broad changes of g_{cr}, and, secondly, for turbulent flow g_{cr} must be larger than in steady-state cases, as controlling g small eddies have lifetimes too short to produce quenching so actual g_{cr} is several times lager then theoretically predicted and experimentally observed in steady-state situations (see Fig. 19 below).

It seems reasonable to use in simulations a simple estimation for the critical velocity gradient obtained by the dimensional analysis $g_{cr} \approx 1/\tau_{ch}$ or

$$g_{cr} = C_{cr} S_L^2 / \chi_u \tag{108}$$

where C_{cr} is an empirical coefficient of order one. We notice that at used log-normal function the stretch effect practically significant for the mode of operation near the boundary of stable combustion that takes place, for example in modern industrial low-emission lean-premixed combustors.

6.2 A Submodel of the Preferential Diffusion Effect

Using submodel for the preferential diffusion effect is practically important in codes aimed for simulations both "light" fuels (in comparison with oxygen – e.g. hydrogen or methane) and "heavy" fuels (for example, propane, vaporized oil fuels) are employed. This effect is associated with the difference in molecular diffusion of fuel and oxygen inside curved and strained flamelets.

The "leading points" concept of premixed combustion (proposed by Zel'dovich), which can describe this preferential diffusion effect, is that turbulent flame velocity is determined by the flamelets most advanced into the unburned mixture. Further propagation of these flamelets are limited by turbulent stretching and critical conditions at such points. The problem of the critical deformation of both spherical and twin flat flames has been theoretically analyzed in [59]. It has been concluded that the mixture composition inside such flames varies owing to different fuel and oxidizer diffusivity coefficients, D_f and D_o respectively. The results derived for twin flat flames was in better agreement with the experimental data for hydrogen-air mixtures, and it has been conjectured that this result is a suitable analytical preferential diffusion submodel:

$$\alpha_{lp} = \frac{\alpha_0(1+C_{st})d + d - 1}{d + C_{st}}, \text{ if } \alpha_{lp} \geq 1; \tag{109}$$

$$\alpha_{lp} = \frac{\alpha_0(1+C_{st})d}{1+\alpha_0 C_{st} + C_{st}(1-\alpha_0 d)}, \text{ if } \alpha_{lp} \leq 1; \tag{110}$$

where $d = (D_0 / D_f)^{0.5}$, C_{st} denotes the mass stoichiometric coefficient, $\alpha = 1/\Phi$ is the air excess coefficient, α_0 and α_{lp} correspond to the initial mixture and the composition at the leading points, respectively.

Our expression for U_t , Eq. (101) contains S_L as a physico-chemical parameter of the mixture. As a result, the leading point concept may be incorporated into this closure, and the simplest way to do so is to use in Eq. (101) $S_L(\alpha_{lp})$ instead of $S_L(\alpha_0)$. Then the direct effect of the preferential diffusion is described by the parameter $S_L(\alpha_{lp})$ alone. This simple method results in reasonable agreement with measurements of mass burning rates in stirred bombs without using additional empirical parameters [43, 44].

7 Validation and Numerical Results

Every premixed combustion model can be recommended for application after testing, i.e. if it produces good agreement with some standard experimental data. Which set of standard experimental data are necessary for validation of the premixed model? This question is still open. Nevertheless it is obvious it is obvious that some one set of experiments with premixed combustion (flames in stirred bombes, or open flames, or flames behind a stabilizer in uniform flows, or impingement flames or so on) would not be sufficient. The reason is that premixed combustion is controlled several physico-chemical and hydrodynamical mechanisms that must be validated independently. So several side of the premixed model must be validated:

- Chemistry dependence of predicted flames, in particular, dependence of the flame speed U_t and width δ_t on used fuel, equivalence ratio of the mixture, and the mixture temperature and pressure, which directly influence on the chemistry, as well as U_t and δ_t dependence on turbulence (u', L, Re_t). These dependences are controlled not only by the chemical kinetics, but also in general case by physico-chemical coupling of chemistry, turbulence and molecular transport processes.

- Developing of the successive stages of the premixed flame including transformation of the scalar transport in general case from the gradient transport through neutral to the counter-gradient one.
- Consistence of used turbulence model with the combustion one especially in the light of existence of the counter-gradient transport phenomenon in the turbulent premixed flame.
- Correspondence of the based for modeling combustion mechanism with real one.
- Possibility to predict boundaries of the unsteady combustion regime (accompanied by large pressure pulsations) and combustion blow off.

Some of these questions we touched above, here we discuss in more details validation of predicted by the model general properties of the turbulent flame such as chemistry and turbulence dependence of the flame speed U_t and general aerodynamics of the flame. One of the most comprehensive correlations of many experimental data was suggested in [61]. It can be reduced for our case of molecular transfer coefficients identical for all species as it follows:

$$U_t \approx u' Da^{0.3} \, \mathrm{Re}_t^{-0.15} . \tag{111}$$

As real premixed flames are transient we should compare experimental data with theoretical predictions for the ISP flames, Eq. (101). These data together with some reliable old results are presented in the Table 2, which demonstrates that predicted theoretically and experimental exponents are very similar. It should be particular emphasized that the correlation by Bradley et al also predicts a negative exponent for the molecular transfer coefficient. We think that a lower exponent for u' in this correlation in comparison with the theoretical prediction, most probably from having included in it experimental data with significant flame-stretch (see about it below).

Table 2

Theoretical results Eq. (101)	$\sim (u')^{0.75}$	$\sim L^{0.25}$	$\sim \chi^{-0.25}$	$\sim T^{0.5*}$	$\sim p^{(0.10\div0.15)**}$
Bradley et al. correlation [59]	$\sim (u')^{0.55}$	$\sim L^{0.15}$	$\sim \chi^{-0.15}$		
Other empirical data	$\sim (u')^{0.7\div0.8}$ Many Russian works	$\sim L^{0.18} \sim L^{0.3}$ Talantov Whol, Shor 1969 [62] 1955 [63]		$\sim T^{0.4}$ Doroshenko 1956 [64]	$\sim p^{0.12}$ Khramtsov 1959 [65]

*) For $u_u \sim T_u^2$, $\chi \sim T_u^2$; **) For $\chi \sim p^{-1}$, $u_n \sim p^{-(0.2\div0.3)}$ (propane)

Comparison of the optical data presented in Fig. 7 with the result of numerical simulation in context of the TFC model, Fig. 8, performed in [6] confirms in general terms the theoretical dependence of the speed U_t on chemistry (through reasonable prediction of the dependence of the flame angle on the equivalence ratio Φ) and a possibility to simulate the transient flame with practically constant in this conditions flame speed and increasing flame brush width.

Testing how correct are predictions of both the chemistry and turbulent dependence of the flame speed and the flame brush width are the key points of every model validations. We touch these questions in more details in the next subsections and also present results of RANS and joint RANS/LES numerical simulations for different flame configurations in the context of our theoretical approach.

7.1 Validation against the Karpov's Standard Data (Spherical Flames in the Stirred Bomb)

The Karpov's data (N.N. Semenov Institute of Chemistry Physics of the Russia Academy of Science) were used for validation of the dependence of the speed U_t, Eq. (101) on chemistry and turbulent velocity pulsation. In these experiments the flame speed follows from the pressure measurements $p(t)$ in spherical stirred bombs with central spark ignition. We believe that such kind experimental data quite necessary for validation. This pressure is directly connected with the mass of burning gas and can be calculated by the combustion model equations. The relative merits of these data for validation are:

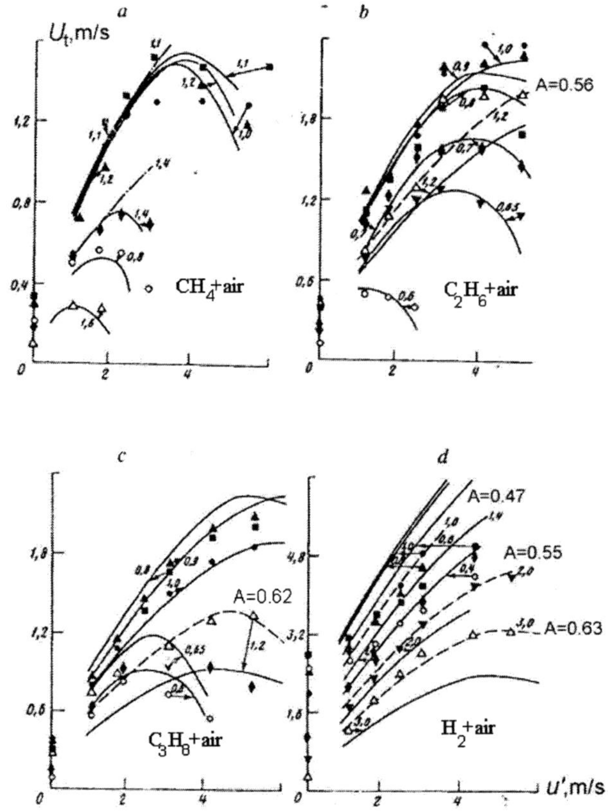

Fig.18 Dependence of U_t on turbulence, type of fuel and the air excess coefficient of the mixture α_0 denoted by numbers on the figure

- Direct determination of the burned mixture mass, therefore of the burning rate;
- Strong fan-created turbulence makes turbulent models not essential in calculations, i.e. these data test precisely
- The combustion model equations, practically without the influence of used turbulent model.
- The possibility to use for validation different fuels, diluted mixtures, special compositions, and so on.

But these data is not sufficient as real combustion takes place in shear flows, so that simulated results depend on the turbulent model used but also on possible additional mechanisms that may be significant to heat release intensity. And these data does not permit to validate a combustion aerodynamics. We pouch this problem in the next subsection.

Here we present some results of validation against the Karpov and Severin experimental data [66-68] performed in [43]. In these experiments calibrated uniform and isotropic turbulence in the central part of the bomb (where combustion was investigated) by several fans situated near the spherical wall of the bomb. During the combustion process the pressure rise was recorded with an oscilloscope. Using the pressure records $p(t)$ and the correlation between the pressure and laminar flames radii, which had been measured for this mixture previously, the evolution of the front (consumption) radius r_p was found. This radius corresponds to the spherical volume, which is equal to the volume occupied by the equilibrium and adiabatic products. The radius $r_p(t)$ looks linear during the phase of the combustion process when (in experiments) the pressure increase was $\Delta p < 0.15\ atm$ for the initial pressure $1\ atm$. The measurements results were presented in terms of the turbulent combustion rate:

$$U_t = (\rho_b / \rho_u)(dr_p / dt). \tag{112}$$

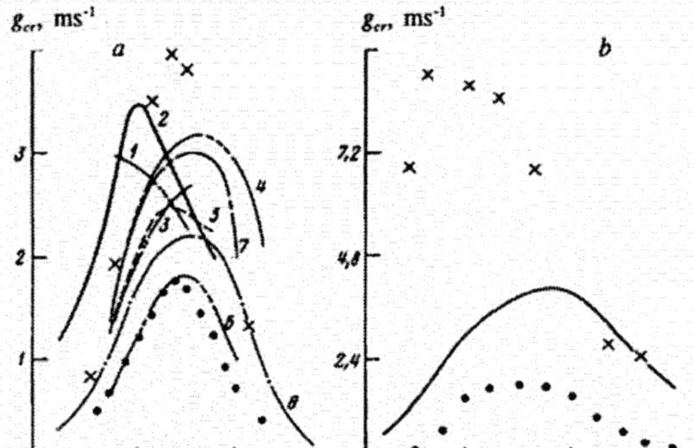

Fig.19 Comparison of the selected g_{cr}(×) parameters with the critical velocity gradient determined in studies of laminar flame extinguishment by hydrodynamic stretching; •experiment [69]; (a) CH_4. asymmetric flow config.: (1) [60], (2) [70]; simmetric config.: (3) [60], (4), (5) and (6) [71], (7) [72], (8)[70]; (b) C_3H_8 solid line,[73], (simmetric config.)

The slope dr_p / dt was the slope averaged in the range $15mm < r_p < 30mm$. The experimental data were obtained with the following initial conditions: $p = 1\ atm$, initial temperature $T = 300K$, integral turbulence scale $L = 1\ cm$, and changed of the r.m.s. velocity pulsation u' were reached by variations of the fan speed.

To simulate the measurements referred to, the set of 1D, unsteady, spherically symmetric equations was used. It includes the equations of continuity, the model combustion equation for the progress variable $\tilde{c}$ and equations of the standard "$k-\varepsilon$" turbulence model. The turbulent combustion velocities in simulations were found with the help of Eq. (112) and of the least square fit of r_p as a function of time during the phase of process when $15mm < r_p < 30mm$, so the calculation of U_t corresponded to the experimental method employed in [66-68].

The comparison between computed and measured U_t is shown in the Fig. 18. The laminar burning velocities S_L measured in the same bomb for the same fuels were used in these calculations. We remind that in [43] was used empirical coefficient A, which is connected with the coefficient A in Eq. (101) by the relation $A = 1/2\ \mathrm{A}$. So used in [43] empirical constant $A = 0.51$ used in Fig. 18 for all solid and dash-dotted curves corresponds to $\mathrm{A} = 1.02$ in Eq. (101). Dashed curves were simulated using changed values of A for better agreement with the experiment, these changed A shown on the figure. The critical velocity gradients were adjusted for each mixture based on the position of the maximum of measured $U_t(u')$ and for two fuels (methane and propane) these g_{cr} are shown in Fig. (18) together with available theoretical and experimental valises for the stationary stretched flames (comprehensive information about this validation performed in fact by A. Lipatnikov is present in [43]). We see that the theoretical predictions give in fact only the order of magnitude and they cannot be used for accurate combustion modeling. Moreover the optimal g_{cr} is larger (especially for the propane) than experimental g_{cr} for the stationary situation from [69]. We explain it by the fact that in the turbulent flame large velocity gradients have short duration and it results in increasing of observed effective g_{cr}.

When u' is moderate, the stretch effect is not significant and the simulated dependences $U_t(u')$ were weakly affected by replacement of the "stretch-factor" G by 1 (compare dashed and solid lines in Fig.18). Thus, the rising branches of the computed curves U_t were in fact affected by the constant A in Eq. (101) (practically $\mathrm{A} \cong 1.0$), which can be assumed the same for all fuels. The fact that these rising brunches, computed without any adjustment, correspond reasonably well to the experimental data for many combustible mixtures is ground to conclude that the combustion theory presented is well tested and that the model equations for premixed combustion (installed in particular in the commercial codes Fluent, CFX and StarCD) may be recommended for l applications. At the same time it should be realized that operation of industrial large-scale lean premixed burner with the thickened flamelet combustion mechanism could correspond to falling branch of the computed curves so using developed in fact for laminar flamelet Eq. (107) could be questinable. So developing the challenging problem is developing a submodel of the stretch-effect for thickened flamelet that also would result in prediction of the blow-off boundary.

7.2 Validation against the Moreau Standard Data (High Speed Combustion in the Channel)

We used the Moreau (ONERA, France) data [74] in a channel for validation of the aerodynamics of the ISP flame. These experiments is closed by parameters to the industrial burners where premixed combustion can takes place at mixture feed velocities of order $(20-50)m/s$ and even higher and combustion takes place in shear flow.

Mix	Case 1 ($\varphi = 0.84$)					Case 2 ($\varphi = 0.8$)				
	T	$\rho\,(Kg/m^3)$	$u\,(m/s)$	$u'\,(m/s)$	$L_t\,(m)$	$T\,(K)$	$\rho\,(Kg/m^3)$	$u\,(m/s)$	$u'\,(m/s)$	$L_t\,(m)$
b	2240	0.1507	108	23	0.008	2192	0.1544	120	23	0.008
u	600	0.573	65	8	0.027	600	0.562	60	8	0.027

Table 3 Inlet flow conditions for burned (b) and unburned (u)mixture.

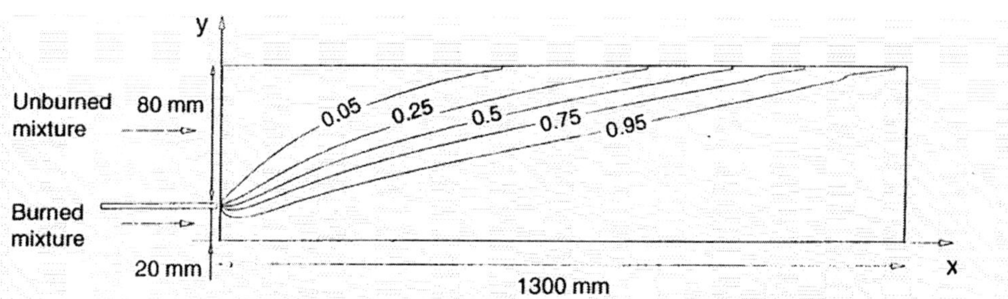

Fig. 20 Sketch of the combustion chamber and progress variable isolines (case1)

Here we present some results of performed in [73] simulations of the turbulent premixed CH_4 – air 2D flame, for two cases presented in Table 3. The combustion regime was far from quenching, so we neglected effects of the flame stretching assuming $G = 1$. The combustion model equations, Eqs. (100, 101), were used together with the standard "$k-\varepsilon$"turbulence model. The pilot flame was simulated as a burned mixture with the same equivalence ratio (and then with the same adiabatic temperature) of the reactants stream. The model is adiabatic so that heat losses were not taken in account. Plug inlet velocity profiles were used. Lacking precise indications about the inlet values of u' and L, the inlet values for k were set on the basis of the experimental profiles in transversal section at 39mm from the inlet [74]. The inlet values of the turbulent length scale (and hence ε) were tuned in order to obtain good agreement for the case 1. The same values were used for case 2.

The isoclines of the progress variable for the case 1, Fig 20, show that the flame is transient with increasing brush width. The correctness of predicted slope and width of the flame is confirmed by good agreement between the velocity and concentration profiles for the case 2 shown in Fig. (21) [75]. For completeness we demonstrate in Fig. 22 together with the temperature also experimental and calculated (using Zel'dovich thermal mechanism) profiles of NO concentrations.

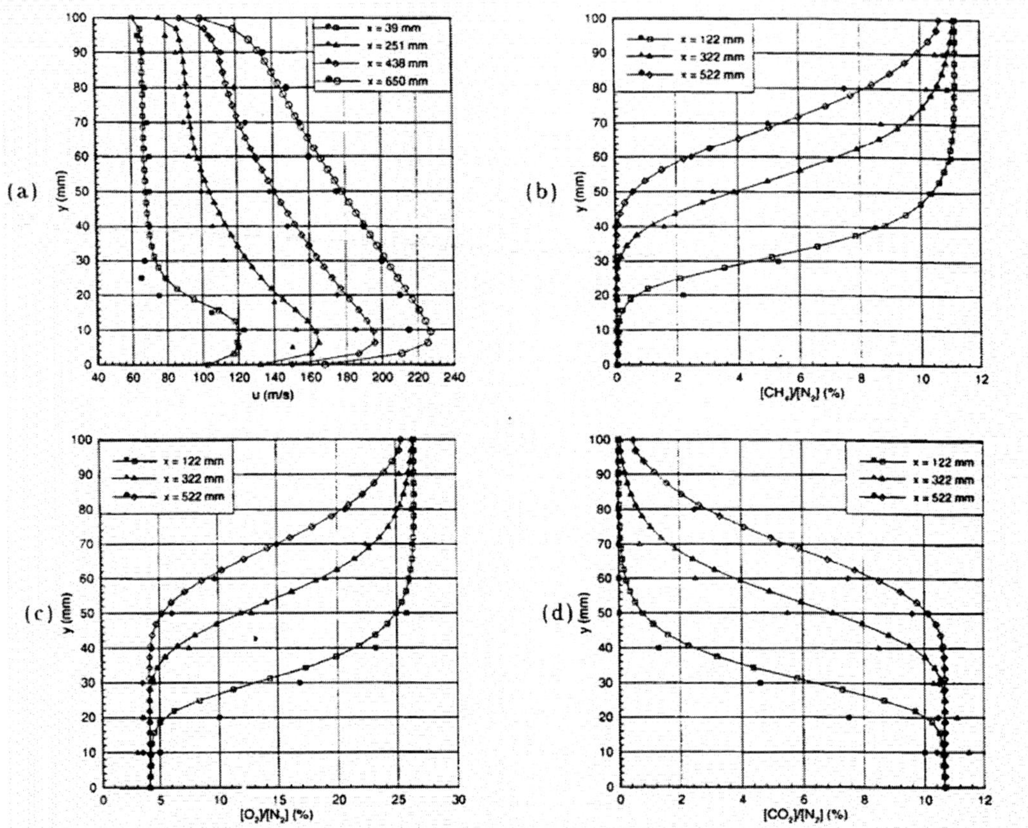

Fig. 21 Experimental (solid symbols) and numerical (outlined symbols) profiles of axial velocity and concentrations (case 1).

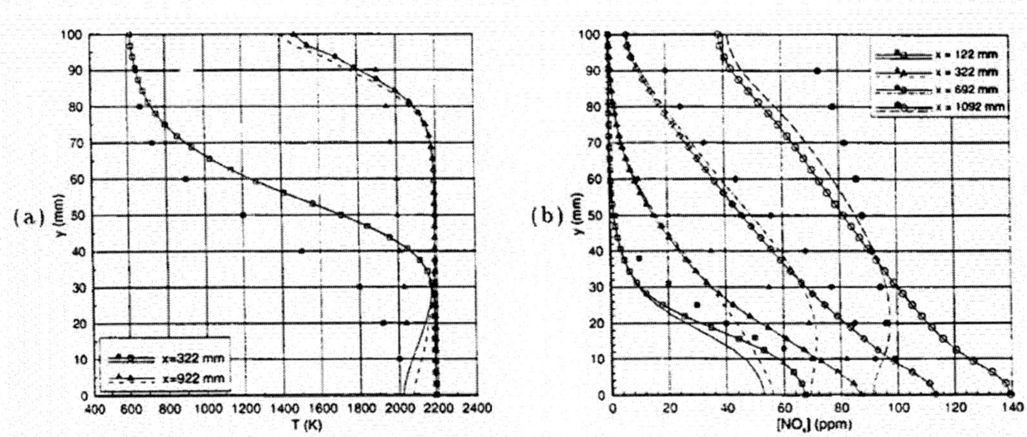

Fig. 22 Experimental (solid symbols) and numerical (1 scalar:outlined symbols: 2: scalars no symbols) profiles of temperature and NO_xconcentrations (case 2).

We see that for small values of y coordinate simulated NO concentrations are significantly larger than experimental. The reason is connected with the fact that NO production strongly depends on the temperature. Experiments in [74] showed that the temperature of the burned mixture was not $\cong 2200K$ as it follows from the thermodynamics but $\approx 2000K$. Using corrected in [75] temperatures (lines in Fig 21) yield better agreement between experimental and calculated profiles of the nitrogen oxide.

Fig. 23 demonstrate the result of joint RANS/LES simulation: a is RANS simulation and b is corresponding LES simulation: the instantaneous image of the flame and averaged LES isosurfaces and profiles of $\widetilde{c}$,, which are very similar to them from a, i.e. RANS and LES in the context of the RANS/LES approach capture well the averaged with and slope of the flame. In this test case LES results are very sensitive to inlet boundaries conditions for turbulence. In simulations we introduced a disturbance into the average inlet velocity with amplitude and length corresponding to the flow r.m.s velocity pulsation u' and the integral scale L.

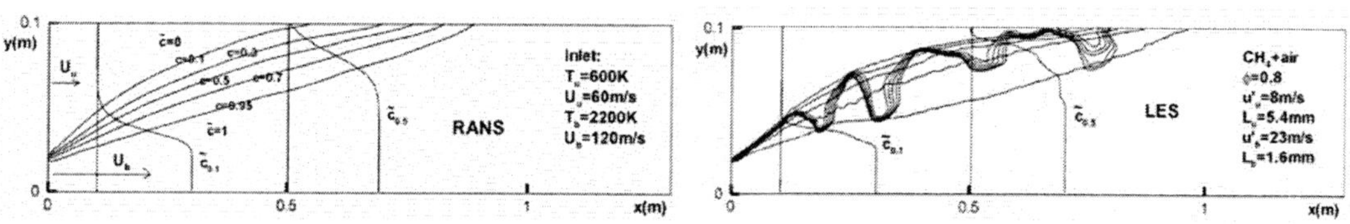

Fig.23 Joint RANS/LES modeling of the Moreau burner.

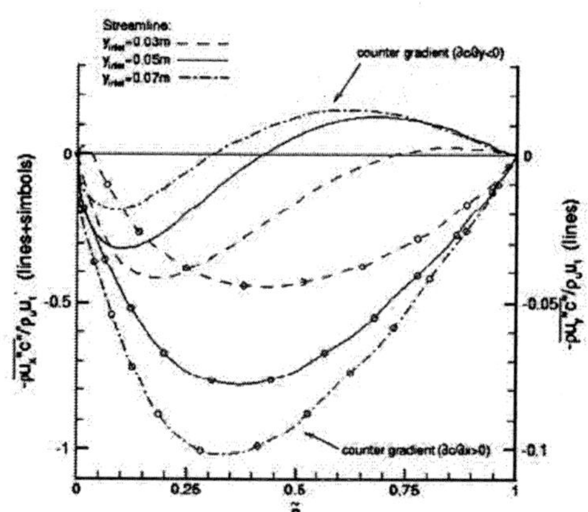

Fig.24 Lateral and longitudinal components of the scalar fluxes for the Moreau experiment (case 2).

Now we present numerical results for the scalar flux $\overline{\rho \vec{u}''c''}$. In our RANS simulations for prediction of this flux in Eq. (99) we added to the gradient turbulent diffusion flux (GTD) from Eq. (100) pressure-driven counter-gradient transport (CGT). The former was described using the "$k-\varepsilon$" turbulence model and the latter by the gasdynamical model described in the subsection 3.3. At the same time LES yields the flux $\overline{\rho \vec{u}''c''}$ directly without modeling. Fig. 24 presents two components of the scalar flux in Eq. (99) calculated on the post-processor stage by adding to the gradient turbulent diffusion component of the flux the counter-gradient gasdynamical one connected with different pressure-driven acceleration of the reactants and products in the flame. We remind that the former component follows from the "$k-\varepsilon$" turbulence model describing the field $D_t(x,y)$ and the combustion model equation Eq. (99) describing the field $\tilde{c}(x,y)$, i.e. the gradient turbulent diffusion component of the scalar transport depends on the constants of the turbulence and combustion model. At the came time our estimation of the latter component in the context of the gasdynamical interpretation of the "counter-gradient" transport phenomenon does not need any additional empirical constants. Fig. 24 show that for the lateral $y-$component of the transport takes place transition from predominantly gradient to predominantly counter-gradient one along the flame, i.e. the turbulent component dominates at the beginning and the pressure-driven one dominates at the end of the flame. At the same time Fig. 24 demonstrate large "counter-gradient" flux of the longitudinal $x-$component of the scalar flux caused by the pressure falling along the combustor. The Moreau data does not contain information about the scalar transport for comparison with simulations, but as we show in 3.2 the gasdynamical model was tested against the data in the Bunsen flames [37], Figs. 13 and 14, and also against the data in impingent flames in [40].

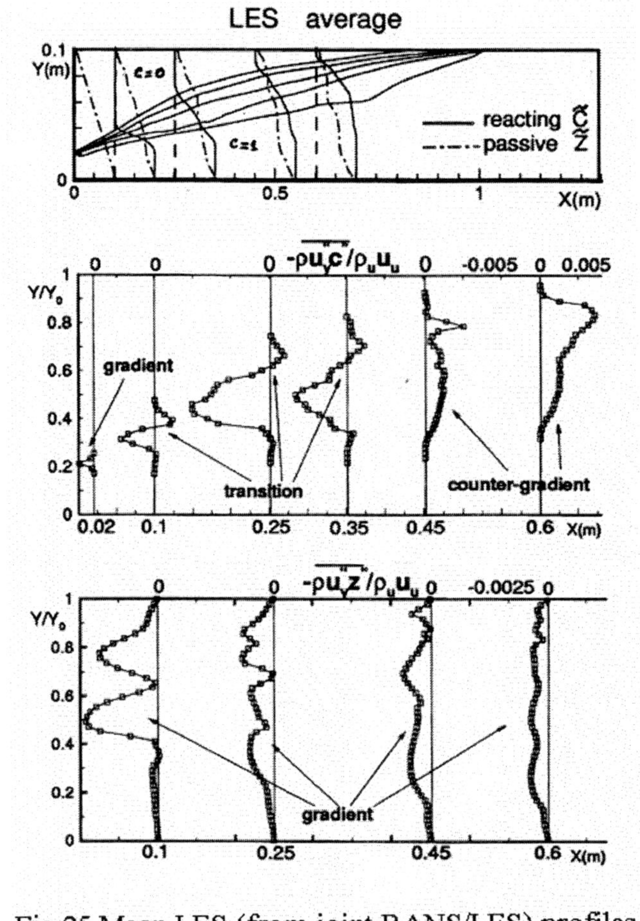

Fig.25 Mean LES (from joint RANS/LES) profiles of $\tilde{c}$ and $\tilde{z}$ and corresponding scalar fluxes.

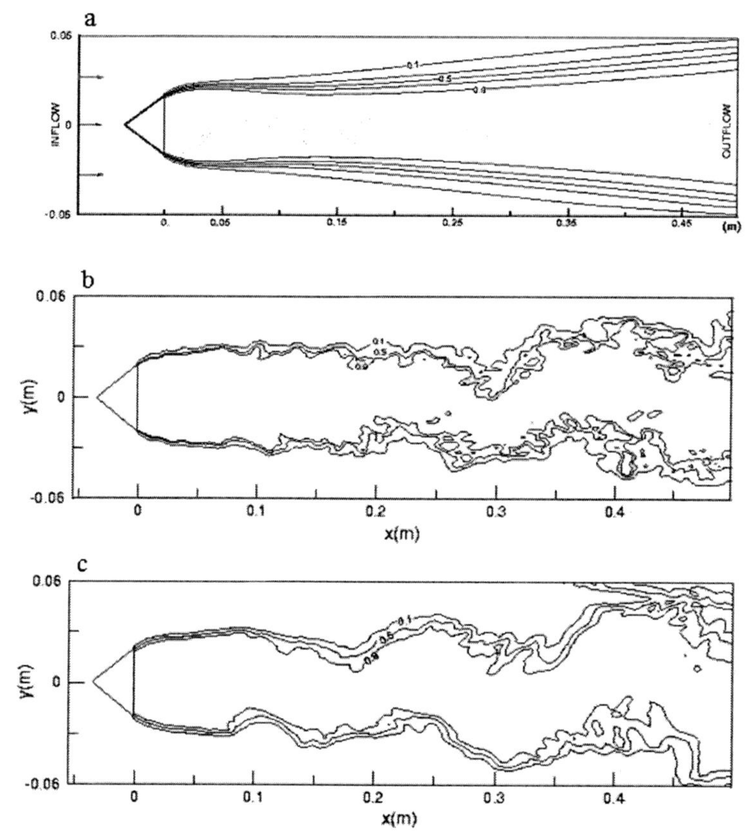

Propane-air: Φ=0.65, u=18m/s, T=288K, u'/u=3%, P=1atm

Fig.26 Contour lines of progress variable for joint RANS/LES (a), (b) and, indipendent on RANS, LES (c).

Our model for the pressure-driven counter-gradient component of the scalar flux is based on strong correlation due to gasdynamics between the progress variable c (as well as between concentrations of a fuel, O_2, H_2O, CO_2 and the temperature) and the velocities of the unburned and burn mixture u_u and u_b and the counter-gradient phenomenon takes place when this gasdynamical correlation prevails over the turbulent correlation between c and the velocities. In this case we have the counter-gradient flux of c and mentioned main species. At the same time in the context of this approach the scalar fluxes of species presented mainly in the flamelet sheet (for example, chemically nonequilibrium OH radical) are always gradient as mentioned gasdynamical part of the correlation between the concentrations and velocities is negligible. Quite similar the flux would be gradient for a nonuniform across the flame passive species as in this case also gasdynamical component could be ignored due to small correlation so the turbulent component dominates. LES of the joint RANS/LES approach, Fig. 25 confirms this conclusion directly (without modeling): we see transition of the progress variable flux and only gradient flux of the passive concentration z assumed linear $z(y) = (1-10y)m$ at $x = 0$.

7.3 Simulation of the Flame behind a Bluff Body (VOLVO Standard Data)

Analyzed Volvo standard data [76] refer to premixed flame inside a rectangular section channel anchored by a triangular section bluff body. The fuel is a propane-air mixture with $\Phi = 0.65$, the width of the channel was 120mm. The flow displays vortex shedding behind the bluff body. In this test case inlet turbulence is less relevant than in the previous case as the main generation of turbulence occurring behind the bluff body, and it make this flame is convenient for premixed combustion aerodynamics validation. Fig. 26a present an example of RANS/LES (a and b) simulation: RANS using Eq. (100) and (101) together with the standard "$k-\varepsilon$" turbulence model and corresponding LES using Eqs. (103), (105) and (101) (instead of Eq. (104), see remark at the end of the paragraph 5.2.2) and the dissipation rate ε from the RANS simulation. Fig. 26c present an example of LES independent on RANS (for comparison with Fig 4b) using the combustion equations Eq. (105) and (101) and the Smagorinsky model for the subgrid turbulence Eq. (106).

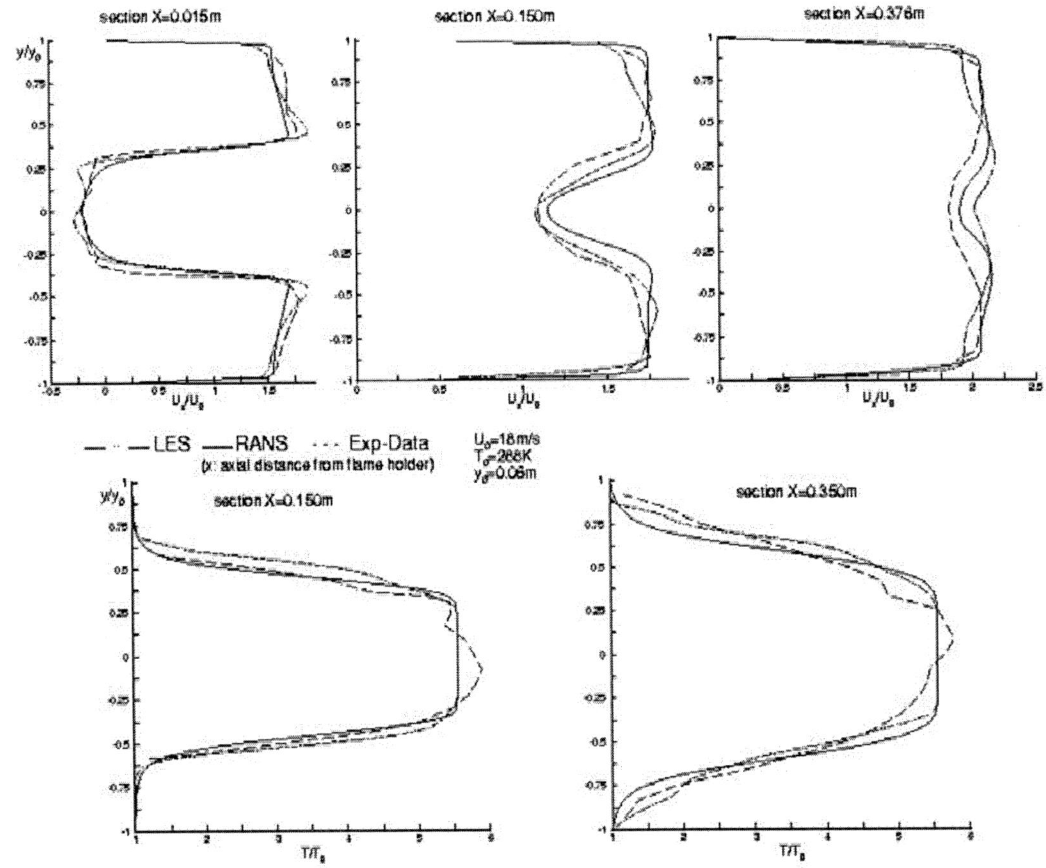

Fig.27 Comparison of the joint RANS/LES with experimental data;
X indicates distance of the sections from the cutoff of the flame holder

We notice that the joint RANS/LES allows to resolve the unsteady phenomena connected with large-scale structures starting from a RANS results more quickly, whilst the LES-Smagorinsky approach requires longer time to develop the starting turbulent field, without the possibility of using previous RANS field. To summarize, the joint RANS/LES presents practical advantages in numerical simulations respect to the standard LES model. The hypothesis of small equilibrium requires a spatial resolution of the mesh larger the required for other subgrid model, this gives the possibility to use a coarser mesh. At the same time the definitions of subgrid parameters do not take into account pulsation; it ensures convergence characteristics with time step compatible with the time scales of the reacting flow, not forcing to use very low time step due to stability of the numerical schemes. Fig.27 shows comparison of the experimental data for the velocity and temperature with RANS simulation and LES in the context of the joint RANS/LES approach. We see that extent of the recirculation zone is predicted also with reasonable accuracy as shown by the profile of axial velocity at sections at 15mm and 150mm. Fig. 28 shows the transport term for progress variable in y-direction $\overline{\rho u''_y c''}$ calculated directly during LES (from RANS/LES) at different section behind the bluff body (x is the distance from the rear of the bluff body. The flux in y-direction undergoes transition from the gradient to counter gradient flux. As for the previous case, the reason for this is that at relatively small width δ_t the gradient turbulent diffusion prevails on the gasdynamics effect while the situation is reversed downstream where the width δ_t becomes larger.

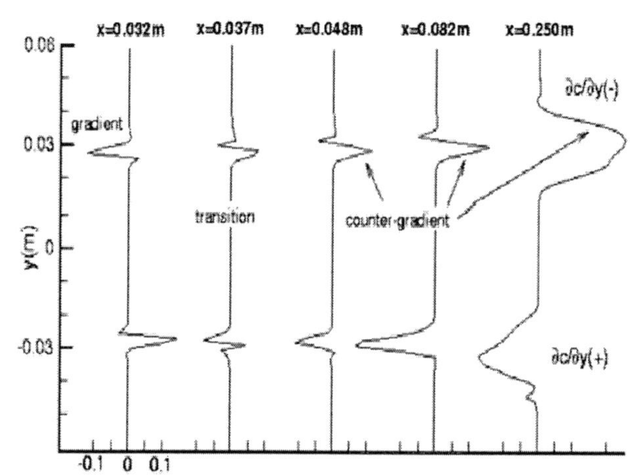

Fig.28 Transition from gradient to counter-gradient of the flux in y-direction of the progress variable

7.4 Impingent Flames and the ISP Flame Concept

Previous TFC-RANS simulations were based on the model of the transient ISP flames and LES modeling in the context of our joint RANS/LES approach and traditional LES modeling directly (without a model) demonstrate existence of this transient regime of combustion.

Demonstrated here TFC-RANS simulations and joint RANS /LES numerical modeling of the impinging flame appear to show that whilst displaying some characteristics of 1D stationary flames, these are in fact also transient ISP flame with corresponding dependent of U_t on chemistry and increasing width due to turbulent diffusion δ_t that is compensated by the divergent flow.

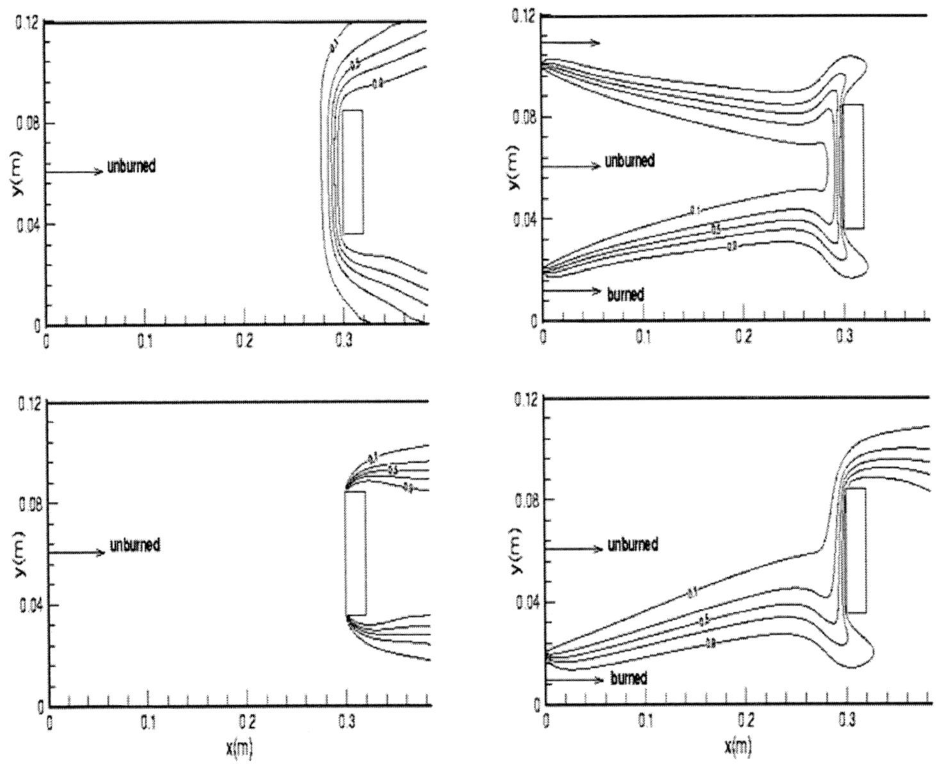

Fig.30 Different configurations of the progress variable simulated with RANS/TFC

Fig. 29 shows results of simulations in the context of the ISP flame using Eq. (100) and (101) together with the "$k-\varepsilon$" model for different configurations of impinging flames characterized by flow and mixture parameters as in the ONERA combustor (T=300K; U=60m/s; p=10^5Pa; $\square_u$=1.1Kg/m^3; $\square_b$=0.157kg/m^3). We notice that the problem of existence of the impingent flame for these conditions is not significant for us as our aim is to show the flame geometry of the ISP flame in the specific for the impingent flame divergent flow. Fig. 29a and 29b demonstrate two possible configurations of the flame, which depend on initial conditions for $\widetilde{c}$ at $t=0$ in iteration solution of the equations (in initial moment $\widetilde{c}=1$ everywhere for Fig. 29a and behind the section of the obstacle front for fig. 29b). Fig. 29c and 29d demonstrate interaction of one and two transient ISP flames with the obstacle.

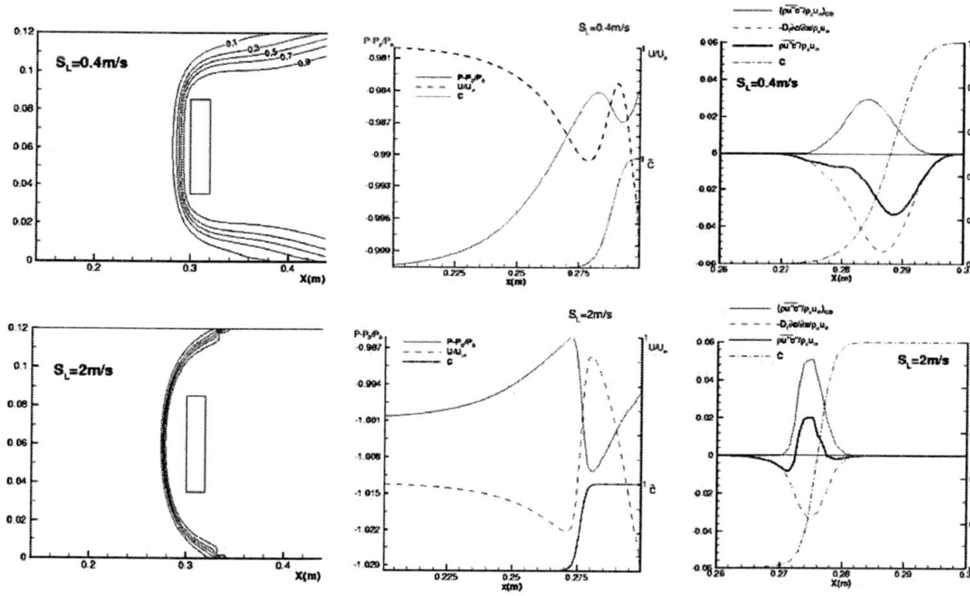

P_{ref}=1atm, T=300K, U_{in}=60m/s, u'=8m/s, L=0.0054m, ρ_u/ρ_b=7

Fig.31 Flame configuration of the 2D plane impingent flames simulated as the transient ISP one (a); parameters (b) and fluxes (c) along the plane of simmetry.

Fig. 31 demonstrates two configurations of the impingent flame for different mixtures characterized by different S_L. We see that for less reactive mixture with $S_L = 0.4m/s$ the flame is situated closer to the obstacle in comparison with the second mixture with $S_L = 2m/s$. It results in that the pressure falling across the flame is suppressed by the obstacle to a greater extend in the first case than in the second and it results in the gradient flux in the former and the counter-gradient (excepting the beginning of the flame) in the latter flame. Some results on validation of the scalar flux prediction in the context of the TFC model against known experimental data for impinging flames are presented in [40].

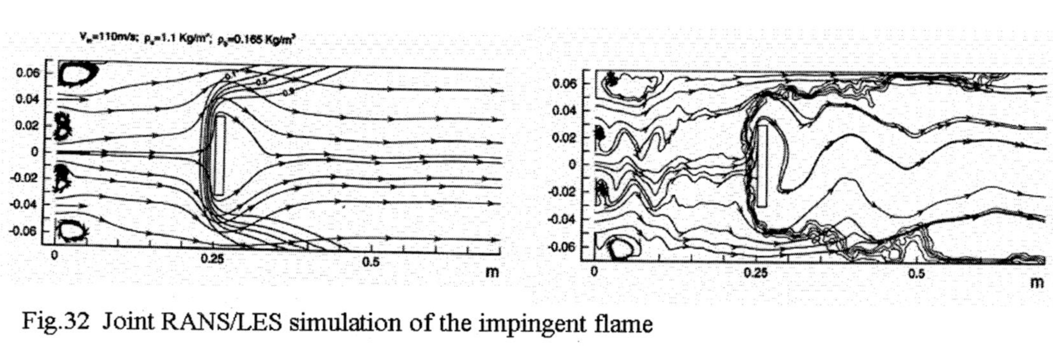

Fig.32 Joint RANS/LES simulation of the impingent flame

Fig. 32 presents as an example joint RANS/LES modeling of the isosurfaces and streamlines: RANS in (a) and corresponding LES in (b). The simulations perform beginning from the large mesh turbulization grid at $x = 0$ that gives an opportunity to avoid setting k

and ε of the turbulent flow for RANS and modeling of the random velocity field for LES parts of the problem.

8 About the Problem of the Partially Premixed Combustion

1. Partially premixed combustion is the most complicated case for modeling as it contains elements both of premixed and non-premixed combustion. The problem connected with small-scale coupling in non-reserved reaction zones refers also to this case. We propose here a generalization combustion model for the case of partially premixed combustion at strong turbulence and fast chemistry that is described by PDF equations in terms of PDF of a passive concentration and the conditional averaged progress variable. These equations are in fact a result of "joint closure" of the exact unclosed PDF equations for the arbitrary combustible system.

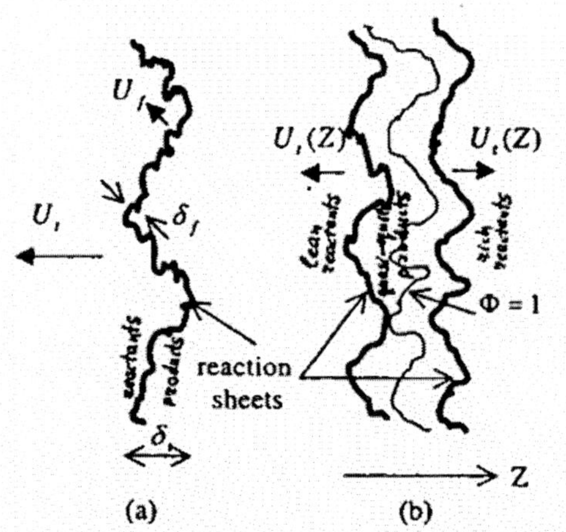

Fig.33 Premixed (a) and partially premixed (b) flames.

In these equations the component of the combustion rate controlling by premixed combustion mechanism is modeled in the context of the TFC model (i. e. assuming hydrodynamics equilibrium of the small-scale structures of the reaction zones) while the component of the combustion rate controlling by non-premixed combustion mechanism is model in assumption of the diffusion mechanism (combustion rate that is controlled by the molecular mixing is expressed in terms of the dissipation rate of the passive concentration.

This reaction zone structure is in fact very similar to the Bilger concept of quasi-equilibrium distributed reaction (QEDR) at non-premixed combustion [77]. In accordance with the QEDR concept the instantaneous reaction zone at non-premixed combustion is more complicated than in the case of perfectly premixed mixture, Fig. 33a and consists of a layer of quasi-equilibrium products in the vicinity of the stoichiometric surface, and a rich and a lean flamelet-like sheets with non-equilibrium chemical reactions inside of them that separate the equilibrium products and frozen mixture, Fig. 33b. The heat release takes place in a narrow vicinity of the stoichiometric surface and it is controlled mainly by mixing (molecular dissipation) as $\tau_{ch} << \tau_{mix}$, in flamelet-like sheet where $\tau_{ch} \sim \tau_{mix}$ reactions are at non-equilibrium, in the frozen mixture where $\tau_{ch} >> \tau_{mix}$ heat release is negligible. At mixing reach and lean reactants flows these wrapping rich and lean sheets propagate through frozen rich and lean mixtures. We assumed that this traveling reaction zones are close to wrinkled and thickened premixed flamelet sheet and could be described by similar formulas in terms of the progress variable that changes from zero (frozen mixture) to unity (quasi-equilibrium mixture). For the extreme case of non-premixed combustion their instantaneous position is controlled by the instantaneous dissipation ($\tau_{ch} \sim \tau_{mix}$).

2. Now we present these equations. At identical molecular transport coefficients $D = \chi$ we can write $C_f = C_f(Z,T)$, $C_o = C_o(Z,T)$, $C_p = C_p(Z,T)$ and $\rho = \rho(Z,T)$ where C_f, C_o and C_p are fuel, oxygen and products concentrations, $Z = C_f + C_p/(1+L_0)$ is the passive concentration, L_0 is the mass stoichiometric coefficient. The general problem reduces to determining the join probability function $p(Z,T)$. The unclosed equation for $p(Z,T,\vec{x},t)$ is as follows [21]:

$$\frac{\partial \rho p}{\partial t} + \frac{\partial \rho p (\overline{u}_k)_{ZT}}{\partial x_k} - \frac{\partial}{\partial x_k}(\rho D \frac{\partial p}{\partial x_k}) + \frac{\partial^2 \rho p \overline{\varepsilon}_T}{\partial T^2} + 2\frac{\partial^2 \rho p \overline{\varepsilon}_{ZT}}{\partial T\, \partial Z} + \frac{\partial^2 \rho p \overline{\varepsilon}_Z}{\partial Z^2} + \frac{q}{c_p}\frac{\partial^2 \rho p W}{\partial T} = 0\,. \quad (112)$$

Here $(\overline{u}_k)_{ZT}$ is the conditional average velocity, $\overline{\varepsilon}_T = \overline{(D(\partial T/\partial x_k)^2)_T}$, $\overline{\varepsilon}_Z = \overline{(D(\partial Z/\partial x_k)^2)_Z}$, $\overline{\varepsilon}_{ZT} = \overline{(D(\partial Z/\partial x_k\, \partial T/\partial x_k)_Z}$ are the conditional average dissipation rates, W is the combustion rate. (The PDF equation in this form at $\rho = const$ was derived using characteristic function technology [59] by V. A.Sabel'nikov in 1977, but was not published.)

It cannot be too highly stressed that in spite of the closed form of the source term chemistry fundamentally cannot be introduced in PDF model equations in closed form as "dissipations" (except $\overline{\varepsilon}_Z$) also directly depend on chemistry.[8] The flamelet mechanism simplifies the problem as in every point we can use instead of the joint PDF $p(Z,T)$ the PDF $p(Z)$ and the conditional probabilities of unburned and burned gases $P_u(Z)$ and $P_b(Z)$ ($P_u(Z) + P_b(Z) = 1$). Hence the conditional averaged and averaged densities are $\overline{\rho}_Z = \rho_u(Z)\, P_u(Z) + \rho_b(Z) P_b(Z)$ and case premixed ISP flames travel through lean and rich reactance and between them takes place diffusion combustion, Fig.33(b))

[8] The statement that the PDF approach gives an opportunity to introduce chemistry in combustion model equations exactly using directly chemical kinetics formulas is valid only at limiting case $Da \to 0$, i. e. in the case of infinitely slow chemistry. In this case there is no coupling between chemistry and turbulence and the dissipation rates is expressed (assuming small-scale equilibrium) in terms of large-scale turbulent parameters.

The generalized TFC model equations for partially premixed combustion (when both premixed and non-premixed combustion mechanisms are presented) in terms of the passive concentration PDF $p(Z,\vec{x},t)$ and the conditional progress variable $\widetilde{c}_Z(\vec{x},t)$ are as it follows:

$$\frac{\partial(\overline{\rho}p(Z,\vec{x},t))}{\partial t}+\frac{\partial(\overline{\rho}\widetilde{u}_k p(Z,\vec{x},t))}{\partial x_k}=\frac{\partial}{\partial x_k}(\overline{\rho}D_t\frac{\partial p(Z,\vec{x},t)}{\partial x_k})-\frac{\partial^2(\overline{\rho}_Z\overline{\varepsilon}_Z p(Z,\vec{x},t))}{\partial Z^2}, \quad (113)$$

$$\frac{\partial(\overline{\rho}\widetilde{c}_Z p)}{\partial t}+\frac{\partial(\overline{\rho}\widetilde{c}_Z\widetilde{u}_k p)}{\partial x_k}=\frac{\partial}{\partial x_k}(\overline{\rho}D_t\frac{\partial\widetilde{c}_Z p}{\partial x_k})-\frac{\partial^2(\overline{\rho}_Z\widetilde{c}_Z p\overline{\varepsilon}_Z)}{\partial Z^2}+p\rho_u(Z)U_t(Z)|\nabla(\widetilde{c}_Z)|, \quad (114)$$

The first term in the left-hand side of Eq.(113) is turbulent diffusion (in this equation the transport is always gradient) while the transport term of Eq.(114) is (quite similar to Eq.(100)) only the turbulent component of the unclosed transport term and the pressure-driven part is included in the model source term. This means that Eq.(114) (and the model) describes ISP flames. In the model equations only the last term in Eq.(114) the model source term depends directly on chemistry. But this term, described at local Z by the expression (101), depends also on turbulence and molecular diffusivity and it contains the counter-gradient component of transport. At $p(Z)=\delta(Z-Z_0)$ (premixed system) Eq.(114) transforms to Eq.(100) with the source term Eq.(101), i. e. the premixed combustion model is a degenerate case of this more general combustion model. Eqs. (113) - (114) correspond to "dispersed" combustion, when the combustion zone is a superposition of independent "conditional" premixed flames (i. e. corresponding to different Z) moving with different combustion velocities $U_t(Z)$. In the case of the chemically equilibrium products (non-premixed combustion controlling by the passive concentration dissipation) we must put at simulation $\widetilde{c}_Z=1$ as $Z=Z_{st}$ (the stoichiometric composition). This approach first formulated in [78] in Russian and in [79] in English. In paper [78] we simulated combustion at mixing rich H_2+air jet with air using presumed method for $p(Z)$ for another limit case: strong interaction between "conditional" flames, when $\widetilde{c}$ does not depend on Z.

9 Conclusiuon

In the chapter was developed first approximation but complete theory of turbulent premixed combustion at developed turbulence $\mathrm{Re}_t >> 1$ and fast chemistry $Da>1$ embracing both laminar and thickened (miscoturbulent) flamelet combustion mechanisms and describing the stages of the turbulent premixed flame that is the following:

(i) initial relatively short ($\tau_i \le \tau_t = L/u'$) stage of forming developed strongly wrinkled laminar or thickened flamelet sheet;

(ii) intermediate (between initial and final) stage with specific for both laminar or thickened flamelet sheet transient flame with nearly steady speed U_t and controlling by turbulent diffusion increasing brush width δ_t (so in fact this flame conceptually is running mixing layer), we called this transient stage (flame) the intermediate steady propagation (ISP) stage (flame) and discriminate initial transient and this transient stages (flames);

(iii) final steady state (SS) stage when flame has constant speed and width, this flame in fact (in contrast to ISP one) is conceptually the running wave, an example of the SS fame is the classical 1D stationary turbulent flame.

Controlling parameters in this theory are two turbulent characteristics: r.m.s. velocity pulsation u' and integral scale L (or the turbulence energy $k = 1.5u'^2$ and the dissipation rate $\varepsilon \approx u'^3 / L$), the kinematical coefficient of the molecular viscosity ν and the characteristic combustion time τ_{ch}. We estimated this time as $\tau_{ch} = \chi / S_L^2$, i.e. in fact we assume equality of the chemical times for the laminar and thickened flamelet sheets. For equal molecular transfer coefficients ($\chi = \nu = D$) we have four controlling parameters of the combustion problem both for the laminar and thickened flamelets namely two large-scale turbulent characteristics u', L and two physico-chemical characteristics of the mixture η, S_L.

Strictly speaking the ratio of reactant and product densities $\gamma = \rho_u / \rho_b$ should be between these parameters but for used theory structure it is not necessary. We deduce all algebraic exponential expressions, which describe the thickened flamelet speed and width, sheet, an effect of the flamelet stretch by turbulence, the flamelet sheet mean area and turbulent flame speed and width, for the hypothetical case of the constant density. Resulting theoretical characteristics of the turbulent flame used directly for designing of the model equations (hence the name of the model "Turbulent Flame Closure" or TFC one) for both cases of constant and variable density and in the latter case the equations contain the ratio γ as a controlling parameters.

The model equations are formulated in terms of the progress variable c, which in fact is the concentration of a generalized reacting species, i.e. the chemical part of the problem is reduces to one reaction with the kinetic characterizing by the chemical time τ_{ch}. This simplification seems justified in the first approximation premixed combustion model as the main mechanisms controlling the chemistry dependence and general picture of the premixed combustion are caused by both large-scale and small-scale turbulence.

The theory contains two key accomplishments and namely they made it possible to develop the complete theory. These accomplishments are the following:

1. It gives the first approximation answer the challenge of the of turbulent combustion (impossibility to resolve by balance equations the small-scale coupling of turbulence and chemistry and hence apparent impossibility to develop a theoretically justified combustion theory in the context the balance equations). We have shown that in

context of developed by us the Kolmogorov ideas of small-scale statistical equilibrium small-scale of the hydrodynamic structures in turbulent flows for turbulent reacting premixed flows combustion, the small-scale coupling of turbulence and chemistry can be expressed in terms of the large-scale turbulence parameters and this result is theoretically justified at least in the fist approximation.

2. Developed approach to describe without empirical constant the counter-gradient transport phenomenon in the context of the gasdynamic nature of it (due to different acceleration of heavy reactants and light products by the generated by the flame pressure gradient). It besides gave an opportunity to combine deduced combustion equations with a two-parametric model in spite of predominantly counter-gradient scalar flux and observed in the experiments transition from the gradient to counter-gradient one as it is used only for modeling of the turbulent component of the scalar transport in the turbulent premixed flame.

In many practical applications the initial stage is not significant and the final one practically unattainable. So presented in the chapter results of numerical simulations refer to the ISP flames and based on one differential combustion $\widetilde{c}-$ equation using "joint closure" when the sum of the actual scalar flux the actual chemical source are approximated by the sum of the turbulent component of the scalar transport and the model source expressed in terms of the ISP flame speed. The developed gasdynamic relations for pressure-driven counter-gradient component of the scalar transport gave an opportunity to restore the actual scalar transport and the actual chemical source.

We paid special attention to the impingent flames and demonstrate results of numerical simulations of them including prediction the counter-gradient scalar transport in the context of the ISP flame where the impingent flames look like 1D steady state flame due to the convergent flow field but is not it by his physical properties (the flame speed dependence on chemistry, much less flame width and so on).

We stress that interpretation of actual flame as the running turbulent mixing layer with nearly constant speed and independent on chemistry brush width controlling by turbulent diffusion (in contrast to interpretations of it as the running combustion front by analogy with laminar premixed combustion (see, for example, [80]) is a key point of our approach to premixed combustion modeling at developed turbulence and fast chemistry.

In conclusion we discuss the problem of the partially premixed combustion that needs to invoke PDF equations. In the PDF equations the chemical sources have the closed form but in our case $Da > 1$ unknown dissipation terms also strongly depend on chemistry and gasdynamic pressure-driven transport component could be significant in the transport (strictly speaking only in the case $Da << 1$ the PDF equations in general contains the chemical kinetics in the closed form, i.e. to treat the dissipation terms as pure turbulent is theoretically justified as well we is justify to ignore completely the gasdynamic pressure-driving contribution in the transport terms). We proposed model PDF equations in terms of the reacting progress variable and a passive concentration based on the joint closure in the context our combustion theory.

References

[1] Pope, S., *Proc. Combust. Inst.* **23** (1990) 591-612.
[2] Bray, K.N.C., *Proc. Combust. Inst.* **26** (1996) 1-26.
[3] Damköhler, G., *NACA Tech. Memo.* **1112** (1947).
[4] Shchelkin, K.I., *NACA Tech. Memo.* **1110** (1947).
[5] Kobayashi, H., Tamura, T., Maruta, K., Niioka, T., Williams, F., *Proc. Combust. Inst.* **26** (1996) 389-396.
[6] Dinkelacker, F., Hölzler, S., *Combust. Sci. and Tech.*, Vol.158 (2000) 321-340.
[7] Moss, J. B., *Combust.Sci. Technol.* **22,** p. 115 (1980).
[8] Frank, J.H., Kalt, P.A.M., and Bilger, R.W., *Combust. Flame* **116** (1999) 220-232.
[9] Chen, Y., and Mansour, M.S., *Proc. Combust. Inst.* **27** (1996) 811-818.
[10] Kolmogorov, A. N., Doklady *AN. SSSR* **30** (1941) 299-30.
[11] Kolmogorov, A. N., Izvestia Academy of Sciences, USSR; *Physics*,**6** (1, 2) (1942) 56-58.
[12] Spalding, D.B., Proc. *Royal Soc. London A* **434** (1991) 211-216.
[13] Spalding, D.B., In: "Turbulence and Stochastic Processes: Kolmogorov's ideas 50 years on", London, The Royal Society, 1991.
[14] Taylor, G. I., *Proc. London Math. Soc. (2),* **20**, pp. 196-211 (1921).
[15] Batchelor, G. K., *Proc. Roy. Soc. Lond.*, **A213,** pp. 349-366 (1952).
[16] Zimont, V.L., *Experimental Thermal and Fluent Science* **21** (2000) 179-186.
[17] Lipatnikov, A., Chomiak J., *Progress in Energy and Combustion Science* **28** (2002)1-74.
[18] Shchelkin, K.I. and Troshin, Ya.K. Gasdynamics of Combustion, *Mono Book Corp.*, 1965.
[19] Zimont, V.L., *Combust. Expl. and Shock Waves* **15** (1979) 305-311.
[20] Monin, A.S. and Yaglom, A.M. *Statistical Fluid Mechanics: Mechanics of turbulence.* Volume II. Edited by John Lumley. The Mit Press (1987).
[21] Zel'dovich, Ya.B., Barenblatt, G.I., Librovich, V.B., and Machviladze, G.M. The mathematical *Theory of Combustion and Explosions*, Plenum Publishing Corporation, New York (1985).
[22] Buschmann, A., Dinkelacker, F., Schafer, T., and Wolfrum, J. *Twenty-Sixth Symposium (International) on Combustion* , The Combustion Institute, Pittsburgh, 1996, pp. 437-445.
[23] Soika, A., Dinkelacker, F., Leipertz, A. *Twenty-Seventh Symposium (International) on Combustion,* The Combustion Institute, Pittsburgh, 1998, pp. 785-792.
[24] Chen, Y-C., Peters, N., Scheemann, G.A., Wruck, N., Renz, U., Mansure, M.S. *Combustion and flame*, **107**: 223-244 (1996).
[25] Chen, Y-C., Mansour M.S. *Twenty-Seventh Symposium (International) on Combustion*, The Combustion Institute, Pittsburgh, 1998, pp. 811-818.
[26] O'Young, F., and Bilger, R.W. *Combustion and Flame*, **109**, (1997) pp. 682-700.
[27] Lipatnikov, A., Chomiak J., *Progress in Energy and Combustion Science* **31** (2005)1-73.
[28] Bray, K.N.C., In: *Complex Chemical Reaction Systems. Mathematical Modelling and Simulation*, Springer, Heidelberg (1987), p. 356.
[29] Zimont, V.L., and Lipatnikov, A.N. *Chem.Phys. Reports*, 1995, Vol. 14(7). pp. 993-1025.

[30] Prudnikov, A.G., Izvestia Academy of Sciences, USSR; *Engineering*, No.1 (1960) 43-54.
[31] Prudnikov, A.G., Volinskii, M.S., and Sagalovich, V.N., *Mixing Processes and Combustion in Jet Engines,* Mashinistroenie, Moscow (1971) (In Russian).
[32] Prudnikov, A.G., *Seventh Symposium (International) on Combustion*, The Combustion Institute, 1959, pp. 575-582.
[33] Dinkelacker, F., Hölzler, S., *Combust. Sci. and Tech*., Vol.158 (2000) 321-340.
[34] Kobayashi, H., Tamura, T., Maruta, K., Niioka, T., Williams, F., Twenty-Sixth Symposium (International) on Combustion, *The Combustion Institute*, 1996, 1996, pp. 389-396.
[35] Starr, V.P., *Physics of negative viscosity phenomenon*, Mc.Graw-Hill Book Company, New York, 1968.
[36] Pope, S.B., *Ann. Rev. Fluid Mech*, **19**, p. 237, 1987.
[37] Zimont, V.L., and Biagioli, F. *Combust. Theory Modelling*, **6** (2002) pp. 79-101.
[38] Bray, K.N.C., and Moss, J.B., *Acta Astronaut*., **4**, 291, 1977.
[39] Moss, J.B., Combust. *Sci. Technol*. **22**:115-129, 1980.
[40] Biagioli F., and Zimont V., Twenty-Nine Symposium (International) on Combustion, the Combustion Institute, (2002) 2087-2096.
[41] Frank, J. H.; Kalt, P.A.M.., Bilger, R.W. *Combustion and Flame*, vol. 116, 1999, pp. 220-232.
[42] Bray, K.N.C., Libby, P.A., Masuya, G., and Moss, J.B., *Combust. Sci. Technol.* **25** (1981), p. 127.
[43] Zimont, V.L., and Lipatnikov, A.N. *Chem. Phys. Reports*, Vol. 14(7), 1995, pp. 993-1025.
[44] Karpov, V., Lipatnikov, A., and Zimont, V., *Twenty-Sixth Symposium (International) on Combustion*, The Combustion Institute, Pittsburgh, 1996, pp. 249-257.
[45] Monin, A.S. and Yaglom, A.M. *Statistical Fluid Mechanics: Mechanics of turbulence*. Volume I. Edited by John Lumley. The Mit Press (1987).
[46] Zimont, V.L., and Sabel'nikov, V.A. Izvestiya Academy of Sciences USSR, *Atmospheric and Oceanic Physics*, Vol. 1, No. 6, (1975), pp. 385-386.
[47] Morse, P. M., and Feshbach. *Methods of Theoretical Physics*, Chap.6. McGraw-Hill, 1953.
[48] Hasegawa, T., and Himeno, R., *RIKEN Rev*. No. **30**, p.16 (2000).
[49] Nishiki, S., Hasegawa, T., and Himeno, R., *RIKEN Rev*. No. **40**, p.12 (2001).
[50] Lipatnikov, A.N. and Chomiak, J. *SAE Transactions*, Vol. 106, Section 3, *Journal of Engines*, pp. 2441-2452, 1997 (SAE Paper 972993).
[51] Barenblatt, G.I., and Zel'dovich, Ja.B., *Ann. Rev. Fluid Mech*., **4**, 285 (1972).
[52] Barenblatt, G. I., "Scaling, Self-Similarity, and Intermediate Asymptotics", Cambridge University Press, 1996
[53] Borghi, R., In: Recent Advances in Aerospace Sciences, Pergamon, Oxford, 1985, pp. 117-138.
[54] L. Kovasznay, L., *Jet Propulsion*, **26**, No. 6, 485 (1956).
[55] Magnussen, B.F., and Hjertager, B.H., *Sixteenth Symposium (International) on Combustion*, The Combustion Institute, 1996, Pittsburgh, 1977, pp.719-729.
[56] Golovitchev, V.I., SAE_NA Paper 2001-01-002, *Internal Combustion Engines, 5th International Conference*, September 23-27 2001, Capri-Italy.

[57] Zimont, V.L., and Lipatnikov, A.N., Doklady *AN. SSSR* **332**(5), 1993, pp. 592-594 (Translation in: Doklady Physical Chemistry, *DKPCAG* **332**(4/5/6/ (1993) pp. 411-472, ISSN0012-5016).

[58] Smagorinsky J., *Mon. Weather Rev.* **91** (1963) 99.

[59] Kuznetsov, V.R., and Sabel'nikov, V.A., *Turbulence and Combustion*, Hemisphere, 1990.

[60] Rogg, B. *Combust. and Flame*, **73** (1988), pp. 45-65.

[61] Bradley, D., Lau, A.K.C., and Lawer, M., *Phil. Trans. Royal Soc., A*, **338** (1992) pp. 359-387.

[62] Talantov, A.V., *Foundation of the combustion theory*, Kazan KAI, 1975 (in Russian).

[63] Whol, K., and Shor, I., *Ind. Eng. Chem.*, **47** (1955), pp. 828-834.

[64] Doroshenko, V.E., and Nikitskii, A.I. Trudi Moskov. *Aviats. Inst.* No. **282**, Moscow (1956).

[65] Khramtsov, V.A. *Seventh Symposium (International) on Combustion.* The Combustion Institute, Pittsburgh, (1959) pp. 609-614.

[66] Karpov, V.P., and Severin, E.S. *Fiz. Gerenia Vzriva* **14**, No.2, 33 (1978).

[67] Karpov, V.P., and Severin, E.S. *Fiz. Gerenia Vzriva* **16**, No.1, 45 (1980).

[68] Karpov, V.P., and Severin, E.S. *Chemical Physics of Combustion and Explosion of Heterogeneous and Gas Systems*, Chernogolovka (1977) p. 74.

[69] Law, C.K., Zhu, D.L., and Yu, G., *Twenty-Second Symposium (International) on Combustion,* The Combustion Institute, Pittsburgh, 1986, p. 1957.

[70] Lebedev, V.N., *Fiz. Gorenia Vzriva* **27**, No.2, 58 (1991).

[71] Kee, R.G., Miller, J.A. et al., *Twenty-Second Symposium (International) on Combustion,* The Combustion Institute, Pittsburgh, 1986, p. 1479.

[72] Stahl, G., and Warnatz, J., *Combustion and Flame* **85**, 285 (1991).

[73] Stahl, G., Warnatz, J., and Rogg, B., *Progr. Aeronaut. Astronaut.***113**, 195 (1988).

[74] Moreau, P., *AIAA paper* **77**/49 and *15th Aerospace Science Meeting* (1977).

[75] Maciocco, L., and Zimont, V. L. /*20-th Annual Meeting of the Italian Section of the Combustion Institute "Frantic '97"*, (1997) X-2.1 - 2.4.

[76] Sjunnesson A., Henrikson P., Lofstrom C., *Proc. 28th Joint Prop. Conf. AIAA*-**92**-3650 (1992).

[77] Bilger, R.W., *Twenty-Sixth Symposium (International) on Combustion*, The Combustion Institute, Pittsburgh, 1996, pp. 475-488.

[78] Zimont, V.L., and Mescheryakov, E.A., In: *Proceedings of International Colloquium "Structure of gas flames" ITPM SO AN USSR*, part 2, 1988, pp. 35-49 (in Russian).

[79] V. Zimont, V.L., *Proceedings of the Third Mediterranean Combustion Symposium*, June 8-13, Marrakech, Morocco, pp. 927-937, 2003

[80] Peters, N., *Turbulent combustion*, Cambridge University Press, 2000.

In: New Devel. in Combustion Research
Editor: William J. Carey, pp. 95-134
ISBN: 1-59454-826-9

Chapter 2

ANALYTICAL FORMULATION OF THE PROBLEM OF TURBULENT PREMIXED COMBUSTION AND MODELING OF THE CHEMICAL SOURCE FOR THE FLAMELET MECHANISM

V. L. Zimont
CRS4 Research Center, POLARIS, 09010 PULA(CA), Italy

Abstract

We derive an original unclosed ensemble averaged equation for the Favre averaged chemical source $\widetilde{W}$ in the progress variable balance equation ($\widetilde{c}-$ equation) describing premixed turbulent combustion. This derived $\widetilde{W}-$equation directly describing the source at the flamelet combustion mechanism can be an alternative one for designing of combustion model to existing unclosed equation for the average surface density function (SDF) Σ which uses the intermediate model presentation $\overline{\rho}\widetilde{W}=\rho_u S_L \Sigma$. The $\widetilde{W}-$ equation is of the same type as any other equation used in CFD for combustion analysis and simulations. However, it has fundamental advantage as it admits to avoid completely a mismatch of a surface averaged SDF $\Sigma-$equation and others Favre averaged CFD equations that inevitably provokes a principal difficulty in designing model equations.

For completeness of analysis we also investigated an intermediate way to describe the chemical source. It which consists in derivation of an unclosed equation for ensemble averaged SDF $\overline{\Sigma}$ which is not used in the classical SDF approach. We show that, in spite of equivalence of deduced $\widetilde{W}-$ and $\overline{\Sigma}-$ equations, the former is preferable as it has logically more simple structure typical for others equations and contains only Favre averaged variables while the latter contains both Favre and Reynolds averaged values. We also address the problem of modeling of the chemical source for premixed combustion in flamelet regime and compare capabilities of the deduced ensemble averaged $\overline{\Sigma}-$ and $\widetilde{W}-$ equations with those of the usually used $\Sigma-$equation.

In conclusion of the chapter we show that that the approaches based on the scalar G and the progress c variables are equivalents. We deduce defined on one particular isosurface kinematical $c-$equation that is completely equivalent to the classical $G-$equation and show that only infinite numbers of such equations both in terms of G or c are sufficient for deducing the Favre average moment equations. Known in the literature speculations about advantage of the scalar G over the progress variable c for resolving of the challenging problems of the turbulent premixed combustion modeling have no any theoretical justification.

1 Introduction

The aim of this chapter is to analyse some conceptual problems connected with the kinematics and moment statistical mathematical description of the premixed flames at flamelet combustion mechanism, i.e. when heat release takes place in thin and strongly wrinkled sheets moving with the speed S_L by reactants. All results below correspond to the case $S_L = const$ that can be assumed equal to the normal laminar combustion velocity. Our analysis is based on universally accepted now simplify description of combustion using instead of actual chemical species a single parameters, so called progress variable c, introduced in [1]. It is, in fact, a generalized reacting species which characterize the combustion completeness ($c = 0$ in reactants and $c = 1$ in products).

The starting equation for our statistical analysis of the turbulent flamelet premixed equations is an instantaneous kinematical $c-$equation which describes an instantaneous flamlet sheet. We will see that it yields not only well known unclosed balance Favre average (weighted by the density ensemble average) $\tilde{c}$ $-$equation, but also unclosed Favre average equations for unknown in it the scalar transport $\overline{\rho c''\vec{u}''} = \overline{\rho} < c''\vec{u}'' >_G$ and mean chemical source usually denoted in the literature as $\overline{\rho W} = \overline{\rho}\tilde{W}$ [1]. The necessity to attract these unclosed $< c''\vec{u}'' >_G$ $-$ and $\tilde{W}$ $-$equations for designing of premixed combustion models is connected with well known fundamental difficulty of direct modeling of the transport and chemical source. It is connected with predominantly counter-gradient scalar transport $\overline{\rho c''\vec{u}''}$ in the premixed flame which cannot be expresses in terms of the turbulent diffusion coefficient and with fundamental difficulty of the mean chemical source modeling. The latter is caused by the fact that the source $\tilde{W}$ strongly depends on an unresolved by model equations small-scale coupling of chemistry and turbulence (so called challenge of turbulent combustion[2]).

We deduce here missing in the literature original unclosed Favre averaged equation describing the mean chemical source ($\tilde{W}$ $-$equation) in the case of the flamelet combustion mechanism and it is one of the main results presented in the chapter. We also state for this combustion mechanism an equation in terms of the scalar transport that is in fact a particular case of known general unclosed scalar transport equation.

Particular attention in this chapter is given to alternative approaches based on using equations which are defined on one-particular isosurface inside the flamelet sheet. They are

---

[1] We use two notations for the Favre averaging namely $\tilde{a}$ and $< ab >_F$.

an unclosed surface average $\Sigma-$equation used for modelling the chemical source[3] and the kinematical $G-$equation used not only for describing instantaneous combustion, but also for designing statistical combustion models[4].

The former equation is weighted by the surface conditional ($c = const$) average one in terms of the average surface density function (FSD) Σ equal to average isosurface area per unite volume. Modelling of the mean chemical source $\widetilde{W}$ is based on the presentation $\overline{\rho}\widetilde{W} = \rho_u S_L \Sigma$. In fact it together with the unclosed $\Sigma-$equation is used for designing model equations which describes the chemical source. The latter equation is used for theoretical analysis and numerical simulations of the instantaneous flamelet sheet kinematics and also for developing RANS and LES combustion model equations.

Though these equation in the case of thin and strongly wrinkled flamelet sheet are theoretically justified and mathematically correct their practical application repeatedly is accompanies by misunderstanding and following from it misconceptions so our aim is to clarify these questions.

We give here only two remarks concerning these points:

1. Defined on one-particular isosurface unclosed $\Sigma-$equation is not consistent with others Favre average unclosed equations of the problem. So using of it for modelling instead of the unclosed Favre average $\widetilde{W}-$ equation yields a know fundamental difficulty in designing of simultaneous model equations. We analyse peculiarities of following model equations and show that they are inevitably in contradiction with some invariance principle.
2. There is a believe that using the scalar G variable instead of the progress variable c has advantage for prediction of the instantaneous combustion as well as for statistical modelling. We show that it is not like this as all results in the context of c variable can be reformulated in the context of G variable and vice versa. In particular the kinematical $G-$equation can be reformulated in terms of c and well know challenge in modelling of the counter-gradient scalar transport and controlled by small-scale processes chemical source in model equations equally inherent to both approaches.

2 Formulation of the Problem

In this chapter we analyze some conceptual problems of the statistical moment description of the turbulent premixed combustion with the flamelet combustion mechanism. The starting point of this analysis is deducing of exact unclosed equations. The premixed flame is described in terms of the progress variable c by a set of first moment unclosed combustion and hydrodynamic equations that are the following:

$$\partial \overline{\rho}\widetilde{c} / \partial t + \nabla \cdot \overline{\rho}\widetilde{\vec{u}}\widetilde{c} = -\nabla \cdot \overline{\rho} < \vec{u}''c'' >_F + \overline{\rho}\widetilde{W}, \tag{1}$$

$$\partial \overline{\rho} / \partial t + \nabla \cdot \overline{\rho}\widetilde{\vec{u}} = 0 \text{ (a)}, \ \partial \overline{\rho}\widetilde{\vec{u}} / \partial t + \nabla \cdot \overline{\rho}\widetilde{\vec{u}}\widetilde{\vec{u}} = -\nabla \overline{p} - \nabla \cdot \overline{\rho} < \vec{u}''\vec{u}'' >_F \text{ (b)}. \tag{2}$$

The progress variable is defined as $c = (\rho_u / \rho - 1)/(\rho_u / \rho_b - 1)$ where ρ_u and ρ_b are the densities of unburned mixture (reactants) and burned gas (products). The set includes the balance combustion $\tilde{c}$ – equation Eq. 1, the hydrodynamic continuity Eq. (2a) and moment Eq. (2b) equations. Eqs. (1) and (2) are Favre (weighted by the density ensemble) average equations. This averaging is generally accepted now instead of the classical Reynolds ensemble averaging as it gives an opportunity to reduce the number of unknown terms. We use here two notation for the Favre averaging: $\tilde{a} = \overline{\rho a} / \overline{\rho}$ and $< ab >_F = \overline{\rho ab} / \overline{\rho}$. Eqs. (1) and (2) are exact up to omitted molecular terms (the average molecular diffusion in Eq. (1) and molecular viscosity stresses in Eq. (2b) as at strong turbulence they are small in comparison with the mean scalar flux $\overline{\rho} < \vec{u}''c'' >_F = \overline{\rho \vec{u}''c''}$ an stress $\overline{\rho} < \vec{u}''\vec{u}'' >_F = \overline{\rho \vec{u}''\vec{u}''}$. At the same time the average chemical source $\overline{\rho}\tilde{W}$ depends on the molecular transport forming together with the chemical kinetics the flamelet structure.

Using of the progress variable c strongly simplify the problem as instead of many equations in terms of actual reacting chemical species concentration we have only single $\tilde{c}$ – equation Eq. (1). In this case average equations of state and heat energy also are not necessary as the mean density $\overline{\rho}$ is expressed in terms of $\tilde{c}$ as it follows:

$$\overline{\rho} = \rho_u / (1 + \tilde{c}(\rho_u / \rho_b - 1)). \tag{3}$$

Eq. (3) follows from the expression for the progress variable taking into account that $\overline{\rho} = \overline{\rho c} / \tilde{c}$.

We stress that this set of unclosed equations Eqs. (1) - (3) is valid irrespective of particular combustion mechanism. Nevertheless we analyze here only the flamelet combustion mechanism when heat release takes place in thin and strongly wrinkled instantaneous flame, so called flamelet sheet. Namely for this particular combustion mechanism we deduce unclosed equations for the chemical source and the scalar transport and analyze peculiarities of modeling this mechanism using the Σ – and G – equations.

Though the temperature and actual chemical species do not appear in our analysis in the case of the flamelet combustion mechanism we can estimate in terms of $\tilde{c}$ the Favre and Reynolds average temperature T and chemical species concentrations Y_i. For this we assume chemical equilibrium of the products and ignore contribution of the flamelet sheet in these parameters (it corresponds to assumption of the bimodal PDF $p(c)$ used, for example, in the BML combustion model [5]). Corresponding expressions are as it follows:

$$\tilde{Q} = Q_u(1 - \tilde{c}) + Q_b \tilde{c}, \ \overline{Q} = [Q_u(1 - \tilde{c}) + Q_b \tilde{c}(\rho_u / \rho_b)][1 + \tilde{c}(\rho_u / \rho_b - 1)]^{-1}, \tag{4}$$

Q denotes ρ, T, Y_i, Q_u and Q_b are parameters of reactants ρ_u, T_u, Y_i^u and products ρ_b, T_b, Y_i^b. Eqs. (4) reasonably predict concentration of the main species and obviously cannot be used for accurate estimation of small concentrations of species presented mainly inside the flamelet such as nonequilibrium radicals and other active particles or significant in application atmospheric contaminants NO_x and CO connected with actual nonequilibrium of the products and so on.

For practical simulation of $\tilde{c}, \tilde{\vec{u}}, \bar{\rho}, \bar{p}$ Eqs. (1) and (2) must be closed by developing some models which would express unknown $<\vec{u}''c''>_F$, $\tilde{W}$ and $<\vec{u}''\vec{u}''>_F$ in terms of known functions described by Eqs. (1) and (2). In our case of thin flamelet sheet the speed of an isosurface with arbitrary $c = c^*$ and corresponding $\rho = \rho^*$ can be put equal to $S_L^{c*} = (\rho_u / \rho_{c*}) S_L$, where S_L is the flamelet speed which is traditionally defined by the reactants with the density ρ_u. At the same time in our analysis we do not assume the bimodal PDF of the progress variable, i. e. we do not assume zero width of the flamelet. It permits to deduce unclosed equations in terms of the scalar flux $<\vec{u}''c''>_F$ and the chemical source $\tilde{W}$ without transferring our main mathematical analysis in the field of the generalized functions. [2]

Necessity to state unclosed Favre average equations in terms of the chemical source $\bar{\rho}\tilde{W}$, scalar flux $\overline{\rho\vec{u}''c''}$ and turbulent stress $\overline{\rho\vec{u}''\vec{u}''}$ for the flamelet combustion mechanism that would be a base for designing of adequate combustion and turbulent model equations. It is connected with the fact that (i) the scalar transport in the general case cannot be approximated in terms of the turbulent diffusion coefficient due to predominantly counter-gradient transport phenomenon connected with different gasdynamical acceleration of cold reactants and hot products in the flame, (ii) the chemical source is depended on the space distribution of the flamelet sheet controlling by turbulence and gasdynamics, (iii) Reynolds stresses also depends on turbulence and this gasdynamical factor that results in strong nonisotropy of turbulence inside the flame.

Strictly speaking direct using of simple two-parametric turbulence models could be questionable in the case of the flamelet combustion mechanism. And though we do not touch this problem here special adequate unclosed $<\vec{u}''\vec{u}''>_F$ – equations must be developed.

We see from Eqs. (1)–(3) that combustion and turbulence problems are coupled: $\tilde{c}$ and $\bar{\rho}$ are described by by Eqs. (1) and (3) which constitute the combustion part of the general problem while the velocity $\tilde{\vec{u}}$ follows from the hydrodynamics part of the problem, Eqs. (2a) and (2b). Besides following from the hydrodynamics problem, the pressure field $\bar{p}$ strongly influences on the flux $\bar{\rho} <\vec{u}''c''>_F$ and source $\bar{\rho}\tilde{W}$ in Eq. (1) due to mention gasdynamical effect.

The most challenging is the unclosed Favre average equation for the source $\bar{\rho}\tilde{W}$ and to state this lacking $\tilde{W}$ –equation is the main aim of this paper. Another aim is the comparative

[2] The generalized functions (δ – function and Heaviside H – function) appear in our analysis only when we derive from the regular instantaneous kinematical c – equation a set of kinematical G – equations defined on particular isosurfaces inside the flamelet sheet and as a particular case the classical G – equation (Eqs. (5), (34) and (35) below).

analysis of this directly deduced Favre average $\widetilde{W}$ − equation with the current indirect approach based on an intermediate presentation of the average source in terms of mean flame surface density (FSD) Σ and unclosed surface average Σ − equation[3]. Inherent to this approach, conditional averaging on one-particular isosurface instead of ensemble averaging yields known fundamental difficulty at designing of model Σ − equations consistent with all other equations of the problem. We analyze the nature of this difficulty and the peculiarities of the model FSD equations that are (as we show in the paper) in contradiction with a general principle.

To gain a better understanding of our idea, we define an analytical (in terms of the moments) formulation of the problem of turbulent combustion that is similar to the analytical formulation of the problem of turbulence stated, for example in [6] as it follows:

"Therefore, the total infinite Friedman-Keller system of equations for all possible moments[3] gives an analytical formulation of the problem of turbulence. But this system of equations is very complicated; any finite subsystem of this system is always unclosed, i.e., it contains more unknown than there are equations in the given subsystem (the impossibility of obtaining a closed system of equations for finite number of moments is a direct consequence of the nonlinearity of the equations of the fluid mechanics). Thus when we study the applications of the Friedmann-Keller method to a finite number of moments, there arises the closure problem of the equations for the moments, there arises the closure problem of the equations for the moments."

In this context the formulation of the turbulent combustion problem is the following:

Instantaneous turbulent combustion is described by a set of instantaneous three-dimensional equations (used, in particular, in DNS) that includes: a) the hydrodynamics continuity and impulse equations; b) the combustion equations containing expressions of the chemical kinetics transport equations in terms of instantaneous chemical species concentration and the enthalpy; c) the thermodynamics equation of state. The ensemble averaging of these instantaneous equations results in the first-order moment unclosed equations in terms of $\overline{\rho}, \tilde{\vec{u}}, \widetilde{Y}_i, \widetilde{T}$ containing unknown functions describing the turbulent stress, the chemical species and heat average scalar fluxes and the mean chemical sources. The second step is the deduction of the second-order unclosed equations for all these unknown functions, yielding in turn a larger number of new unknown functions. The third step yields some third-order unclosed equations for these new unknown functions and so on. Obviously, at every step we will have a set of finite number of connected turbulent and combustion unclosed equations describing the turbulent combustion. Finally as a result of an infinitely large number of such steps we will have the infinitely large set of moment equations. Namely this total system of the infinite number of equations gives an analytical formulation of the problem. It is in principle closed, i. e. it does not need modeling. Every subsystem of this general system, corresponding to some finite number of steps, contains a larger number of unknown functions than the number of equations, i. e. it is unclosed. So here the problem of closure inevitably arises and its practical resolving is to develop a turbulence

[3] The authors mean a system results from using "a general method of obtaining (using the equation of motion of a real fluid) the differential equations for the moments of arbitrary order of fluid mechanical fields of turbulent flow" first proposed in the paper: Keller, L. V., and Fridman, A. A., *Proc. 1st Intern. Congr. Appl. Mech.*, Delft, 395-405 (1924).

model for the hydrodynamics part of the problem and a turbulent combustion models for the thermodynamics part of the problem.

It is pertinent to note here that in contrast to the transport equations constituting the combustion part of the problem the dynamical equations do not contain the chemistry kinetics (though indirectly they are coupled through dependence of the instantaneous density, composition and temperature on combustion). So, conventionally the general problem is divided into the turbulent combustion and turbulence hydrodynamics parts. It practically takes the form of independent developing of the turbulence and turbulent combustion models. We should keep in mind that these turbulence and combustion model must be consistent. We notice also that there is a fundamental difficulty in deducing of the set of unclosed hydrodynamics equations connected with the pressure in them as it is controlled by the velocity distribution in all space. So in general case we can speak about the set of infinite moment equations of the hydrodynamical part of the problem only hypothetically.

In our case of turbulent premixed combustion, the first-order system is presented by Eqs. (1) and (2). The second-order system must include unclosed differential equations describing the unknown $\overline{\rho\vec{u}''c''}$, $\overline{\rho}\widetilde{W}$ and $\overline{\rho\vec{u}''\vec{u}''}$. The model differential equations would follow from them by semi-empirical closure using some algebraic approximations of the new unknown functions in terms of known variable.

Speaking about unclosed second-order equations, the current state of the art is the following:

A. The Favre average equations describing the stress $\overline{\rho}<\vec{u}''\vec{u}''>_F$ for the case of variable density is known [7]). This general unclosed equation is valid also in the case of turbulent combustion. As we analyze here only the combustion part of the problem we do not touch this equation referring to the turbulence modeling;
B. The Favre average equation for the flux $\overline{\rho}<\vec{u}''c''>_F$ is also known and discussed in many publications (we use it below in a form presented in [8]). This general unclosed equation is valid irrespective of the specific combustion mechanism, so we present one particular form of it corresponding to the flamelet combustion mechanism analyzed here;
C. As we mentioned above, the $\widetilde{W}$ −equation necessary for the second order system has not been stated in the literature. Instead modeling of the chemical source is commonly based on the flame surface density (FSD) concept. The average FSD Σ is a local characteristic of an isosurface $c = const$ inside the flame and it is defined as the mean isosurface area per unit volume.[4] Using of this concept includes two parts[3]:

The ensemble average source is expressed in terms of the surface average FSD Σ (equal to ratio area-to-volume of one particular isosurface) using the model expression $\overline{\rho}\widetilde{W} = \rho_u S_L \Sigma$. This Σ is defined on one particular isosurface $c = const$ and the value of this c for thin flamelet sheet in fact can be arbitrary.

[4] It is pertinent to remind that application of this characteristic for for description of the turbulent mixing in nonreactive systems was suggested by Corrsin in: Corrsin, S. A measure of the area of a homogeneous ramdom surface in space, Quart. Appl. Math., **12**, 404-408 (1954).

Defined on one-particular isosurface, the unclosed surface averaged $\Sigma-$equation together with Eq.(21) describes the source $\overline{\rho}\widetilde{W}$ appearing in Eq. (1).

Though the FSD concept is physically justified and the $\Sigma-$equation is mathematically correct, the latter is inconsistent (due to surface averaging) with the other ensemble averaged equations, i. e. combination of the $\Sigma-$equation with the other Favre average equations does not yield simultaneous equations. It is the main reason of the well known fundamental difficulty in using the FSD concept for modeling. At the same time, using our unclosed $\widetilde{W}-$equation consistent with the others for designing combustion models basically free of this difficulty.

Known attempts to transform, at the stage of model designing, the surface average FSD equation into the ensemble average form (to be consistent with the other equations of the problem) forced (as we show in the paper) to use unjustified approximations and inevitably yield model $\Sigma-$equations in contradiction with the general principles. We analyze it using as an example one of the existing models.

Our statement about possibility of consistency of the FSD concept with other ensemble average equations of the problem on the stage of developing of unclosed equations is the following: it can be consistent if we attract for modeling not a single $\Sigma-$equation (corresponding to one-particular isosurface) but an infinitely large number of them corresponding to all isosurfaces from the range $0<c<1$, bringing together consistent ensemble average variables in the analysis. This infinitely large number of $\Sigma-$equations together with the corresponding densities constitute a Favre average FSD equation in the field of ensemble average functions and we also state it.

In fact modeling of the ensemble average chemical source using as a base the FSD equation inevitably results in the violation of two principles of statistical modeling. Namely it is a real reason of arising difficulty of such kind of premixed flamelet combustion modeling. In the conclusive part of the paper we formulate them as the principle of compatibility for modeling of the basic unclosed equations (that is violated when combining Favre and surface averaged equations) and the principle of invariance of the flamelet speed definition in the model equations (that is violated in the known model FSD equations).

The main results of this paper were preprinted in [9].

3 Unclosed Equations for the Chemical Source and Scalar Transport

3.1 Instantaneous Kinematical Equation for the Flamelet Sheet

We start from the kinematical equation for the progress variable c describing a flamelet sheet propagating through reactants with the speed S_L:

$$\rho\partial c/\partial t+\rho\vec{u}\nabla c=\rho_u S_L\,|\nabla c|. \qquad (5)$$

We show below that to be the same in all equations of the problem the flamelet speed inevitably must be constant. We can assume that $S_L = const$ is the speed of the normal laminar flame. In general case this speed depends on the local flamelet curvature, stretch and so on, we will touch this question later. We notice also that strictly speaking Eq.(5) represents only approximately the flamelet sheet dynamics described by the equation

$$\rho\partial c/\partial t + \rho\vec{u}\nabla c = \nabla\cdot(\rho D\nabla c) + \rho\dot{\omega}_c \tag{6}$$

where D is the molecular diffusion coefficient and $\dot{\omega}_c$ is the instantaneous chemical source for c.

Now we show that Eq. (5) yields not only the unclosed $\widetilde{c}-$ equation Eq. (1) but also an unclosed equation in terms of $\widetilde{W}$ (Eq. (14) below) directly describing the chemical source term in Eq. (1). But first we rewrite Eq. (5) in following more appropriate form:

$$\partial c/\partial t + \vec{u}\nabla c = S\,|\,\nabla c\,|. \tag{7}$$

where S is the flamelet speed by the local isosurface. Kinematics Eq. (7) describes a particular isosurface inside the flamelet sheet with $c = const$ moving with the speed $S(c) = \rho_u S_L / \rho(c)$.

Multiplying Eq. (7) by ρ and adding in the l. h. s. instantaneous continuity equation

$$\partial\rho/\partial t + \nabla\cdot\rho\vec{u} = 0 \tag{8}$$

multiplied by c, Eq. (6) is transformed to the divergent form

$$\partial(\rho c)/\partial t + \nabla\cdot(\rho\vec{u}c) = \rho W \tag{9}$$

In Eq. (9) the "kinematical" source expresses the generation rate of products (the combustion rate) due to the traveling of the flamelet sheet through the reactants. It is the following:

$$\rho W = \rho S\,|\,\nabla c\,| = \rho_u S_L\,|\,\nabla c\,|\;[=\nabla\cdot(\rho D\nabla c) + \rho\dot{\omega}_c]. \tag{10}$$

The unclosed equation for the chemical source (-equation)

The expression in square brackets in Eq. (10) demonstrates the origin of the chemical source term in the kinematical equation which is controlled both by the molecular transport and chemical kinetics. So the chemical source in the unclosed $\widetilde{c}$ - equation Eq. (1) which directly follows from Eq. (10) using the Favre decomposition $a = \widetilde{a} + a''$ and averaging is the following:

$$\bar{\rho}\widetilde{W} = \overline{\rho W} = \overline{\rho S\,|\nabla c|} = \rho_u S_L \overline{|\nabla c|}\ [= (\nabla \cdot \overline{(\rho D \nabla c)} + \overline{\rho \dot{\omega}_c})]. \qquad (11)$$

Using the obvious identity laws

$$\partial(|\nabla c(\vec{x},t)|)/\partial t = -\vec{n}\cdot\nabla(\partial c(\vec{x},t)/\partial t) \text{ and } \partial(|\nabla c(\vec{x},t)|)/\partial x_i = -\vec{n}\cdot\nabla(\partial c(\vec{x},t)/\partial x_i)$$

(where $\vec{n} = -\nabla c(\vec{x},t)/|\nabla c(\vec{x},t)|$ is the unit vector normal to an isosurface pointing towards reactants) from Eq. (5) follows the equation (we adopted the mathematics from [10])

$$\partial(|\nabla c|)/\partial t + \vec{u}\cdot\nabla(|\nabla c|) = -|\nabla c|(\vec{n}\vec{n}:\nabla u) - \vec{n}\cdot(\nabla(S|\nabla c|) \qquad (12)$$

where $\vec{n}\vec{n}:\nabla\vec{u} = n_i n_j \partial u_i/\partial x_j$. Introducing the rate of strain acting in the isosurface tangent plane $\nabla_t \cdot \vec{u} = \nabla\cdot\vec{u} - \vec{n}\vec{n}:\nabla\vec{u}$, expressing $|\nabla c|$ in terms of W from Eq. (10) and using $\rho S = \rho_u S_L$ finally results in an equation for the instantaneous reaction rate

$$\partial(\rho W)/\partial t + \nabla\cdot(\rho\vec{u}W) + \rho_u S_L \nabla\cdot(\vec{n}W) = \rho(\nabla_t\cdot\vec{u})W + \rho_u S_L(\nabla\cdot\vec{n})W. \qquad (13)$$

Ensemble averaging of Eq. (13) gives the final desired result:

$$\begin{aligned}&\partial(\bar{\rho}\widetilde{W})/\partial t + \nabla\cdot(\bar{\rho}\tilde{\vec{u}}\widetilde{W}) + \nabla\cdot(\bar{\rho} < \vec{u}''W'' >_F) + \rho_u S_L \nabla\cdot(\overline{\vec{n}W}) = \\ &\bar{\rho}\widetilde{W} < \nabla_t\cdot\vec{u} >_F + \bar{\rho} < W''(\nabla_t\cdot\vec{u})'' >_F + 2\rho_u S_L \overline{KW}\end{aligned} \qquad (14)$$

where $K = 0.5\nabla\cdot\vec{n}$ is the curvature of the local isosurface.

The l. h. s. of Eq. (14) contains one nonstationary term and three transport terms describing convection by (i) mean velocity transport, (ii) turbulent diffusion type transport due to correlation of velocity and combustion rate pulsations and (iii) transport caused by the flamelet movement.

Three r. h. s terms describe effects of different physical mechanisms (sources and sinks) controlling $\widetilde{W}$: (i) the flame stretch by the mean velocity field, (ii) the flamelet stretch by the pulsation component of the velocity, (iii) the effect of the wrinkled flamelet propagation on the sheet area.

It should be realized that the unclosed Eq. (14) does not contain all mechanisms. The reason is that it follows from pure kinematical analysis with the implicit assumption of the existance of a one-sheet smooth flamelet surface. It results that Eq. (14) does not contain at least two factors:

1. An effect of the cusps, self-intersections of isosurfaces and formation of islands of unburned gas in products. There inevitably arise also at flamelet with nonzero width and result in the reduction of the flamelet area that could yield additional sink in Eq. (14).

2. Effect of possible Darrhieus-Landau flame instability that generates mainly small-scale flamelet wrinkles increasing the sheet area that could yield additional source.

3.2 The Unclosed Equation for the Scalar Transport ($\overline{\rho \vec{u}''c''}$ – Equation) for the Flamelet Mechanism

As we mentioned in the introduction, the unclosed equation for the progress variable flux $\overline{\rho \vec{u}''c''} = \overline{\rho} < \vec{u}''c'' >_F$ is well known in the literature and is the following[8] (τ below is the tensor of the molecular stresses):

$$\begin{aligned}\partial \overline{\rho} < \vec{u}''c'' >_F / \partial t + \nabla \cdot (\overline{\rho} \tilde{\vec{u}} < \vec{u}''c'' >_F) = -\nabla \cdot (\overline{\rho} < \vec{u}''\vec{u}''c'' >_F) - \overline{\rho} < \vec{u}''\vec{u}'' >_F \nabla \tilde{c} - \\ \overline{\rho} < \vec{u}''c'' >_F \nabla \cdot \tilde{\vec{u}} - \overline{c''}\nabla \overline{p} - \overline{c''\nabla p'} + \overline{c''\nabla \cdot \tau} + \overline{\vec{u}''\nabla \cdot (\rho D(\nabla c)} + \overline{\vec{u}''\dot{\omega}_c}\end{aligned} \quad (15)$$

This equation follows from the $c -$equation Eq. (6) and the Navier-Stokes density and moment equations. In our case of flamelet combustion mechanism, the two last terms (describing molecular diffusion and the chemical source) are coupled and they control the flamelet speed S_L, see Eq. (10). Additionally in our case of thin flamelet width we can use with satisfactory accuracy the expression $\overline{c''} = (\rho_u / \rho_b - 1)\tilde{c}$ strongly valid for bimodal PDF. So we can rewrite Eq. (15) as it follows:

$$\begin{aligned}&\partial \overline{\rho} < \vec{u}''c'' >_F / \partial t + \nabla \cdot (\overline{\rho} \tilde{\vec{u}} < \vec{u}''c'' >_F) = \underset{(a)}{-\nabla \cdot (\overline{\rho} < \vec{u}''\vec{u}''c'' >_F)} \underset{(b)}{- \overline{\rho} < \vec{u}''\vec{u}'' >_F \nabla \tilde{c}} - \\ &\underset{(c)}{(\overline{\rho} < \vec{u}''c'' >_F \cdot \nabla)\tilde{\vec{u}}} \underset{(d)}{- \tilde{c}(\rho_u / \rho_b - 1)\nabla \overline{p}} \underset{(e)}{- \overline{c''\nabla p'}} \underset{(f)}{+ \overline{c''\nabla \cdot \tau}} \underset{(g)}{+ \rho_u S_L \overline{\vec{u}'' |\nabla c|}}.\end{aligned} \quad (16)$$

We also can use also for interpretation of the terms of Eq. (16) the algebraic expression for the scalar transport corresponding to the BML[5]

$$\overline{\rho \vec{u}''c''} = \overline{\rho} < \vec{u}''c'' >_F = \overline{\rho}\tilde{c}(1 - \tilde{c})(\overline{\vec{u}}_b - \overline{\vec{u}}_u), \quad (17)$$

where $\overline{\vec{u}}_b$ and $\overline{\vec{u}}_u$ are conditional mean velocities of products and reactants. For thin flamelet Eq. (17) in fact is approximate integral of Eq. (16) expressed in terms of the conditional averaged velocities. As an example we analyze a1D stationary flame perpendicular to the axis x and with velocity direction along x so $\tilde{u} > 0$, $d\tilde{c} / dx > 0$, $d\tilde{u} / dx > 0$, $d\overline{p} / dx < 0$. In this case, if $\overline{u}_b > \overline{u}_u$ we have the counter-gradient transport ($\tilde{c} -$ flux points from reactants with $\tilde{c} = 0$ toward products with $\tilde{c} = 1$). When $\overline{u}_b < \overline{u}_u$ the scalar transport has the opposite direction coincident with the normal gradient diffusion direction. Using Eq. (17) a simplified interpretation of the different terms (a) - (g) of Eq. (16) is:

(a) It is a diffusion type term (in contrast with other terms that are source or sink terms) as it only redistribute $< u''c''>_F$ across the flame;

(b) It is a sink term as $d\tilde{c} / dx > 0$ and $< u''^2 >_F > 0$;

(c) This term is a source at gradient and a sink at counter-gradient transport;

(d) This term is the source as in the flame $d\overline{p} / dx < 0$, this term describes effect of different acceleration of light products and heavy reactants by falling across the flame pressure, more strong acceleration of products over reactance contributes to the counter-gradient tendency of the transport;

(e) This term is in fact connected with the pressure fall in the instantaneous flamelet, it results in lower conditionally averaged pressure in reactance over the pressure in products $\overline{p}_b < \overline{p}_u$, and it contributes additionally to the counter-gradient tendency of the transport;

(f) This term is controlled in fact by molecular viscous dissipation inside the flamelet that additionally to heat release increases the entropy (decreases the total pressure of the products) that reduces $\overline{u}_b$ and so contributes to the gradient tendency;

(g) This term describes the direct effect of combustion. We can expect that term to be small in comparison with the terms that describe indirect combustion effects connected with the generation of instantaneous and mean pressure nonuniformity, resulting in different acceleration of reactants and products in Eq. 17.

Some DNS numerical results confirming this qualitative analysis presented in [8].

4 General Properties of the Unclosed $\tilde{W}$ – and $\overline{\rho \vec{u}''c''}$ – Equations

In this section we stress several general properties of the stated equations and a specific inherent to the flamelet combustion mechanism invariance principle they obey in contrast to known FSD model equations which we discuss in the next section.

4.1 The Flamelet Speed in the Unclosed Equations

The laminar flamelet sheet differs from the normal 1D laminar flame due to effects of the flamelet stretch and curvature in turbulent flows. So, generaly speaking, S_L is not constant. Assume that α and β are some parameters which characterize the stretch and the curvature and $S_L(\alpha, \beta)$ is a known function. So, in the general case ensemble averaged flamelet speed $\overline{S}_L$ is variable in space and time and can be presented as follows

$$\overline{S}_L(\vec{x},t) = \int S_L(\vec{x},t)\ p(\alpha,\beta,\vec{x},t)\ d\alpha\ d\beta \tag{18}$$

where $p(\alpha, \beta, \vec{x}, t)$ is the joint PDF.

This rises the question how to generalize the unclosed transport and source equations Eqs. (14) and (16) for this situation deduced in the assumption $S_L = const$. Is it possible to replace the constant S_L by a variable mean $\overline{S}_L(\vec{x},t)$ or at least a mean $\overline{S}_L = const$ in space and time? The answer is "NO" as mean $\overline{S}_L$ would be different in different equations: in the $\widetilde{W}-$ equation Eq. (14) we have two different speeds S_L^I, S_L^{II} and in the transport equation Eq. (16) the third speed $\overline{S}_L^{III}$ defined as it follows:

$$\overline{S}_L^I = \overline{S_L \nabla \cdot \vec{n} W} / \overline{\nabla \cdot \vec{n} W}\,,\ \overline{S}_L^{II} = \overline{S_L K W} / \overline{K W}\,,\ \overline{S}_L^{III} = \overline{S_L \vec{u}'' \,|\nabla c|} / \overline{\vec{u}'' \,|\nabla c|}\,.$$

Our statement is the following: *if we want to use the same flamelet speed in all the unclosed equations of the problem, it inevitably must be constant ($S_L = const$) and cannot be some averaged speed variable in the field ($\vec{x},t$) or even constant $\overline{S}_L$.*

We believe that it is more practical to introduce inconstancy of the flamelet speed on the stage of model equations designing, using some approximate theoretical or empirical correlations for the mean $\overline{S}_L$, than to try to deduce a more general unclosed equations including effects of the flamelet stretch and curvature.

4.2 Invariance of the Equations by Definition of the Flamelet Speed

S_L is the flamelet speed conventionally defined as the flame speed in reactants at $c = 0$ and $\rho = \rho_u$. But quite equally the flamelet speed could be defined by products at $c = 1$ and $\rho = \rho_b$, the value of this speed S_L^b is different ($S_L^b > S_L$). In the general case, the flamelet speed S_L^c can be defined by an arbitrary isosurface with some value of the progress variable c and density ρ_c. In all these cases, the physical content of Eqs. (14) and (16) would be the same as the speed appears in the product $\rho_u S_L$ and using other definitions of the speed, this product must be replaced by an equal product $\rho_b S_L^b$ or $\rho_c S_L^c$. The terms where S_L does not appear contain the mean density $\overline{\rho}$ that ensure the same dimension of all terms. We specially stress this point because the FSD model equations discussed in the next section (see Eq. (28)) contain S_L but do not contain neither ρ_u nor $\overline{\rho}$. It means that these equations are not invariant by definition of the flamelet speed. We explain below the consequence of this imperfection on the model equations.

We can state the following: *Obeying to the condition of invariance by definition of the flamelet speed is necessary for correctness of every equations containing this speed as a parameter.*

4.3 Agreement of the Equations with the Mass Conservation Law

Nonstationary and convective terms of Eqs. (1), (2), (14) and (16) analyzed above have the form:

$$\partial(\overline{\rho}\widetilde{Q})/\partial t + \nabla\cdot(\overline{\rho}\tilde{\vec{u}}\widetilde{Q}) \text{ (a)},\ \overline{\rho}\ (\partial\widetilde{Q}/\partial t + \tilde{\vec{u}}\ \nabla\cdot\widetilde{Q}) = \overline{\rho}\ D\widetilde{Q}/Dt \text{ (b)}, \qquad (19)$$

where $\widetilde{Q}$ can be either $\tilde{c}, \tilde{\vec{u}}, <\vec{u}''c''>_F, \widetilde{W}$ or $<\vec{u}''\vec{u}''>_F$. Expression Eq. (19a) is in agreement with the mass conservation law, it obviously contains the continuity equation. Subtraction of Eq. (2a) results the l. h. s. of Eq. (19b) which is expressed in terms of the total derivative, the r. h. s. of Eq. (19b). The total derivative denotes the rate of change of $\widetilde{Q}$ along Favre average streamline so others terms of the equations describe the effect of different mechanisms on this rate of change.

We note this typical structure of the equations as the structure of the model $\Sigma-$equations analyzed in the next section is atypical (see Eq. (28)) and does not correspondent to Eq. (19a). That is why the l. h. s. of Eq. (28) cannot be expressed in terms of the total derivative presented by the r. h. s. of Eq. (19b). It means that this model $\Sigma-$equation (as well as others) in fact do not describe exactly the rate of change of mean FSD along Favre average streamline as it takes in all other equations. Below we explain this peculiarity of model $\Sigma-$equations.

4.4 Consistency of the Ensemble Averaged Equations

The balance $\tilde{c}-$equation Eq. (1) together with the source and scalar transport equations Eqs. (14), (16) constitute the second order unclosed set of simultaneous moment equations describing the combustion part (the fields $\tilde{c}, \overline{\rho}, \overline{\rho\vec{u}''c''}, \widetilde{W}$) of the general turbulent combustion problem. The turbulent hydrodynamics part of this general problem includes Eqs. (2) and the unclosed equation for the mean stress $\overline{\rho\vec{u}''\vec{u}''}$ non-analyzed here. They also constitute simultaneous equations describing the fields $\tilde{\vec{u}}, \overline{p}, \overline{\rho\vec{u}''\vec{u}''}$. All these equations are formulated in the fields of the ensemble average function (Favre version), so when designing model equations for the source, scalar transport and mean stresses they would naturally be ensemble average.

The unclosed equations contains both Favre and Reynolds average parameters. From a conceptual point of view, it is interesting to note that Reynolds averaging could be presented as the Favre averaged terms, i. e. the problem can be formulated completely in the field of Favre average functions. Really, transformation of Reynolds average terms containing S_L in Eq. (14) gives the following: $\rho_u S_L \nabla\cdot(\overline{\vec{n}W}) = \overline{\rho} < S\nabla\cdot(\vec{n}W) >_F$ and $2\rho_u S_L \overline{KW} = 2\overline{\rho} < SKW >_F$ where S is the flamelet speed by the isosurface with the density ρ equal to $S = (\rho_u/\rho)S_L$. So we can present the Reynolds averaged terms through Favre averaged as it follows:

$$\rho_u S_L \nabla \cdot (\overline{\vec{n}W}) = \overline{\rho}\rho_u S_L < \rho^{-1}\nabla \cdot (\vec{n}W) >_F \ , \ 2\rho_u S_L \overline{KW} = 2\overline{\rho}\rho_u S_L < SKW / \rho >_F$$

and keeping in mind Eq. (3) ($\overline{\rho} = f(\widetilde{c}, \rho_u, \rho_b)$ formulate equation Eq. (14) completely in terms of the Favre average functions as well as the scalar transport equation Eq. (16) which is the following:

$$\begin{aligned}
&\partial\overline{\rho} < \vec{u}''c'' >_F / \partial t + \nabla \cdot (\overline{\rho}\widetilde{\vec{u}} < \vec{u}''c'' >_F) = -\nabla \cdot (\overline{\rho} < \vec{u}''\vec{u}''c'' >_F) - \overline{\rho} < \vec{u}''\vec{u}'' >_F \nabla\widetilde{c} - \\
&(\overline{\rho} < \vec{u}''c'' >_F \cdot \nabla)\widetilde{\vec{u}} - \widetilde{c}(\rho_u / \rho_b - 1)\nabla(\overline{\rho} < p\rho^{-1} >_F) - \overline{\rho} < c''\rho^{-1} >_F \nabla\overline{p} + \\
&\overline{\rho} < c''\rho^{-1} >_F \nabla(\overline{\rho} < p\rho^{-1} >_F) - \overline{\rho} < c''\rho^{-1}\nabla p'' >_F + \overline{\rho} < c''\rho^{-1} >_F \nabla\widetilde{\tau} + \\
&\overline{\rho} < c''\rho^{-1}\nabla\tau'' >_F + (\rho_u S_L)\overline{\rho} < \vec{u}''\rho^{-1} | \nabla(\widetilde{c} + c'') | >_F .
\end{aligned} \tag{20}$$

The unclosed $< \vec{u}''\vec{u}'' >_F$ – equation describing the stress of the impulse equation Eq. (2b) can be presented in the similar form. The Reynolds average pressure is presented as $\overline{p} = \overline{\rho} < p\rho^{-1} >_F$).

It demonstrates that in fact the unclosed problem as well as model equation can be formulated in the field of Favre average functions so we will call below our problem as defined in the field of Favre average functions.

So we can state that combination the Favre average equations of the premixed combustion problem with a FSD surface average one yields inconsistence of them and it is a main reason of the difficulty in designing of the FSD combustion models.

4.5 A Remark about the Turbulent Combustion Modeling

These two combustion and hydrodynamics subproblems are connected through common for both fields $\overline{\rho}(\vec{x},t)$ and $\widetilde{\vec{u}}(\vec{x},t)$. Predicted by the hydrodynamics subproblem, the fields $\widetilde{\vec{u}}$ and $\overline{p}$ strongly depend on controlled by combustion field $\overline{\rho}$. At the same time, nonuniform $\overline{p}$ causes a different acceleration of reactants and products resulting in the well known counter-gradient scalar transport but not only it. Different acceleration also deforms the field of mean chemical source and yields strong anisotropy of the velocity pulsation in the flame that is much larger in the direction across the flame. So designing model equations for the source, scalar transport and mean stresses must be in agreement. Strictly speaking combustion and hydrodynamics subproblems cannot be modeled independently to guarantee that model equations are not only mathematically simultaneous but also consistent from the physical point of view i.e. both submodels contain the same controlling physico-chemical mechanisms. Nevertheless combustion and turbulence modeling traditionally refer to different disciplines (turbulent combustion and turbulence theories) and are developing as a rule independently. In this paper we follow this tradition that is justified here as we analyze only a particular aspects of the combustion part of the general problem.

5 Flame Surface Density (FSD) Concept and Chemical Source Modeling

In this part we analyze the approach for modeling of the ensemble average chemical source in terms of the flame surface density and known surface average $\Sigma -$equation. In the literature the $\Sigma -$equation is used instead of the $\widetilde{W}-$ equation Eq. (14) and it yields known and in fact irresistible difficulty at designing on this base model equations for the chemical source. As we see below this difficulty forced designers of the model $\Sigma -$equations to use inadmissible approximations that not only are theoretically unjustified but moreover change the physical meaning of the unknown terms of the unclosed $\Sigma -$equation. It was necessary for designing model $\Sigma -$equations consistent with the other Favre average ones. But even with these tricks resulting model equations are inevitably ambiguous: we show that they basically depend on the chosen isosurface where they are defined.

To gain a better understanding our statements, we give short preliminaries where we first shortly formulate the nature of this basic difficulty.

5.1 The Nature of the Fundamental Difficulty of the FSD Concept for Modeling

The nature of the difficulty is connected with the following:

1. Using for modeling of the Favre average chemical source the unclosed surface averaged $\Sigma -$equation yields a system based on unclosed equations of the problem which is not any more a set of simultaneous equations. The reason is that unclosed surface average $\Sigma -$equation defined on one particular isosurface is inconsistent with other Favre average equations of the problem.
2. It should be particular emphasized that this difficulty is not connected with some fallacy of the FSD concept. It itself is physically justified and mathematically correct . The difficulty is connected exclusively with the fact that unclosed Favre average equations and unclosed $\Sigma -$equation defined on one particular isosuface are basically inconsistent. It is obvious that for practical simulations a model system must contain only simultaneous equations, i. e. it must contain only Favre average equation. So the current strategy of the FSD applying is to convert the inconsistent unclosed FSD equation into a consistent one on the stage of designing model $\Sigma -$equations, several such consistent FSD models are analyzed, for example, in [3].
3. Below we show that a conversion of the unclosed $\Sigma -$equation into a consistent model one is impossible using theoretically justified approximations. The reason of it is that the surface average unclosed $\Sigma -$equation is defined on one-particular isosurface and basically cannot contain mechanisms obligatory in a consistent with others model $\Sigma -$equation, in particular the turbulent type transport (appearing in the ensemble average $\widetilde{W}-$equation Eq. (14) as well as in other Favere average ones Eq. (1), (2) and (16)). Besides the surface average speed $<\vec{u}>_S$ (referring to one

particular isosurface) cannot basically be expressed in terms of the Favere average velocity $\tilde{\vec{u}}$ appearing in all other Favre average equations.

4. We show, using as an example one of these model, that even using theoretically unjustified approximations (cardinally changing it physical meaning of the terms and bringing in the converted Σ – equations turbulent type transport and Favre average velocity) resulting model equations contain the flamelet speed as a parameter but do not contain the density, i. e. the product $\rho_u S_L$ do not appear in them. We show that in the model Σ – equations take place violation of an invariance principle (stated below) and it yields inherent ambiguity of them.
5. We demonstrate in conclusion of the section that consistent FSD concept is possible but must contain an infinitely large number of the Σ – equations corresponding to all isosurfaces and these unclosed surface average Σ – equations are equivalent to a single unclosed Favre average $\tilde{\Sigma}$ – equation consistent with the other Favre average equations of the problem.

5.2 Exact Equations of the FSD Concept

The current surface area concept is based on model representation of the ensemble average source term of the $\tilde{c}$ – equation Eq. (1) in terms of the average surface-to-volume ratio of one-particular isosurface Σ corresponding to some $c = c_f$, i. e. $\Sigma = \Sigma_{c_f}$

$$\bar{\rho}\tilde{W} = \rho_u S_L \overline{|\nabla c|} = \rho_u S_L \int_0^1 \overline{|\nabla c|}_c p(c)dc \cong \rho_u S_L \overline{|\nabla c|}_{c=c_f} p(c_f) = \rho_u S_L \Sigma_{c_f} \quad (21)$$

as $\overline{|\nabla c|}_c p(c)$ practically does not depend on c (from a physical viewpoint it is clear as the smaller is $\overline{|\nabla c|}_c$, the greater is $p(c)$ and vice versa). As for thin flamelet Σ_{c_f} practically does not depends on chosen isosurface inside the flamelet sheet the index c_f is usually omitted.

The origin of the analyzed fundamental difficulty of the traditional FSD approach is caused by a peculiarity of Eq. (21), namely that its l. h. s. part $\bar{\rho}\tilde{W}$ is Favre ensemble averaged (it is the source term of the $\tilde{c}$ – equation) while its r. h. s. part $\rho_u S_L \Sigma_{c_f}$ expresses the source in terms of the surface averaged characteristic corresponding to one-particular isosurface. The instantaneous and averaged equations of the FSD approach in original notation are the following [11] - [14]:

$$\partial \Sigma' / \partial t + \nabla \cdot ((\vec{u} + w\vec{n})\Sigma') = (\nabla_t \cdot \vec{u} + w\nabla \cdot \vec{n})\Sigma', \quad (22)$$

$$\partial \Sigma / \partial t + \nabla \cdot (< \vec{u} + w\vec{n} >_S \Sigma) = < \nabla_t \cdot \vec{u} + w\nabla \cdot \vec{n} >_S \Sigma \quad (23)$$

$$\partial\Sigma/\partial t + \nabla\cdot\tilde{\vec{U}}(\vec{x},t)\Sigma + \nabla\cdot\langle\vec{u}''\rangle_S\Sigma + \nabla\cdot\langle w\vec{n}\rangle_S\Sigma =$$
$$\langle a_T\rangle_S\Sigma + \langle A_T\rangle_S\Sigma + \langle w\nabla\cdot\vec{n}\rangle_S\Sigma, \quad (24)$$

where $\Sigma' = |\nabla c(\vec{x},t)|\,\delta(c_f - c(\vec{x},t))$ and $\Sigma = \overline{|\nabla c|}_{c=c_f}\, p(c_f)$ (see, for example, Eq. (27) in [13]), w is the local flamelet speed (that in contrast with our local speed $S = (\rho_u/\rho)S_L$, assumed to be constant at an isosurface) can depend on the flamelet curvature, stretch and so on), $< a_T >_S = < \nabla\cdot\vec{u}'' - \vec{n}\vec{n} : \nabla\vec{u}'' >_S$ and $< A_T >_S = \nabla\cdot\tilde{\vec{U}} - < \vec{n}\vec{n} >_S : \nabla\tilde{\vec{U}}$. We remind that Eq. (24) follows from Eq. (23) using the decomposition

$$< \vec{u} >_S = \tilde{\vec{u}} + (< \vec{u} >_S - \tilde{\vec{u}}) = \tilde{\vec{u}} + < \vec{u}'' >_S$$

and namely its r. h. s. was used for designing of known FSD models.

We remind that Eqs. (22)-(24) are valid in every point $(\vec{x},t)$: Eq. (22) is instantaneous while Eqs. (23) and (24) are surface averaged and defined on one particular isosurcace $c = const$.

The expression $\bar{\rho}\tilde{W} = \rho_u S_L \Sigma$ (Eq (21)) is used usually together with Eq. (23) or Eq. (24). We attract attention to a contradiction between them. In Eq. (21) it was implicitly assumed that the flamelet speed is constant. To be consistent with Eq. (22) and (23) containing the variable flamelet speed w (depending on instantaneous stretch, curvature and so on) Eq. (21) $c^* = const$ must be replaced by the following

$$\bar{\rho}\tilde{W} = \rho_{c*} < w\Sigma' >_S^{c*} = \rho_c(< w >_S^{c*} \Sigma_{c_f} + < w\Sigma''' >_S^{c*}$$

where Σ''' is (keeping original notation: Σ' and Σ are instantaneous and mean FSD) the FSD pulsation $\Sigma''' = \Sigma' - \Sigma$. To avoid this complication we put (as in our previous analysis) $S_L = const$, i. e. $w_{c*} = S_L^{c*} = S_L(\rho_u/\rho_{c*}) = const$.[5]

For designing of FSD model is used the $\Sigma-$equation defined on one particular isosurface $c = c^* = const$. To avoid confusion we rewrite the basic equations for FSD modeling Eqs. (23) -(24) and the expression for the source Eq. (21) as it follows:

$$\partial\Sigma_{c*}/\partial t + \nabla\cdot(< \vec{u} + S_L^{c*}\vec{n} >_S^{c*} \Sigma_{c*}) = < \nabla_t\cdot\vec{u} >_S^{c*} + S_S^{c*} < \nabla\cdot\vec{n} >_S^{c*} \Sigma_{c*} \quad (25)$$

$$\partial\Sigma_{c*}/\partial t + \nabla\cdot\tilde{\vec{u}}(\vec{x},t)\Sigma_{c*} + \nabla\cdot\langle\vec{u}''\rangle_S^{c*}\Sigma_c + \nabla\cdot\langle S_S^{c*}\vec{n}\rangle_S^{c*}\Sigma_{c*} =$$
$$\langle a_T\rangle_S^{c*}\Sigma_{c*} + \langle A_T\rangle_S^{c*}\Sigma_{c*} + S_L^{c*} < \nabla\cdot\vec{n}\rangle_S^{c*}\Sigma_{c*}. \quad (26)$$

[5] We ignore a hypothetical case when w is variable but the correlation $< w\Sigma''' >_S^{c*} \equiv 0$.

$$\bar{\rho}\widetilde{W} = \rho_{c*} S_L^{c*} \Sigma_{c*} \tag{27}$$

Choosing $c*$ for modeling is not significant as for thin flamelet Σ_{c*} is practically the same for all $0 < c* < 1$. We note the specific peculiarity of Eqs. (25) (and hence of Eq. (26)) connected with the fact that while Σ_{c*} practically does not depend on the chosen isosurface at the same time $<\vec{u}>_S^{c*}, S_L^{c*}$ and so on strongly depend on $c*$. It means that there is mutual compensation of these influences to yield $\Sigma_{c*} \cong \Sigma$ where Σ is independent on $c*$.

For combustion simulation in fact namely Eq. (26) is used, written using more detailed notation in contrast with the original Eq. (24) to avoid any embarrassment in further analysis of the FSD model equations.[6]

5.3 The Problem of the FSD Modeling

The difficulty arising at designing of a model FSD equation is connected with the fact that while equation Eq. (25) based unclosed surface average FSD is inconsistent with other Favre average equations of the problem designed for practical applications model, the $\Sigma-$equation must be consistent with them, i. e. be presented in ensemble average parameters. As we mentioned above, it can be achieved only using inadmissible approximations. But even in this case, the model FSD equation is inevitably in contradiction with a general invariance principle. A consequence for modeling is discussed below.

The problem of designing is connected with the fact that though for a thin flamelet sheet Σ_{c*} practically does not depend on $c*$, all other surface average parameters have a strong dependence, for example, the variation of $<\vec{u}>_S^{c*}$ or S_L^{c*} across thin flamelet at real combustion could be by a factor 5-7 times. For current modeling, the surface averaged parameters are approximated with ensemble averaged values and this dependence on $c*$ is ignored. It is, as we will see below, a compensating error that is inevitably necessary for transforming the exact unclosed surface averaged equation Eq. (26) into a model ensemble averaged equation:

1. The expression $\langle\vec{u}''\rangle_S^{c*}\Sigma_{c*}$ in the third r. h. s. term of Eq. (26) has the physical meaning of the difference between the actual $<\vec{u}>_S^{c*}\Sigma_{c*}$ and dummy $\tilde{\vec{u}}\Sigma_{c*}$ convective transport. For the bimodal PDF $\overline{\vec{u}''}$ could be expressed in terms of the progress variable flux using mathematics of the BML model[5] that results $\overline{\vec{u}''} = \overline{\vec{u}} - \tilde{\vec{u}} = ((\rho_u - \rho_b)/\rho_u\rho_b)\overline{\rho\vec{u}''c''}$. It means that $\langle\vec{u}''\rangle_S^{c*}\Sigma_{c*}$ is not a turbulent transport of the flame surface density as it is usually assumed. So the

[6] We note that Eqs. (25)-(27) are valid only in the case of thin flamelet sheet while Eqs. (22)-(23) are valid in general case of combustion. Obviously the latter equations are valid also for isosurfaces of every scalar field in moving liquid (independently on his nature) self-propelled moving with the speed W, which is zero for passive isosurfaces.

approximation $<\vec{u}''>_S^{c*}\Sigma_{c*} = -(\nu_t/\sigma_\Sigma)\nabla\Sigma_{c*}$ (where l. h. s. strongly depends on $c*$ while r. h. s. practically does not depend on it) is in fact a compensating error that resolve two fundamental problems of the surface averaged equation Eq. (26) when it is applied to combustion modeling coupled together with others equations:

a. It eliminates from the equation the surface average velocity $<\vec{u}>_S^c$;
b. It introduces turbulent type transport through questionable (due to the counter-gradient tramsport phenomenon in the premixed flame) gradient diffusion approximation. The turbulent type transport fundamentally is not presented in Eq. (26) but naturally presented in the $\widetilde{W}$ − equation Eq. (14) and other ensemble averaged combustion and CFD equations.

2. Usual approximations of other $<...>_S^{c*}$ − type terms using ensemble averaged parameters that do not contain dependency on $c*$ also bring in the analysis the compensation errors.
3. Only using these in fact inadmissible approximations gives an opportunity to transform the surface averaged unclosed equation Eq. (26) into a form consistent with other Favre average equations of the problem.

As an illustration let us consider one of the simplest models, the coherent flame model (SFD1), discussed among other models in [3]. The model equation based on several reasonable approximations (that we do not discuss here) is, in original notations of [3], the following:

$$\partial\Sigma/\partial t + \nabla\cdot\tilde{u}\Sigma = \nabla\cdot(\nu_t/\sigma_\Sigma)\nabla\Sigma + A_{ik}\partial\bar{u}_k/\partial x_i\Sigma + \alpha_0(\varepsilon/\kappa)\Sigma + \beta_0(S_L + C\sqrt{k})\Sigma^2/(1-\bar{c}) \quad (28)$$

First we rewrite this equation with a more detailed notation in a form consistent with other Favre average equations. This equation with Σ_c defined on the isosurface $c* = const$ is the following:

$$\begin{aligned}&\partial\Sigma_{c*}/\partial t + \nabla\cdot\tilde{u}\Sigma_{c*} = \\ &\nabla\cdot(\nu_t/\sigma_\Sigma\nabla\Sigma_c) + A_{ik}\partial\tilde{u}_k/\partial x_i\Sigma_{c*} + \alpha_0(\tilde{\varepsilon}/\tilde{k})\Sigma_{c*} + \beta_0(S_L^{c*} + C\sqrt{\tilde{k}})\Sigma_{c*}^2/(1-\tilde{c})\end{aligned} \quad (29)$$

and this equation must be used with Eq. (27). We see that both Eqs. (27) and (29) contain the same S_L^{c*} as they both are defined on the same one-particular isosurface $c* = const$.

Remarks below are very significant:

1. As the model Eq. (29) follows from the unclosed Eq. (26) using some approximations, it must inevitably contain only the speed S_L^{c*} of the chosen isosurface $c = c*$ as the base unclosed equation “knows” only this speed S_L^{c*} and

"knows" nothing about the speed by reactants S_L. If we follow the Bray idea that the isosurface with ρ_m and S_L^m corresponding to the maximum reaction rate is physically justified [2] (where $S_L^m = (\rho_u / \rho_m) S_L \approx (\rho_u / \rho_b) S_L = (5-7) S_L$), in this case the flamelet speed in Eq. (28) must be several times larger than S_L. Selection of S_L in Eq. (28) between basically equivalent definition of the flamelet speed by reactants, product or in general case by arbitrary isosurface is based in fact on tradition. A general agreement is to use the speed S_L defined by reactants (by the way it is the reason why notation S_L is traditionally used instead of detailed S_L^u). Preferring namely S_L has no any theoretical justification and tradition cannot be considered as a rational argument.

2. The reason of the basic ambiguity of this model $\Sigma -$ equation (as well as all others, discussed in [6, p. 232]) is connected with the following:

   a. While all surface average variables $<...>_S^{c*}$ and S_L^{c*} in unclosed Eqs. (25) and (26) strongly depend on $c*$, the FSD Σ_{c*} does not depend on it resulting in the same $\overline{\rho}\widetilde{W}$ independently on the chosen isosurface $c* = const$. This independence is connected with mutual compensation of the $c*$ dependences in these variables in the framework of the equation. Using approximation of $<...>_S^{c*}$ in terms of independent on $c*$ ensemble average variables (to be consistent with others equations) obviously breaks this compensation structure and leaves in the model FSD equations a single S_L^{c*} whose strong dependence on $c*$ cannot be compensated.
   b. As the unclosed Eqs. (25) and (26) basically do not contain the density, the model equation can contain the flamelet speed which is inevitably only S_L^{c*}, in contrast to the invariant (by the flamelet speed definition) form as the product $\rho_{c*} S_L^{c*} = \rho_u S_L = const$.
   c. We state that the reason of ambiguity of the known model FSD equations (in contrast to the based unclosed FSD equation) is due to their inevitable non-invariance by definition of the flamelet speed.[7]

We note that the l. h. s. of Eqs. (21) and (27) is ensemble averaged while the r. h. s. is surface averaged. From the viewpoint of mathematical statistic, it is a strange equation (usually in both sides is used the same type of averaging). Mixture of these averaging types in the l. h. s. of Eqs. (28) and (29) results that its standard form (see Eq. 20a) is broken. Using expression $\Sigma_c = \overline{\rho}\widetilde{W} / (\rho_c S_L^c)$ following from Eq. (27), we can rewrite Eq. (29) as follows:

[7] We note that a quite similar ambiguity takes place in the known FSD models developed for LES but we do not touch this question here.

$$\partial \bar{\rho}\widetilde{W} / \partial t + \nabla \cdot \overline{\rho \widetilde{u}} \widetilde{W} = \nabla \cdot (\nu_t / \sigma_\Sigma) \nabla(\bar{\rho}\widetilde{W}) + A_{ik}\bar{\rho}\widetilde{W}(\partial \widetilde{u}_k / \partial x_i) + \alpha_0 \bar{\rho}(\widetilde{\varepsilon} / \widetilde{k})\widetilde{W} + \beta_0 \bar{\rho}^2 (S_L^{c*} + C\sqrt{\widetilde{k}})\widetilde{W}^2 / ((1-\widetilde{c})\rho_u S_L) \tag{30}$$

That is formulated in terms of Favre average variables (i. e. it is consistent with other Favre average equations of the problem) but reminds ambiguous due to the dependence of S_L^c on the chosen isosurface.

We do not touch in this paper the problem of designing model equations and make only several remark following from comparison of the model equation Eq. (30) with our unclosed $\widetilde{W}-$ equation Eq. (14).

First: In contrast to our equation Eq. (30) is not invariant be definition of the flamelet speed. For this invariance it must contain S_L^{c*} in combination with the density on the chosen isosurface ρ_{c*} in the form of the product $\rho_{c*}S_L^{c*}$ (equal to $\rho_u S_L$). As we mention above, usual use of S_L instead of S_L^{c*} is based on tradition but in accordance with a theoretical justification, and in this case the equation remains non-invariant .

Second: In accordance with the unclosed $\widetilde{W}-$equation Eq. (14) the source term connected with S_L is equal to $2\rho_u S_L \overline{KW}$. If we assume that $\overline{KW}$ does not depend on S_L (we call it the first extreme case) the source is $2\rho_u S_L \overline{KW} \sim S_L \widetilde{W}$. Real curvature K depends on S_L and reduces when S_L increasing (the flamelet sheet becomes more smooth) so we can expect that S_L dependence is weaker than in the first extreme case. The second extreme case corresponds to the situation when effect of S_L increasing is completely compensated by smoothing of the sheet, $2\rho_u S_L \overline{KW} \sim S_L^0 \widetilde{W}$, i. e. $2\rho_u S_L \overline{KW}$ in fact does not depend on chemical kinetics presented by S_L. So using the power dependence $2\rho_u\, S_L \overline{KW} \sim S_L^n$ the reasonable range of n is $0 \le n \le 1$. The value $n = -1$ in Eq. (30) seems appear due to using squared relationship Σ^2 in Eq. (28), but we do not analyze arguments for it. At the same time an idea to present the source as a sum of two terms (in our case depending and independent on S_L) seems reasonable for description of the intermediate situation namely using instead of the power dependence $2\rho_u\, S_L \overline{KW} \sim S_L^n$ with unknown exponent n the presentation $2\rho_u\, S_L \overline{KW} \sim S_L A + B$. Here A and B are some algebraic expressions which are composed of variables known from the equations of the problem.

Third: Likely there is correspondence between the other terms of the model equation Eq. (30) and the terms of the unclosed equation Eq. (14):

(a) $A_{ik}\bar{\rho}\widetilde{W}(\partial \widetilde{u}_k / \partial x_i)$ in Eq. (30) corresponds to $\bar{\rho}\widetilde{W} < \nabla_t \cdot \vec{u} >_F$ which in Eq. (14) describes an effect of the flame stretch by the mean velocity field.

(b) $\alpha_0 \overline{\rho}(\tilde{\varepsilon}/\tilde{k})\widetilde{W}$ in Eq. (30) corresponds to $\overline{\rho} < W''(\nabla_t \cdot \vec{u})'' >_F$ which in Eq. (14) describes an effect of the flamelet stretch by the pulsation component of the velocity.

(c) $\nabla \cdot (\nu_t / \sigma_\Sigma)\nabla(\overline{\rho}\widetilde{W})$ in Eq. (30) is a distorted gradient transport term (correct one is, obviously, $\nabla \cdot (\overline{\rho}\nu_t / \sigma_\Sigma)\nabla(\widetilde{W})$). It corresponds to the transport term $\nabla \cdot (\overline{\rho} < \vec{u}'' W'' >_F)$ of Eq. (14). We notice that gradient approximation of the transport term seems questionable due to known counter-gradient transport phenomenon in premixed flames.

(d) An effect of the term $\rho_u S_L \nabla \cdot (\overline{\vec{n}W})$ in Eq. (14) is omitted in Eq. (30). This term describes the transport caused by the flamelet movement and in the case of relatively small flamelet speed it can be ignored in the model equation.

5.4 An Ensemble Average Formulation of the FSD Concept

As we mentioned above the FSD concept is physically justified and instantaneous and surface average FSD equations are mathematically correct. Difficulty in using this concept, as analyzed above, for the chemical source $\overline{\rho}\widetilde{W}$ description is connected with the fundamental inconsistence of the surface averaging of the $\Sigma -$equation with the Favre averaging used for the other equations of the problem. It does not mean that the FSD concept cannot be applied to the problem of the premixed flames modeling. Our statements on this question are the following:

1. Using only one surface average $\Sigma -$equation defined on one particular isosurface $c^* = const$ fundamentally does not permit to design in a conceptually correct manner a model equation (i. e. based on justified approximations keeping the physical meaning of the approximated terms, without inadmissible assumption, resulting in equations in invariant form and so on). In fact an attempt to do it is a misconception.
2. Nevertheless FSD concept gives an opportunity to correctly resolve this fundamental difficulty if it was possible to evoke an infinitely large number of $\Sigma -$equations corresponding to all isosurfaces from the interval $0 < c^* < 1$. All these equations bring in the analysis all densities ρ_{c^*} and conditionally average velocities $< \vec{u} >_S^{c^*}$ that gives an opportunity to form Favre average variables consistent with other Favre average variables of the problem.
3. We state below an equivalent Favre average unclosed $\widetilde{\Sigma} -$equation consistent with the others and equivalent to an infinitely large number of unclosed surface average Σ_{c^*} $-$ equations so that their actual involving is not necessary.

Eqs. (23, 24) refer to one-particular isosurface with $c = c^*$ where

$$\Sigma_{c*} = \Sigma(c^*;\vec{x},t) = \overline{|\nabla c|\,\delta(c^* - c(\vec{x},t)} = \overline{|\nabla c|}_{c*} p(c^*) \tag{31}$$

and contribution of all Σ_{c*} form the ensemble average FSD. So we will have that

$$\overline{\Sigma} = \int_0^1 \overline{|\nabla c|}_{c*} p(c^*)dc^* = \int_0^1 \Sigma_{c*} dc^* = \overline{|\nabla c|} \text{ (a) and } \overline{\rho}\widetilde{W} = \rho_u S_L \overline{\Sigma} \text{ (b).} \tag{32}$$

Using Eq. (32b) we can present our Eq, (14) in terms of $\overline{\Sigma}$ as it follows:

$$\begin{gathered} \partial(\overline{\rho}\overline{\Sigma})/\partial t + \nabla\cdot(\overline{\rho}\widetilde{\vec{u}}\overline{\Sigma}) + \nabla\cdot(\overline{\rho} < \vec{u}''\Sigma'' >_F) + \rho_u S_L \nabla\cdot(\overline{\vec{n}\Sigma'}) = \\ \overline{\rho}\,\overline{\Sigma} < \nabla_t\cdot\vec{u} >_F + \overline{\rho} < \Sigma''(\nabla_t\cdot\vec{u})'' >_F + 2\rho_u S_L \overline{K\Sigma'}, \end{gathered} \tag{33}$$

where (keeping notation of the original works) Σ' is instantaneous FSD and $\Sigma'' = \Sigma' - \overline{\Sigma}$. We remind that Reynolds average functions could be expressed in terms of Favre average $(\overline{\vec{n}\Sigma'}) = \overline{\rho} < \vec{n}\Sigma'/\rho >_F$ and $(\overline{K\Sigma'}) = \overline{\rho} < K\Sigma'/\rho >_F$.

L. Vervish was probably first to realize that using the FSD on one particular isosurface is not sufficient and has introduced in [6 p. 217] a 'generalized flame surface density' $\overline{\Sigma}$. But his statement [4, p. 218] that in this case "all balance equations remain formally the same" and that "$\overline{|\nabla c|}$ follows a balance equation similar to the $\Sigma(c^*)$ in balance equation (Eq. 23 in this paper), where surface averages $< Q >_S$ are replaced by 'generalized surface averages' $\overline{< Q >_S}$ which do not depend on c^*" is not quite accurate: $\overline{|\nabla c|}$ directly follows a Favre average balance $\widetilde{W}-$equation without evoking the FSD concept and superposition of all Σ_{c*}-equations yields a basically different equation in comparison with every single Σ_{c*}-equation.

We state the logic structure of our analysis of the ensemble average FSD concept which is the following:

1. Defining Σ_{c*} by Eq. (31), we obtain the Σ_{c*}-equation Eq. (26);
2. Superposition of all Σ_{c*} yields $\overline{\Sigma} = \overline{|\nabla c|}$ in Eq. (32a);
3. At the same time from $\overline{|\nabla c|}$ follows our $\widetilde{W}-$equation Eq. (14);
4. Finally $\widetilde{W}-$equation Eq. (14) directly yields (using Eq. (32b)) the $\overline{\Sigma}-$ equation Eq. (33) which together with Eq. (32b) constitute the Favre average formulation of the FSD concept;
5. $\overline{\Sigma}-$ equation Eq. (33) could be deduced directly using more complicated mathematical treatment combining all Σ_{c*}-equations Eqs. (26) and corresponding

densities from the range $0 < c^* < 1$, but we did not perform it as the result, obviously, would be the same.[8]

It is present only for completeness as our results demonstrate that the FSD approach is in fact superfluous for modeling of the source $\bar{\rho}\tilde{W}$ in Eq. (1) . It only brings in the analysis unjustified complication in mathematics due to the unnecessary intermediate transition of the problem into the field of the surface average functions and as analyzed above leads to inevitable misconception at using one-particular $\Sigma_{c^*}-$ equation for designing model equations. It does not mean that we repudiate the FSD concept, not at all. FSD is a justified variable and the surface average Σ is an indispensable local statistical characteristic of the scalar isosurface area as well as the balance unclosed $\Sigma-$equation describing the mechanisms controlling average ratio area-to-volume of the isosurface not only in the premixed flame ($w \neq 0$) but in others systems (for passive isoscalars $w \equiv 0$ in Eq. (26)).

Finally, unclosed Eqs. (31b) and (32) represent the Favre average formulation of the FSD concept consistent with the other unclosed equations of the turbulent premixed combustion at flamelet mechanism .

6 A Graphic Illustration of Different Approaches to $\tilde{W}$ −Modeling

We present here for clearness the graphic presentation of the analyzed approaches to modeling of the mean source term $\overline{\rho W} = \bar{\rho}\tilde{W}$ of Eq. (1):

1.Our approach (deducing of the Favre average unclosed $\tilde{W}$ −equation):

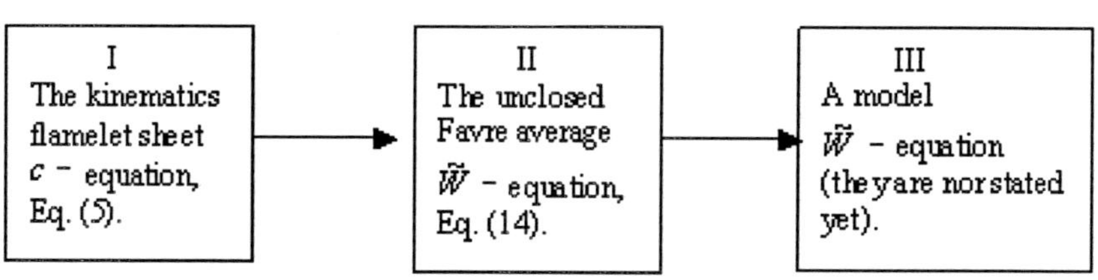

This is the direct way for modeling of the chemical source without evoking the FSD concept.

We state it as a conceptually correct approach to the problem.

2.The FSD concept in the field of the surface average functions:

[8] In our preprint [7] we attempted to deduce both Reynolds and Favre unclosed FSD equation taking, as the starting point, instantaneous equation in terms of $|\nabla c|$, Eq. (12). Though these equations (Eqs. (12) and (16) in [7]) have different form, we think that in essence they are equivalent to the unclosed $\overline{\Sigma}-$ equation Eq. (33).

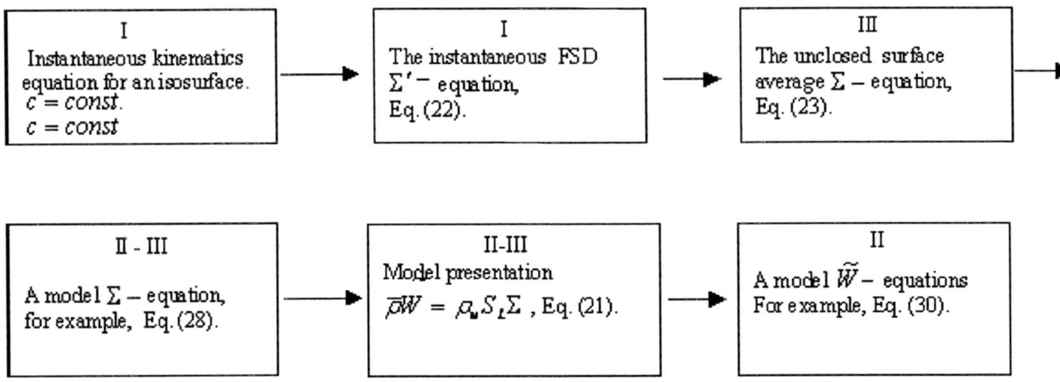

This approach is not conceptually correct as the FSD equation defined on one particular isosurface is fundamentally insufficient for the design of Favre average model equations as the surface average equation basically does not contain all physical mechanism inherent to Favre average equations. Using this approach, we basically cannot include the unclosed surface average $\Sigma-$equation into the basic set of unclosed simultaneous equations and it is a reason of misconception of this approach.

3.The FSD concept in terms of the ensemble average $\overline{\Sigma}$:

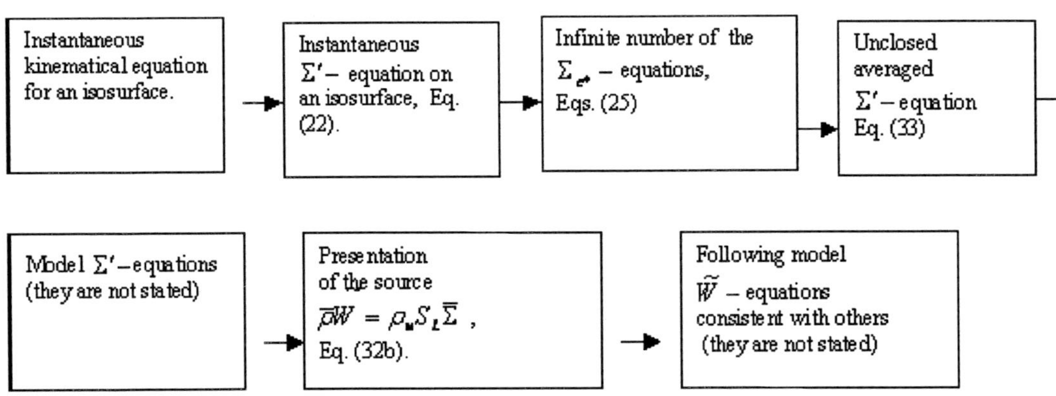

This approach gives an opportunity to include the equation in a set of simultaneous equations and from this viewpoint it is conceptually correct. But this way is indirect (in contrast to ours) and more complicated that is why it is not practical.

So we can state the following: *evoking the FSD approach is not obligatory for modeling the chemical source in premixed flames with flamelet combustion mechanism as there is a direct and mathematically more simple way based on the unclosed* $\widetilde{W}$ *−equation Eq. (14). Nevertheless the FSD concept also can be used but only in the ensemble average formulation. Only unclosed* $\widetilde{\Sigma}$ *− equation is conceptually consistent with unclosed others of the problem. At the same time using a single surface average* Σ *− equation is in fact a misconception and this inconsistence is a reason of the basic difficulty in using the FSD concept for turbulent premixed combustion modeling.*

7 Two Principles of Turbulence Modeling

Summarizing the results of our analysis of the challenging problem of the chemical source term modeling in the premixed flame with the flamelet combustion mechanism urged us to formulate two rules obligatory for conceptually correct modeling of turbulent premixed combustion in the context of the statistical moment description. Though these rules seems obvious, existence of a considerable body of publications where they are violated forced us to formulate them as principles that we called "the principle of compatibility" and "the principle of invariance".

7.1 Principle of Compatibility of Unclosed Equations Starting for Modeling

We formulate the principle of compatibility as follows:

Basic unclosed equations for turbulent combustion modeling including both hydrodynamics and thermodynamics equations must be consistent. The unclosed equations must constitute a set of simultaneous equations naturally following the consistent model equations. A necessary condition is that all equations must be formulated in terms of variables using the same type of averaging. Physical mechanisms used for the closure of combustion and turbulent equations must be common.

This principle is more general, i. e. it refers not only to premixed flamelet combustion.

Several examples below based on previous analysis illustrate this principle:

1. The system of equations of the premixed combustion problem including the first moment combustion $\tilde{c}$ – equation Eq. (1) and hydrodynamics Eqs. (2) together with the equations for their unknown variables Eqs. (14) and (16) and with the equation for the turbulent stresses presented in the literature obey this principle of compatibility. This system is defined in the field of Favre average equations that is why following model systems naturally also are defined in this field. Here we note, that in the context of this principle, the model equations must be also consistent from the physical viewpoint: modeling of the combustion and turbulent part of the problem must be used on similar or compatible physical mechanisms. For example, if we include the different pressure-driven acceleration of reactants and products in the scalar transport model equation for prediction of the counter-gradient transport phenomenon, we must include this mechanism also in the model equation for the Favre average stress tensor $<\vec{u}\vec{u}>_F$ as this mechanism is responsible also for strong anisotropy of the velocity pulsations in the premixed flame.
2. Replacement of the Favre average $\widetilde{W}$ – equation Eq. (14) by the unclosed equations in terms of the surface average FSD defined on one-particular isosurface, Eqs. (21) and (23) (or in more detailed notation Eqs. (25), (27)) violates this principle and makes these equations incompatible. We show in the paper that in this case it is impossible to develop a conceptually correct model (based on theoretically justified approximations and obeying to the second principle of invariance).
3. At the same time, replacement of the $\widetilde{W}$ – equation by the unclosed Favre average SDF equations, Eqs. (31b) and (32) does no violate this principle. It is quite another

matter that, in fact, it is unnecessary to use these compatible equations for modeling of the mean source as they are equivalent to the $\widetilde{W}$ −equation and using of FSD only brings in the analysis needless mathematical complexity.

4. We note that a system of equations in the field of the ensemble average function does not obligatory obey this principle. For example, let us replace the continuity equation Eq. (2a) in the subsystem of the simultaneous Favre average equations Eqs. (1) and (2) by the Reynolds average continuity equation $\partial \overline{\rho}/\partial t + \nabla \cdot (\overline{\rho}\ \overline{\vec{u}} + \overline{\rho' \vec{u}'}) = 0$. The resulting equations (containing unknown $\widetilde{Q}$, $<\vec{u}''c''>_F$, $<\vec{u}''\vec{u}''>_F$ and $\overline{\rho'\vec{u}'}$, obviously, do not obey this principle. They do not constitute a system of simultaneous equations as in the context of the moment approach there is no compatibility between the velocities $\widetilde{\vec{u}}$ and $\overline{\vec{u}}$. We remind that the velocities $\widetilde{\vec{u}}$ and $\overline{\vec{u}}$ are connected only in terms of the joint PDF $p(\rho,\vec{u})$, i. e. in the context of a more profound probability approach to the problem these both velocities can appear in the analysis without violating the principle of compatibility.
5. We specially analyzed the previous example to demonstrate this subtle point connected with the inevitable appearance of Favre and Reynolds average functions together as it took place in the Eqs. (14) and (16). We stress that in our case all equations obey this principle as they do not contain incompatible variables and even formally they can be represented in terms of only Favre average functions as we showed above.

7.2 The Principle of Invariance by Definition of the Flamelet Speed

We formulate the principle of compatibility as follows:

Unclosed and model equations must contain the flamelet speed (used as a parameter) in a form such that the physical content of the equations does not depend on the definition of this speed.

This principle states that the physical content must be the same for the flamelet speed traditionally defined by reactants $S_L = S_L^u$ or by products S_L^b and in general case by an arbitrary isosurface S_L^{c*}. To obey this principle the equation must contain the flamelet speed in the form of the product $\rho_u S_L = \rho_b S_L^b = \rho_{c*} S_L^{c*}$ that inevitably makes $\overline{\rho}$ obligatory appear in the terms which do not contain this speed.

All unclosed equations stated in the paper obey this invariance principle. So, the model equations naturally following for the scalar flux and the chemical source will also obey this principle. At the same time, in known FSD model this principle is violated and it yield the basic ambiguity of these model equations.

8 Correlation between Descriptions of the Turbulent Premixed Flame in the Context of the Progress Variable and the Scalar G

In the literature presently there are two main variables to description of the turbulent premixed flames with the flamelet combustion mechanism. They are the progress variable c that is in fact a generalized reacting chemical species and treated as nonreacting scalar variable G that is a marking parameter used for identification of the thin flamelet sheet. These variables are associated with two fundamentally different approaches to premixed combustion modeling:

1. The progress variable usually is used for statistical (in terms of averaged parameters) approach to combustion modeling with the basic moment $\tilde{c}-$ equation Eq. (1). In the context of this equation there are two main difficulties of modeling connected with existence of the counter-gradient scalar transport phenomenon and small-scale coupling of turbulence and chemistry controlling the chemical source. The latter difficulty is known in the literature as the "challenge of turbulent combustion" [Bray]. Though this approach was developed by several authors we for simplicity will call it as the Bray approach.

2. The variable G is associated usually with description of the premixed flame by analysis and numerical simulation the instantaneous kinematics of one particular isosurface inside the flamelet sheet described by so called G - equation. Obviously this approach does not yield mentioned above difficulties which arise at statistical description. There is much speculations that using the G - equation instead of the instantaneous equations in terms of the progress variable has significant technological advantage as there is no necessity to resolve the flamelet structure and it is sufficient to have in simulations much smoother distributions of G in comparison with actual distribution of the progress variable.

3. There is the third in fact intermediate approach that is statistical but based on using the G variable instead of the progress variable[Peters]. There is a belief that this approach is free from mentioned above difficulties of modeling: "An alternative model for premixed turbulent combustion, based on the nonreacting scalar G rather than on the progress variable, has been developed in recent years. It avoids complications associated with counter-gradient diffusion and, since G in nonreacting, there is no need for a source term closure"[4]. We will call it the Peters approach.

Similar to previous section in this one we analyze these three approaches from a conceptual point of view so we do not touch here particular combustion models and deal only with exact instantaneous and statistical unclosed moment equations based on both c and G variables. The results of this sections were announced in one-page publication [/].

8.1 The Instantaneous Kinematical $c-$ and $G-$Equations

The instantaneous kinematical $c-$equation was stated above, Eq. (5). In this equation $c(\vec{x},t)$, $\rho(\vec{x},t)$ and $\vec{u}(\vec{x},t)$ are functions defined in all space$(\vec{x},t)$. The classical G-equation is the following[15] (see also [4]):

$$\partial G/\partial t + \vec{u}_f \nabla G = S_L \mid \nabla G \mid, \tag{34}$$

which is defined (in contrast to the kinematical $c-$equation) only on the one-particular isosurface and hence the velocity $\vec{u}_f$ is also defined only on it. The constant G on the isosurface usually is denoted as G_0 and the value of it could be arbitrary. It is traditionally assumed that outside of the surface $G < G_0$ in the direction of the reacts and $G > G_0$ in the direction of the products. As Eq. (34) is valid only on the one-particular isosurface , the distribution of G outside of it is not defined by the equation and can be arbitrary.

First we notice qualitative difference of the l.h.s. terms of the c - equation Eq. (5) and the G - equation Eq. (34). The source term in Eq. (5) contains the density of the reactants ρ_u. This explicit "asymmetry" of the c- equation is connected with asymmetry of the conventional definition of the speed S_L by the reactants. We have the product $\rho_u S_L$ in Eq. (5) using traditional definition of the flame speed S_L or equal products $\rho_b S_L^b$ and $\rho_c S_L^c$ using equally possible definitions of the speed by product or an isosurface with arbitrary $c = const$. It means that the kinematical c- equation Eq. (5) is invariant by the definition of the flame speed, we discussed this point in the section 3. At the same time the G- equation is not invariant. The fact that the classical G- equation Eq. (34) contains S_L and does not contains the density means that it is valid only on the "cold" isosurface with $c = 0$ (the front edge of the flamelet sheet) and is not valid for other isosurfaces. If we use replace S_L in Eq. (34) by S_L^b or S_L^{c*} these new G- equations would be valid on "hot" $(c = 1)$ or corresponding intermediate isosurface with $c = c^* = const$ and the speed $\vec{u}_f$ would be defended only on them. So in principle there are infinitely large number of the G- equations corresponding to all isosurfaces with $c^* = const$ $(0 \le c^* \le 1)$ with different for different c^* values of the marking parameter G_0, i.e. $G_0 = G_0(c^*)$. If we denote the speed $\vec{u}_f$ and marking parameter G_0 corresponding to $c = c^* = const$ as $\vec{u}_f^*$ and G^* (different for different c^*, i. e. $G^* = G_0(c^*)$) the general form an equation of this set would be the following

$$\partial G/\partial t + \vec{u}_f^* \nabla G = S_L^{c*} \mid \nabla G \mid, \tag{35}$$

Outside the isosurface $c^* = const$ with $\vec{u}_f^*$ and G_* the scalar G is arbitrary and in accordance with the tradition we can assume $G < G_*$ in the direction of the reacts and $G > G_*$ in the direction of the products. But we must remember that Eqs. (35) are not define scalar G outside of their particular isosurfacres $c^* = const$. We can rewrite Eqs. (35) in terms of the traditionally defined flamelet speed S_L which, obviously, would be as it follows:

$$\rho^* \partial G/\partial t + \rho^* \vec{u}_f^* \nabla G = \rho_u S_L \,|\nabla G|. \tag{36}$$

It should be stressed that the classical G - equation Eq. (34) as well as Eqs. (35) and (36) are in fact phenomenological and they are not connected with kinematical c - equation. We will show below that they rigorously follows from the kinematical c - equation Eq. (5). Moreover we show that only the infinite number of G - equations yielded by the Eq. (5) is equivalent to it but not the single classical G - equation.

8.2 Deduction of the Set of the G - Equations

We start with the kinematical c - equation Eq. (5) that is defined on all isosurfaces $c = c^* = const$ inside the flamelet sheet while a single G - equation is defined on one-particular isosurface with some $c^* = const$. As G outside this isosurface is arbitrary we assume that outside of it $G = 0$ $(G < G_*)$ at $0 \le c < c^*$ and $G = 2G_*$ $(G > G_*)$ at $c^* < c \le 1$, i.e. $G(c) = 2G_* H(c - c^*)$ where $H(\alpha)$ is the special Heaviside function ($H(\alpha) = 0$ at $\alpha < 0$, $H(0) = 1/2$ and $(H(\alpha) = 1$ at $\alpha > 0$). For sake of simplicity we assume that $G_* = 1/2$ on this particular isosurface with $c(\vec{x}, t) = c^*$. In this case

$$G(\vec{x}, t) = G(c(\vec{x}, t)) = H(c(\vec{x}, t) - c^*). \tag{37}$$

It means that

$$G(c) = \int_0^c \delta(c - c^*) dc \text{ (a)}, \; dG/dc = \delta(c - c^*) \text{ (b)}, \tag{38}$$

where δ is the special $\delta -$ function

Writing symbolically the mathematical treatment of special functions as normal functions (this procedure is standard and provides correct final result [\cite {GS58}]) we can write $dG/dc = (\partial G/\partial t) \cdot (\partial t/\partial c)$, hence using Eq. (38) we have $\partial c/\partial t = (\partial G/\partial t)/\delta(c - c^*)$. Similar $dG/dc = (dG/dn) \cdot (dn/dc)$, hence $|\nabla c| = |\nabla G|/\delta(c - c^*)$. Inserting these expressions in Eq. (5) results in the following:

$$\delta(c-c^*)(\rho\partial c/\partial t+\rho\vec{u}\nabla c-\rho_u S_L\,|\nabla c|)=\rho\partial G/\partial t+\rho\vec{u}\nabla G-\rho_u S_L\,|\nabla G|=0 \quad (39)$$

The special function $\delta(c(\vec{x},t)-c^*)$ is defined in the infinitely small vicinity of the isosurface $c(\vec{x},t)=c^*$ so the central (with G) part of Eq. (39) also defined only in it, whence it follows that

$$\rho^*\partial G/\partial t+\rho^*\vec{u}_f^*\nabla G=\rho_u S_L\,|\nabla G|. \quad (40)$$

In Eq. (40) ρ^* and $\vec{u}_f^*$ are parameters that refer to the isosurface $c(\vec{x},t)=c^*$ where only this equation is defined (and is not defined outside of it). We see that deduced Eq.(40) is the same as the phenomenal Eq. (36) . We can rewrite Eq. (40) in the form that does not contain the density using the expression $S_L^{c^*}=(\rho_u/\rho^*)S_L$ corresponding to the isosurface $c(\vec{x},t)=c^*$

$$\partial G/\partial t+\vec{u}_f^*\nabla G=S_L^{c^*}\,|\nabla G| \quad (41)$$

that is the same as the phenomenal Eq. (35).

For the analyzed particular isosurface we assumed $G_*=1/2$, but due to the linearity of the equation the value of the marking parameter G_* could be arbitrary but, obviously, different for different isosurfaces. We will assume below that G_* is a single-valued continuous function of c^*, so every isosurface with a particular value of c^* is characterized by a particular value of the marking parameter G_*.

Eqs. (40) or (41) represent a set of infinite number of the kinematical $G-$equations describing all isosurfaces $0\le c^*\le 1$ inside the instantaneous flame with and the classical $G-$equation Eq. (3) is one of them corresponding the isosurface $c^*=0$ (the front edge of the thin flamel6et sheet) with the marking parameter $G_*(0)=G_0$. In this connection it is pertinent to note that G. H. Markstein (who probably was first who proposed to use kinematical equations for description of the premixed flame) understood that one kinematical equation is not sufficient and he stated in the book [16] two kinematical equations referring to the front and back edges of the instantaneous flame corresponding to the cases $\vec{u}_f^*=\vec{u}_f^{c=0}$, $S_L^{c^*}=S_L^{c=0}$ and $\vec{u}_f^*=\vec{u}_f^{c=1}$, $S_L^{c^*}=S_L^{c=1}$ in Eq. (41) and the classical $G-$equation Eq. (34) is in fact one of them describing the front edge of the instantaneous flame. At the same time Eqs. (40) or (41) describe all isosurfaces, i.e. they describe the instantaneous flamelet sheet completely. As all Eqs. (40) or (41) follows from the kinematical $c-$equation Eq. (5) they, obviously, are equivalent to it.

We stress that Eq. (5) is a regular equation defined in the $3D$ field $(\vec{x},t)$ while Eqs. (40) or (41) are valid on the $2D$ surfaces $c(\vec{x},t)=c^*=const$. Infinite number of the latter equations define the $3D$ field of the marking parameters $G_*(\vec{x},t)$. The field $G_*(\vec{x},t)$ is obeyed the regular equation

$$\rho \partial G_* / \partial t + \rho \vec{u} \nabla G_* = \rho_u S_L \mid \nabla G_* \mid \tag{42}$$

defined in the $3D$ field $(\vec{x},t)$. This equation directly follows from Eq. (5) taking into account that $G_* = G_*(c)$.

As a particular case we can assume that $G_*(c^*) = c^*$, i.e. the marking parameter is the progress variable itself. In this case Eqs. (40) and (41) become

$$\rho^* \partial c / \partial t + \rho^* \vec{u}_f^* \nabla c = \rho_u S_L \mid \nabla c \mid . \tag{43}$$

and

$$\partial c / \partial t + \vec{u}_f^* \nabla c = S_L^{c^*} \mid \nabla c \mid . \tag{44}$$

where c (marketing parameter here) is equal to c^* on the isosurface and is arbitrary outside of it but in accordance with the tradition we can assume $c < c^*$ in the direction of the reacts and $c > c^*$ in the direction of the products (we notice that the value of this arbitrary c could be less than zero or larger then unite as in this case it is not actual progress variable).

A kinematical equation in the context of c which is equivalent to the classical $G-$equation (i.e. describing the "cold" isosurface) is the following:

$$\partial c / \partial t + \vec{u}_f \nabla c = \rho_u S_L \mid \nabla c \mid . \tag{45}$$

On this isosurface the marking $c = c_0 = 0$ and following to the concept of the $G-$equation we can assume that $c < 0$ in reactants and $c > 0$ in the direction of the products (and the fact that $c < 0$ would not be a mistake as c here is not the progress variable but a formal parameter).

The physical meaning of the presented derivation is that described by the kinematical $c-$ equation Eq. (5) thin flamelet sheet breaks up for infinite number of the isosurfaces described by Eqs. (42) or (43). We stress that the Eq. (5) express the conservation law in all space (the rate of the product generation is equal to the rate of the reactants consumption by the thin flamelet sheet moving with the speed S_L by reactants) while Eqs. (42) and (43) (as well as Eqs. (41) and (42)) express the same law but only on one-particular isosurface $c(\vec{x},t) = c^*$ (or $G^* = const$).

From these results follows three consequences:

1. The marking parameter scalar G cannot be interpreted as a nonreacting parameter in contrast to the reacting progress variable c. They both are constant on one-particular

isosurface (both are “nonreactiong” on it) and both are variable (“reacting”) across the flamelet sheet.

2. The scalar G is in fact superfluous parameter as the kinematics of the flamelet sheet in general as well as the kinematics of a paricular isosurface can be in the context of the progress variable c. Evoking the scalar G means in face using a change of variables.
3. The kinematical $c-$equation Eq. (5) is defined in the field of the regular functions and describes $3D$ fields $c(\vec{x},t)$ at the same time the kinematical equations describing one-particular isosurface both in the context of the scalar G (Eqs. (40) and (41)) and the progress variable c (Eqs. (42) and (43)) refers to the class of equations in the field of the special functions and describe only the $2D$ solution (the surface $c(\vec{x},t) = c^*$) that is fact is the section of the $3D$ solution $c(\vec{x},t)$ in terms of G or c. Keeping in mind this difference is significant for correct deriving statistical moment combustion equations.

8.3 The Unclosed Balance Moment Favre Average Equation in Terms of the Scalar G Function

Now we show that the balance moment equations in terms of the Favre average progress variable and scalar G are equivalent so known in the context of the $c-$approach fundamental difficulty in modeling in the scalar flux and chemical source is inherent of the $G-$approach.

First remark is that we cannot deduce the moment equation using only the classical $G-$equation Eq. (34) as it is not define the $G-$function outside one-particular (“cold”) isisurface. At the same time all isosurfaces define marking $G_*(\vec{x},t)$ everywhere together with corresponding densities that yield the Favre average $\widetilde{G}^*(\vec{x},t)$. They are equivalent to the regular G_*-equation Eq. (42) so the instantaneous marking G^* in it (but not G in Eq. (34)!) can be presented as

$$G_*(\vec{x},t) = \widetilde{G}_*(\vec{x},t) + G_*''(\vec{x},t) \quad (43)$$

that yields after installation, using the continuity equation and averaging the moment equation

$$\partial \bar{\rho}\widetilde{G}_* / \partial t + \nabla \cdot \bar{\rho}\tilde{\vec{u}}\widetilde{G}_* = -\nabla \cdot \bar{\rho} < \vec{u}''G_*'' >_F + \bar{\rho}\widetilde{W}_G , \quad (44)$$

where the chemical source equal to $\bar{\rho}\widetilde{W}_G = \rho_u S_L \overline{|\nabla G_*|}$. It is obvious that Eq. (44) is equivalent to the moment $\tilde{c}-$equation Eq. (1) together with inherent to this problem challenge in description of the scalar transport and chemical source. Obviously, the unclosed equations in terms of unknown chemical source $\overline{\rho W_G} = \bar{\rho}\widetilde{W}_G$ and the scalar flux

$\overline{\rho\vec{u}''G_*''} = \overline{\rho} < \vec{u}''G_*'' >_F$ would be similar to the equations Eqs. (14) and (16) deduced in the context of the progress variable.

Three main consequence follows from the analysis which is presented in this section:

1. Opportunity to describe the kinematics of one-particular isosurface inside the instantaneous flame is not a prerogative of the scalar G function. It equally can be done in terms of c and advantage in using the latter is that the progress variable has clear physical meaning.
2. Using in the moment equations the scalar G instead of the progress variable c fundamentally does not offer theoretically justified possibilities to avoid the challenging problems of the scalar transport and the chemical source modeling.
3. Due to equivalence of the moment Eqs. (1) and (44) every combustion model formulated in terms of c can be reformulated in terms of G and vice versa.

In spite of stated above equivalence of the both approaches there is a believe in the literature that the combustion models in the context of the scalar G could be different. For example, the guiding idea in [4] at designing of a combustion model equation in terms of the scalar G was to obtain the hyperbolic type differential model equation (i.e. in fact without the transport term) similar to the instantaneous $G-$equation describing in the running combustion front kinematics. To understand what does it means we rewrite this developed in [4] equation (Eq. 2.151 in [4]) in our notations

$$\overline{\rho}\partial \widetilde{G}_* / \partial t + \overline{\rho}\widetilde{\vec{u}}\nabla \widetilde{G}_* = \rho_u U_t \mid \nabla \widetilde{G}_* \mid, \tag{45}$$

(where U_t is the turbulent flame velocity) and present by comparing it with Eq. (44) the r.h.s. of Eq. (45) as it follows:

$$\rho_u U_t \mid \nabla \widetilde{G}_* \mid = \overline{\rho}\widetilde{W}_G - \nabla \cdot \overline{\rho} < \vec{u}''G_*'' >_F . \tag{46}$$

We see that the "kinematical" source (r.h.s. term in Eq. (45)) is controlled both by the actual chemical source and the scalar transport, i.e. possible counter-gradient transport phenomenon is "recondite" in the kinematical source. Absence of the transport term in Eq. (45) does not mean that the complication connected with the counter-gradient transport phenomenon is overcome. Eq. (45) itself is not enough to simulate the flame as it alone cannot predict the flame brush width that is controlled in general case both by the scalar transport and chemistry.

It must be emphasize that we cannot attract an equation in terms of the second moment $< G_*'' >_F$ as independent for estimation of the flame brush width as in our case of thin and strongly wrinkles flamelet sheet the first and the second moments are not independent and are connected by the expression $< G_*''^2 >_F = \widetilde{G}(1-\widetilde{G})$ that is valid in the second order. It means that corresponding $< G_*'' >_F$ − equation is not independent on Eq. (45). Using in [4] as independent the $< G_*''^2 >_F$ – equation which in fact is formulated for the case of the distributed combustion does not seem theoretically justified.

It should be also realized that in general case the turbulent flame speed U_t in Eq. (45) depends on the small-scale coupling of the chemistry and turbulence so a theoretical estimation of U_t as a function of composition and turbulence is a challenging problem.

We stress that all ideas and following equations expressed in terms of the scalar G can be equally reformulated in terms of the progress variable c where existing of mentioned above challenging problems is common knowledge.

8.4 Two Concept of the Turbulent Premixed Flame

Eq. (45) is the kinematical equation describing the turbulent flame as a moving wave. In fact it is based on assumption that the turbulent combustion front very soon reach equilibrium steady state structure, i.e. the turbulent flame has constant speed U_t^{st} and width δ_t^{st}. We rewrite it in more general form as it follows

$$\overline{\rho}\partial \widetilde{Q}/\partial t + \overline{\rho}\widetilde{\vec{u}}\nabla\widetilde{Q} = \rho_u U_t^{st} \,|\nabla\widetilde{Q}\,|, \tag{47}$$

where $\widetilde{Q}$ could be equally $\widetilde{G}_*$ or $\widetilde{c}$.

In accordance with reasoning of the pioneers of the turbulent premixed combustion theory [17], [18] the turbulent front speed at strong turbulence $u' >> S_L$ does not depend on chemistry and is $U_t^{st} \approx u'$. A large body of data shows that real premixed flames are different:

1. The premixed flame structure is transient, the flame brush increases and steady state structure practically unattainable in real experiments;
2. The turbulent flame velocity depends on the chemical kinetics of the mixture even at strong turbulence.

These peculiarities become more clear if we remember behavior of the plane surface in nonreactive turbulent medium. This surface increases its area and spread in space due to stretch and bending that results in turbulent mixing of the gases disposed bilaterally along it. The averaged width of the mixture layer δ_m is in terms the turbulent diffusion coefficient $\delta_m = (2D_t t)^{1/2}$. At the same time the dimensionless are of the surface increases in accordance with the Batchelor analysis [19] exponentially $(\overline{A}/A_o) \sim e^{at}$ (at the initial moment $(\overline{A}/A_o) = 1$). The reason of this so different behavior of the turbulent mixing layer width and the boundary surface area connected with the fact that the width is controlled mainly by the large-scale wrinkles of the surface while the area is controlled mainly by the small-scale wrinkles with rapidly developing structure at strong turbulence.

At the same time we do not observe exponential increase of the flame speed $U_t = S_L(\overline{A}/A_0)$ while the increasing of the flame brush similar to turbulent mixing is well documented [20]. The physical reason of it is that the flamelet moving with relatively small speed ($S_L << u'$) suppress developing of the small-scale structure of the wrinkles, to consume them. So very soon in the case of combustion very soon we will have some quasi-equilibrium structure of the small-wrinkled and close to the turbulent mixing nonequilibrium structure of the large wrinkles. The flames in this case have nearly constant U_t and increasing δ_t. Namely such flames are observe in simple hydrodynamics situations (the flames with increasing brush with and nearly constant inclination to the flow[20]). And only at much larger time we will have the flames with equilibrium both small-scale and large-scale wrinkles structure corresponding to the steady state flames.

We called in [21] the former flames "the intermediate steady propagation flames" (ISP flames) as they correspondent some intermediate asymptotic when small-scale wrinkles are already quasi-equilibrium and at the same time the large-scale wrinkles are practically nonequilibrium. A kinematical equation describing the ISP flame (that equally can be formulated in term of both G or c variables is the following:

$$\overline{\rho}\partial\tilde{c}/\partial t + \overline{\rho}\tilde{\tilde{u}}\nabla\tilde{c} = \nabla\cdot(\overline{\rho}D_t)\nabla\tilde{c} + \rho_u U_t\,|\nabla\tilde{c}\,|, \tag{48}$$

where as $\tilde{Q}$ equally could be $\tilde{G}_*$ or $\tilde{c}$.

It is obvious, that the believe about opportunity using the scalar G *instead of the progress variable* c *gives an opportunity to propose different formulation of the problem has no a theoretical justification: they inevitably would be equivalent in the both variables.*

Simple estimation in [21] demonstrates that ISP flames take place at $t << \tau_t(u'/S_L)^2$ (practically at $t < \tau_t(u'/S_L)^2$) while the steady state flames take place asymptotically at $t >> \tau_t(u'/S)^2$ ($\tau_t = L/u'$ is the turbulent time). This estimation and experimental data show that the latter flames are practically unattainable: at real turbulence and chemistry the residence time in the flame τ_r is relatively small $\tau_r < \tau_t(u'/S_L)^2$. Practically it means that the completion of combustion (the flame reach a wall in a channel or the brushes crosses in open flames) long before forming the steady state flame structure.

It is significant that while for described by Eq. (47) steady state flames the speed U_t^{st} does not depend on chemistry (U_t^{st} is controlled only by turbulence) and the width δ_t^{st} directly depends on it (δ_t^{st} is controlled both by turbulence and chemistry), at the same time for described by Eq. (48) ISP flames situation is quite the reverse: the speed U_t depends on chemistry (U_t is controlled both by turbulence and chemistry) and δ_t does not depend directly on chemistry (δ_t is controlled only by turbulence).

In the next chapter we explain the reason of it and present theoretical estimations of these parameters. Arising fundamental difficulty connected with existing of small-scale coupling of turbulence and chemistry (that is one of the main controlling mechanism) we overcome in the context of Kolmogorov ideas, i. e. assuming statistically equilibrium structure of the reaction zones. We demonstrate also how this ideas can be applied for combustion modeling and demonstrate numerical results in the context of RANS and LES approaches.

9 Conclusion

1. The main result of the paper is the development of the unclosed $\widetilde{W}-$ equation Eq. (14) describing for the case of the premixed flamelet combustion mechanism the chemical source term $\overline{\rho}\widetilde{W}$ of the unclosed $\widetilde{c}-$ equation Eq. (1). These equations, together with the unclosed equation for the scalar transport, constitute the simultaneous equations completely consistent with the other CFD Favre average equations formulated in the field of Favre average functions.
2. The set of equations is in fact a second-order subsystem of a general system of infinitely large number of connected unclosed equations constituting the analytical moment formulation of the turbulent combustion problem (which is similar to the known analytical moment formulation of the turbulence problem). They offer a justified base for developing a model equation for the scalar transport and chemical source for premixed combustion with flamelet mechanism.
3. We analyze the nature of well known difficulty in using the flame surface density (FSD) approach for modeling the chemical source $\overline{\rho}\widetilde{W}$ based on using the FSD $\Sigma-$equation which is defined on one-particular isosurface. We show that this difficulty is not connected with some fallacy of this concept (FSD concept is physically justified and the unclosed $\Sigma-$equation is mathematically correct) but it is due to fundamental inconsistence of all other unclosed Favre average equations of the problem with the unclosed FSD $\Sigma-$equation defined on one-particular isosurface.
4. We show that known attempts to transform the inconsistent unclosed $\Sigma-$equation into a consistent model $\Sigma-$equation (i. e. its transition from the field of surface average into the field of Favre average) are based on using in fact inadmissible approximation changing the physical meaning of the unknown terms. And it is inevitable for designing physically justified model equations because obligatory Favre average variables and turbulent type transport mechanism basically do not follows from one-particular unclosed $\Sigma-$equation. We demonstrate that the model FSD $\Sigma-$equations inevitably are not invariant by definition of the flamelet speed contained in the equations as it appears solo (in invariant form it must appear as the product $\rho_{c*}S_L^{c*} = \rho_u S_L$). It results in the basic ambiguity of the model FSD equations.
5. Our statement is that the FSD concept, correct in essence, would be consistent with other Favre average equations if we evoke for the analysis an infinitely large number

of unclosed Σ_{c*}- equations defined on all isosurfaces from the range $(0 < c^* < 1)$. These equations bring in the analysis all densities ρ_{c*}, velocities $<\vec{u}>_S^{c*}$ and isosurface speeds S_L^{c*} necessary for the formation of Favre average invariant model equations. Practically this infinitely large number of unclosed equations is equivalent to one Favre average FSD $\overline{\Sigma}-$equation Eq. (33) that is equivalent to our $\widetilde{W}-$equation.

6. We show that that approach based on the scalar G and progress c variables are equivalents and it is only a tradition to use c for statistical description of the flame and G for kinematical description of the flame through description of one-particular isosurface inside the flamelet sheet. We deduce defined on one particular isosurface kinematical $c-$equation that is completely equivalent to the classical $G-$equation and show that only infinite numbers of such equations both in terms of G or c are sufficient for deducing the Favre average moment equations. Known in the literature speculations about advantage of the scalar G over the progress variable c for resolving of the challenging problems of the turbulent premixed combustion modeling have no any theoretical justification.
7. During this work we arrived at the following conclusions:

   a. Using as a basis the straightforwardly deduced unclosed $\widetilde{W}-$ equation Eq. (14) (together with unclosed scalar flux Eq. (16)) is the simplest and conceptually correct way to describe the unknown terms of the $\widetilde{c}-$equation in the context of moment differential equations modeling;
   b. Using the FSD concept based the $\Sigma-$equation for modeling of the chemical source $\overline{\rho}\widetilde{W}$ in the context of the Favre average equations of the problem is in fact a misconception due to a fundamental disagreement between it and others based unclosed equations of the problem. The reason is that theoretically justified and mathematically correct $\Sigma-$equation defined on one-particular isosurface refers to different formulation of the combustion problem namely in the fields of conditional averaged functions. We show that attempts to overcome this disagreement on the stage of designing of combustion model equations basically impossible. We formulate two principles (compatibility and invariance) obligatory for conceptually correct model equations and show that using a single $\Sigma-$equation inevitably results in violation of them.
   c. The FSD can be consistent with others ensemble averaged equations of the problem if to evoke infinite number of the $\Sigma-$equations (corresponding to all $0 \le c^* \le 1$) which are together equivalent to one unclosed ensemble average $\overline{\Sigma}-$equation Eq. (33) that it was in fact stated in the paper but not deduced from the basic surface averaged Eq. (26). And though it gives the same result as direct $\widetilde{W}-$equation Eq. (14), we believe that evoking of the FSD concept (useful elsewhere) is not necessary in this problem.

d. Though the approaches based on the scalar G and progress variable c are equivalent for description of the flamelet sheet kinematics in general or by using one-particular isosurface as well as for statistical moment description of the premixed combustion, the progress variable is certainly preferable as it has clear physical meaning. We believe that evoking the G variable additionally to the universally accepted progress variable c and developing an approach on this base is in fact a superfluous strategy that can bring confusion and unjustified hopes to avoid the basic difficulty of the turbulent combustion modeling by changing variables.

References

[1] Bray, K. N. C., and Moss, J. B., *Acta Astronaut.* **4**:291 (1977).

[2] Bray, K. N. C., *Twenty-Sixth Symposium (International) on Combustion,* The Combustion Institute, Pittsburgh, 1996, p. 1-26.

[3] Veynante, D., Vervisch, L., *Progress in Energy and Combustion Science* **23**: 193 (2002).

[4] Peters, N. *Turbulent Combustion*, Cambridge University Press, 2000.

[5] *Turbulent Reacting Flows*, Edited by P. A. Libby and F. A. Williams, Springer, N.Y., 1980.

[6] S. Monin, A. M. Yaglom, *STATISTICAL FLUID MECHANICS: Mechanics of turbulence*, The MIT Press, 1987, p. 9.

[7] Wilcox, D. C. *Turbulence Modeling for CFD*, DCW Industries, Inc. La Canada, California, 1994, p. 179.

[8] Hasegava, T. and Himeno, R., "Transport properties in turbulent premixed flames", *RINEN Review* No. **30**, September, 2000).

[9] Zimont, V., *Proceedings of the Third Mediterranean Combustion Symposium,* June8-13, Marrakech, Morocco, pp. 1115-1126, 2003.

[10] Vervisch, L., Bidaux, E., Bray, K. N. C., and Kollmann, W., *Phys. Fluids* **7**(10): 2496 (1995).

[11] Pope, S., *Int. J. Engng. Sci*, **26**(5):445 (1988).

[12] Candel, S. M. and Poison, T. J., *Combust. Sci. Technology.* **70**:1 (1990).

[13] Cant, R. S., Pope, S. B., and Bray, K. N. C., *Twenty-Third Symposium (International) on Combustion,* The Combustion Institute, Pittsburgh, 1990, p. 809.

[14] Trouve, A., and Poinsot, T., *J. Fluid Mech.* **278**:1 (1994).

[15] Williams, F. A. in: *The Mathematics in Combustion* (Buckmaster J., Ed.) *SIAM*, Philadelphia, 1985, pp. 97-131.

[16] *Nonsteady Flame Propagation*, Edited by G. H. Markstein, Pergamon Press, 1964.

[17] Damköhler, G., *NACA Tech. Memo.* **1112** (1947).

[18] Shchelkin, K.I., *NACA Tech. Memo.* **1110** (1947).

[19] Batchelor, G. K. *Proc. Roy. Soc. Lond.,* **A213**, 349-366 (1952).

[20] Lipatnikov, A., Chomiak J., *Progress in Energy and Combustion Science* **28** (2002)1-74.

[21] Zimont, V. L., *Experimental Thermal and Fluent Science* **21** (2000) 179-186.

In: New Devel. in Combustion Research
Editor: William J. Carey, pp. 135-162
ISBN 1-59454-826-9

Chapter 3

APPLICATION OF REYNOLDS-AVERAGED NAVIER-STOKES TURBULENCE MODELS IN SIMULATIONS OF NON-PREMIXED FLAMES

Bart Merci*
Ghent University - UGent, Ghent, Belgium
Dept. of Flow, Heat and Combustion Mechanics
Web site: http://www.FloHeaCom.UGent.be/

Abstract

Recent developments in Reynolds-Averaged Navier-Stokes (RANS) turbulence models are discussed in the context of applications in numerical simulations of turbulent non-premixed flames. Both two-equation ($k-\varepsilon$ type) models and second moment closure models are considered. Some results are discussed with both the pre-assumed PDF (probability density function) approach and the transported PDF methodology to model interaction between turbulence and chemistry. In the pre-assumed PDF simulations the conserved scalar approach is followed, with mixture fraction as conserved scalar. Combustion is modelled by assumption of (constrained) chemical equilibrium or by means of a steady laminar flamelet model in the results discussed. In the transported PDF approach detailed chemistry with finite rate kinetics is applied in the form of skeletal schemes or augmented reduced mechanisms. Results are discussed for different turbulent non-premixed jet type flames: piloted jet flames ('Sandia Flame D-F' and 'Delft Flame III') and a bluff-body stabilized flame ('Sydney Flame HM1-3').

*E-mail address: Bart.Merci@UGent.be

1 Nomenclature

D	molecular diffusivity
f	probability density function
F	mass density function
g	mixture fraction variance / gravity
G_b	buoyancy production rate of turbulent kinetic energy
k	turbulent kinetic energy
p	pressure
P_k	production rate of turbulent kinetic energy by turbulent stresses
$\mathbf{S}$	strain rate tensor
$\overline{S}$	chemical source term vector
v_i	velocity component
V_i	velocity component
W	molecular weight
x_i	coordinate direction
y	normal distance from nearest solid boundary
Z	mass fraction
α	blending function
δ_{ij}	Kronecker delta
ε	turbulent dissipation rate
μ	molecular dynamic viscosity
μ_t	turbulent or 'eddy' viscosity
ν	molecular kinematic viscosity
ψ	sample space variable
ρ	density
τ	molecular stress tensor
$\tau^{\mathbf{t}}$	turbulent stress tensor
τ_t	turbulence time scale
ξ	mixture fraction
■	vorticity tensor
"	Favre fluctuation
$\sim$	Favre average
LES	Large-Eddy Simulation
MDF	mass density function
PDF	probability density function
RANS	Reynolds-Averaged Navier-Stokes
RSM	Reynolds stress model
SMC	second moment closure

Subscripts

C	carbon
F	fuel
H	hydrogen
O	oxygen / oxidizer

2 Introduction

Before effectively starting this chapter, I want to apologise to any person who feels that their work should have been referred to, while it has not been included in the list of references. I do not intend to include every valuable paper ever written, and as such many articles or chapters in books are not covered here. Instead I walked through the literature and picked some papers that I consider relevant for the current chapter. By no means any form of completeness is aimed at.

Turbulent mixing is the key process in turbulent non-premixed combustion. Indeed, fuel and oxidizer are only able to react as they mix. As such, the quality of the turbulence model is of utmost importance in numerical simulations of turbulent non-premixed flames. This chapter aims at giving an impression of the application of classical and advanced RANS turbulence models in the context of reacting flows.

In the 70s, major steps forward have been taken in the field of turbulence modelling. Firstly the model that is now known as the standard $k-\varepsilon$ turbulence model, has been formulated[1]. In the same decade, the 'basic' second moment closure (SMC) or Reynolds stress model (RSM) was presented[2]. Inevitably, shortcomings have been discovered after applying these models and studying them more thouroughly during the past thirty years. This has led to improvements over the years. In the next section we discuss some of these improvements, basically restricting ourselves to models that have already been applied in turbulent combustion simulations.

Although this chapter primarily focuses on turbulence modelling, we briefly address the aspect of combustion modelling, too. We consider two strategies to model turbulence - chemistry interaction, within the context of probability density functions (PDF). In the pre-assumed PDF simulations combustion is typically modelled with the conserved scalar approach and the chemistry model provides the relation between the thermo-chemical quantities and the mixture fraction. In the transported PDF method chemistry appears in closed form in the transport equation and finite rate kinetics can be accounted for. Combustion modelling is discussed in section 4.

A brief discussion is devoted to the modelling of turbulence - chemistry interaction in section 5. The importance of the turbulence model in the transport equations is highlighted.

Finally some results are discussed for three turbulent non-premixed flame types in section 6. The first series of piloted flames is 'Sandia Flames D-F'[3]. These flames have a strong pilot annulus and have been very popular in the literature (in particular Flame D). The second flame is again a piloted flame: 'Delft Flame III'[4, 5, 6]. This flame has a much weaker pilot in the form of twelve small holes, rather than a complete annulus. The flame has not yet been discussed in the literature as extensively as the Sandia flame series. Finally the bluff-body burner flame series 'Sydney HM1-3'[7, 8] is considered. These flames do not have a pilot: the actual jet type flame is stabilized by a recirculation region behind the bluff body. This flame series combines a higher degree of flow complexity (which is in a sense more realistic in burner applications) with well-defined boundary conditions, which is an appealing feature for model validation. All flames mentioned have been target test cases in the international workshop series on 'Turbulent Non-Premixed Flames' ('TNF'). The experiments are very well documented and the data are available on the web[9].

We deliberately avoid the discussion about swirling flames. The reason is that simulations with steady RANS turbulence models are not capable of predicting such flows with a sufficient degree of accuracy. As such unsteady simulations are necessary and in the author's opinion LES ('Large-Eddy Simulations') type simulations are best suited. This falls beyond the scope of the current chapter. Completeness obliges to mention, however, that recently promising results have been reported for unsteady simulations with second moment closure RANS turbulence modelling for swirling flames [10]. This is briefly discussed later.

A concluding discussion is presented in section 7.

3 RANS Turbulence Modelling

3.1 Two-Equation Models

We restrict ourselves here to $k-\varepsilon$ type turbulence models. The general remarks are equally valid for $k-\omega$ type models (e.g. [11, 12]) or the v^2-f model[13]. To the best of the author's knowledge these models have not yet been applied very often to non-premixed combustion simulations, so that they are not considered further in this chapter.

In [14] it is explained that two-equation RANS turbulence models rely on an intrinsic hypothesis that the turbulent (or 'Reynolds') stresses can be expressed as a function of the local mean strain rate tensor components, mean vorticity components and turbulent time scale. This is a major difference with the Reynolds stress models (or 'second moment closure' models). The specific hypothesis (still in the terminology used in [14]), which determines the functional expression of the intrinsic hypothesis, is discussed next.

3.1.1 Constitutive Law

In the classical $k-\varepsilon$ model the specific hypothesis is a direct linear relationship between the turbulent stress tensor and the mean strain rate tensor:

$$\tau^t_{ij} = -\rho\widetilde{v''_i v''_j} = \mu_t\tilde{S}_{ij} - \frac{2}{3}\rho k\delta_{ij} , \qquad (1)$$

with the mean strain rate tensor (omitting averaging symbols from now on):

$$S_{ij} = \frac{1}{2}(\frac{\partial v_i}{\partial x_j} + \frac{\partial v_j}{\partial x_i}) - \frac{1}{3}\delta_{ij}\frac{\partial v_l}{\partial x_l} , \qquad (2)$$

In (1) the turbulent (or 'eddy') viscosity must be introduced. The classical expression is:

$$\mu_t = \rho c_\mu k\tau_t , \qquad (3)$$

with τ_t the turbulent time scale. In the standard $k-\varepsilon$ model the expression for this quantity is $\tau_t = \frac{k}{\varepsilon}$. A low-Reynolds number expression ($\tau_t = \frac{k}{\varepsilon} + \sqrt{\frac{\nu}{\varepsilon}}$) is applied in [15, 16].

More important is the expression for c_μ. In the standard $k-\varepsilon$ model its value is kept constant, equal to $c_\mu = 0.09$. This value has been determined from a flat plate boundary layer flow, but it is widely known[14] that this must not be interpreted as a universally valid value. Instead c_μ can be made locally mean flow field dependent. If the expression is chosen appropriately, effects from streamline curvature can be taken into account, as explained in [15].

Regardless of the possible complexity of the expression for the (scalar) quantity μ_t, the linear expression (1) implies alignment of the turbulent stress tensor with the mean strain rate tensor. It is well-known that, even in a simple shear flow, this is not true in reality[14].

A second immediate observation from (1) is its inability to reproduce the physical anisotropy in turbulent normal stresses in e.g. the simple fully developed two-dimensional channel flow.

Both shortcomings mentioned reduce to the fact that expression (1) relies on an isotropic eddy viscosity assumption. One manner to obtain more general expressions for the constitutive law for the Reynolds stress tensor is to use a basis of independent tensors, combining the local mean strain rate and vorticity tensors. In [17] it is explained, relying on the Cayley-Hamilton theorem, that there are only ten linearly independent tensor combinations. In principle, the highest order possible is five. In practice, combinations up to third order are encountered (e.g. [15, 18, 19]). The coefficients in the expression for the Reynolds stresses as a function of these basis tensors must be in terms of tensor invariants.

Another way of arriving at higher order expressions for the turbulent stress tensor is to start from the weak equilibrium assumption in order to reduce the set of transport equations for the Reynolds stresses to a set of algebraic equations[20]. Making this set of implicit equations explicit by imposing a functional dependence of the turbulent stresses on the local mean strain rate and vorticity tensors, leads again to a higher order expression. One of the first models derived in this manner is described in [21], reporting a second order model. In [22] a fourth order expression is given, but in many simple flow conditions it reduces to a second order formulation, too.

Relatively few simulations results have been reported in the literature for this type of higher order models in the context of turbulent non-premixed combustion. Some examples are found in [16, 23]. Several reasons may be found for the moderate success of this type of models in the context mentioned.

A first reason is the poor ratio of added value to added cost of the higher order terms in the constitutive law. Indeed in practically all non-premixed turbulent flames the anisotropy of turbulent normal stresses does not affect the mixing of fuel and oxidizer (and as such the mean combustion process). The reason is that the resulting force from these stresses in the momentum equations is negligible compared to the force associated with the turbulent shear stresses. One could argue that cubic terms have their merits in flows containing swirl. Arguments against this point of view are that the major effects can already be incorporated in a two-equation turbulence model with the linear expression (1): the dynamic balance of static pressure and centrifugal forces associated with swirl is automatically accounted for through the momentum equations, while the primary streamline curvature kinematic effect can be captured in the first order term, provided that the eddy viscosity is formulated so that it becomes sensitive to streamline curvature [24].

A second reason, closely related to the first one, is that the flow field for many well-documented test cases is simple, namely a jet flow. In such a configuration it does not pay off to apply a more complex constitutive law than (1), because the cost of the calculations increases while the accuracy remains unchanged.

A third reason is that, when the flow field really is complex such as in strongly swirling flows, two-equation RANS turbulence models are not suited whatsoever, regardless of the complexity of the expression for the turbulent stress tensor. The representation of turbulence by a single time and length scale involves too much simplification. Some encouraging results have been obtained recently with unsteady second moment closure modelling[10], which is briefly discussed in the next section. Probably LES type simulations are the most reliable for such complex flow configurations (with, as stated, the possible exception of unsteady second moment closure RANS simulations).

A fourth reason may be a time delay between developments in the turbulence modelling research community and application in the combustion community. It is too early to make a strong statement with respect hereto at the moment of writing.

To conclude this section we mention that the cases presented in the current chapter involve some complexity, but remain sufficiently simple to be modelled with a two-equation turbulence model. In the Sandia Flame series there is a central jet, mixing with a low speed co-flow air stream. In 'Delft Flame III' a central jet mixes with a high speed annular flow and a low-speed secondary air co-flow. In the bluff-body stabilized 'Sydney HM1-3' flames there is a recirculation region behind the bluff body, stabilizing the actual jet-like flame behind. A more extensive discussion is given later in this chapter, but we already point out that, although a cubic model is applied in [16], for all cases considered the first order expression (1) is sufficient, provided that the eddy viscosity accounts for streamline curvature.

3.1.2 Transport Equations

Even more important than the expression for the turbulent stress tensor are the transport equations for the turbulent quantities (here the turbulent kinetic energy k and the dissipation rate ε). They enter the constitutive law through the eddy viscosity and, in more complex expressions than (1), through the turbulent time scale in the coefficients of the higher order tensor combinations.

In the literature there is little discussion on the transport equation for the turbulent kinetic energy, although there are some model assumptions involved. For incompressible flows direct term by term modelling of the exact transport equation derived from the exact equations only requires a model for the turbulent dissipation rate (which is provided by the second transport equation) and the transport by velocity and pressure fluctuations. The latter are modelled with a simple gradient diffusion hypothesis, introducing a model constant σ_k. The value of σ_k is usually kept constant ($\sigma_k = 1$), although this is certainly an approximation (e.g. [25]). For reacting flows, a term (G_b) is normally added to account for variable density effects on turbulence. This is not of importance for the test cases under study in this chapter, because the momentum of the flows is much too high for buoyancy to have a significant effect, but it is important in buoyancy driven flows (e.g. in fires). The resulting transport equation reads:

$$\frac{\partial(\rho k)}{\partial t} + \frac{\partial}{\partial x_l}(\rho k v_l) = P_k - \rho\varepsilon + G_b + \frac{\partial}{\partial x_l}[(\mu + \frac{\mu_t}{\sigma_k})\frac{\partial k}{\partial x_l}] \ , \tag{4}$$

with the production rate by turbulent stresses given by:

$$P_k = \tau^t_{ij}\frac{\partial v_j}{\partial x_i} \ . \tag{5}$$

In sharp contrast there has been ongoing discussion on the dissipation rate transport equation. The major difficulty is that term by term modelling of the exact equation is not straightforward. The exact equation focuses on small scale phenomena, while the dissipation rate is mainly governed by large scale processes. The way out in the standard $k-\varepsilon$ model is the construction of a transport equation very similar to the turbulent kinetic energy transport equation, introducing additional model constants and dividing by the turbulent time scale (mainly for dimensional reasons):

$$\frac{\partial(\rho\varepsilon)}{\partial t}+\frac{\partial}{\partial x_l}(\rho\varepsilon v_l)=(c_{\varepsilon 1}P_k-c_{\varepsilon 2}\rho\varepsilon+c_{\varepsilon 3}G_b)\frac{1}{\tau_t}+\frac{\partial}{\partial x_l}[(\mu+\frac{\mu_t}{\sigma_\varepsilon})\frac{\partial\varepsilon}{\partial x_l}] \ . \tag{6}$$

Despite the large amount of empiricism in its derivation, this expression was supported later by renormalization group theory[26] and a time scale interpretation[27].

In a first attempt to overcome limitations of equation (6) the model 'constants' $c_{\varepsilon 1}$ and $c_{\varepsilon 2}$ have been reformulated in terms of mean strain rate and vorticity, so that certain theoretical consistency conditions could be fulfilled (e.g. [28]). Such an expression is used in [15, 16]. The major deficiency with respect to non-premixed turbulent flames, however, is the plane jet - round jet anomaly: the spreading rates of plane and round jets cannot be predicted correctly with a single set of values for the model constants $c_{\varepsilon 1}$ and $c_{\varepsilon 2}$. Their values can be tuned for one type (the standard values are suited for plane jets), but not for both at the same time. Even local flow field dependence does not cure this shortcoming. It is an inherent deficiency of the ε transport equation, formulated as in (6).

This was recognized in [29], where a new transport equation for the dissipation rate was suggested, based on a modelled transport equation for enstrophy (i.e. mean square of the vorticity fluctuations). They combine this transport equation with an advanced expression for c_μ in the eddy viscosity so that the most important realizability conditions (i.e. conditions in order not to produce physically impossible solutions) are fulfilled. As a consequence this model is sometimes referred to as the 'realizable' $k-\varepsilon$ model. In [30] the dissipation rate transport equation presented in [29], that was only valid for high Reynolds incompressible flows, has been extended for use in turbulent non-premixed flames. The result is:

$$\frac{\partial(\rho\varepsilon)}{\partial t}+\frac{\partial}{\partial x_l}(\rho\varepsilon v_l) \quad = \quad C_1 S\rho\varepsilon-c_{\varepsilon 2}\frac{\rho\varepsilon}{\tau_t}+c_{\varepsilon 3}\frac{G_b}{\tau_t}+\frac{\partial}{\partial x_l}[(\mu+\frac{\mu_t}{\sigma_\varepsilon})\frac{\partial\varepsilon}{\partial x_l}] \ . \tag{7}$$

The major difference with equation (6) is the 'production' rate of turbulent dissipation rate: in eq. (7) this is no longer proportional to the production rate of turbulent kinetic energy, but proportional to the strain rate $S=(2S_{ij}S_{ij})^{1/2}$. As such the Reynolds stresses do not appear in the generation term of turbulent dissipation rate. The large benefit is the correct prediction of the spreading rates of both the planar and the round jet, so that the model becomes more generally valid without test case dependent parameter tuning. Some draw-backs must be kept in mind, though. Firstly some 'self-regulation', present in the standard $k-\varepsilon$ model, is lost when equation (7) is used. This is easily understood: if in the standard $k-\varepsilon$ model the production term P_k for turbulent kinetic energy is over-estimated (or under-estimated), the production term for the dissipation rate is over-estimated (under-estimated) similarly, so that the effect is weakened. In eq. (7) the behaviour is no longer linear. As a result the model with eq. (7) is much more sensitive to the expression for c_μ

than when the classical transport equation (6) is used, so that care must be taken if one would like to modify this expression. Secondly it is not straightforward to apply equation (7) in the neighbourhood of solid boundaries, because it relies on a relationship between the turbulent dissipation rate and enstrophy that is only valid in high-Reynolds number flow regions. Extension for use near walls is not straightforward. Finally, the model with eq. (7) is more sensitive to inlet boundary conditions than the standard $k-\varepsilon$ model, which can again be explained through the absence of a self-regulative mechanism.

In [31] a solution has been suggested to remove to some extent the sensitivity to inlet boundary conditions and to make the 'realizable' model applicable in the neighbourhood of solid boundaries. The idea is to blend equations (6) and (7) with a blending function α going from zero near solid boundaries to one sufficiently far away from walls. The exact expression of α is not crucial. In [31] $\alpha = 0$ if $R_y = \frac{\rho k^{1/2} y}{\mu} < 1000$ and $\alpha = 1$ if $R_y > 2000$, while α evolves sinusoidally in between. More important is the final expression for the dissipation rate transport equation:

$$\frac{\partial(\rho\varepsilon)}{\partial t} + \frac{\partial}{\partial x_l}(\rho\varepsilon v_l) = (1-\alpha)c_{\varepsilon 1}\frac{P_k}{\tau_t} + \alpha C_1 S\rho\varepsilon - c_{\varepsilon 2}\frac{\rho\varepsilon}{\tau_t} + c_{\varepsilon 3}\frac{G_b}{\tau_t} + \frac{\partial}{\partial x_l}[(\mu + \frac{\mu_t}{\sigma_\varepsilon})\frac{\partial\varepsilon}{\partial x_l}] \quad (8)$$

To conclude this section it is recalled that the standard set of model constants for the classical $k-\varepsilon$ model is suited for the spreading rate of planar jets, while the spreading rate of round jets is overpredicted. In general this is remedied by either increasing $c_{\varepsilon 1}$ from $c_{\varepsilon 1} = 1.44$ to $c_{\varepsilon 1} = 1.6$ or by decreasing $c_{\varepsilon 2}$ from $c_{\varepsilon 2} = 1.92$ to $c_{\varepsilon 2} = 1.83$. While this is justified for configurations where the flow field is relatively well known a priori, such practice does not serve as great support for a model's general validity. A major strength of the turbulence model, consisting of transport equation (8) and constitutive law (1), where the eddy viscosity automatically accounts for streamline curvature, is the wide range of applicability without test case dependent model parameter tuning. As such it becomes possible to predict flows to a certain extent rather than just to reproduce experimental measurements.

3.2 Reynolds Stress Models

In this class of RANS turbulence models an individual model transport equation is solved for each turbulent stress. Due to tensor symmetry this implies (in a three-dimensional simulation) six transport equations instead of one (for turbulent kinetic energy) in two-equation turbulence models. This immediately reveals one major disadvantage of second moment closure models: they are by far more time consuming. The situation is even worse: simulations with Reynolds stress models are numerically in general less stable, so that convergence is not always easy to achieve. This implies a large CPU time increase.

Different Reynolds stress models are applied in non-premixed combustion applications in the literature. Three examples are the basic LRR-IP ('Launder, Reece, Rodi - Isotropization of Production') model[2], the SSG ('Speziale, Sarkar, Gatski') model[32] and the JM ('Jones, Musonge') model[33]. These models mainly differ in the pressure - rate-of-strain modelling. In the LRR-IP model the classical linear return-to-isotropy model of Rotta[34] is used to model the slow part of the pressure - rate-of-strain tensor. This is combined with a result from RDT (rapid distortion theory) that expresses the isotropization of production by the rapid part of the tensor. In the SSG model the slow part contains a non-linear return to isotropy, in better agreement with reality. The JM model was specifically designed for reacting flows, paying attention to variable density effects. For the slow part of the pressure - rate-of-strain tensor the Rotta model is applied. In [35] this model is extensively discussed. In [14] the most important features and shortcomings are discussed for the different Reynolds stress models. No existing models satisfies all possible theoretical constraints, but from a practical point of view it may be more important that good agreement of results with experimental data is obtained for realistic flows. In general the LRR-IP has a good over-all performance for a wide variety of flow types.

It is important to appreciate that the length scale determing transport equation remains necessary for second moment closure models. The potential to reproduce the turbulent stresses more accurately than with two-equation turbulence models does not remove the need for a turbulent length scale to close the individual transport equations.

In the context of non-premixed combustion, Reynolds stress models have not been applied as often as two-equation turbulence models. The major reason is the increase in computing time, while there is often relatively little gain in accuracy. As explained in the previous section this is mainly due to the fact that, in the mixing process of fuel and oxidizer, the detail of individual turbulent stresses is by far of less importance than the resulting force from the turbulent shear stress. The author sees, however, some areas where second moment closure may prevail over (advanced) two-equation turbulence models in non-premixed combustion simulations: unsteady flows, mainly because the assumption of alignment of the turbulent stress tensor with the mean strain rate tensor is not incorporated (see also next section); swirling flows, although definitely further research is required[36]; and transported PDF simulations (see later). In the latter simulations most of the total computing time is put into the solution of transport equations for particles and inm the chemistry part, so that the disadvantage of longer computing times with Reynolds stress models compared to two-equation turbulence models disappears to a large extent. Moreover there is always the possibility to derive a second moment closure model from the turbulence model applied in the PDF transport equation (when velocity components are included as PDF density variables)[37], so that for consistency reasons the use of second moment closure models may be required.

3.3 Order of Importance

In this section we give a personal impression on the importance of the different aspects of a turbulence model, based on personal experience and literature review.

The most important aspect seems to be the transport equation for the turbulent length scale determining quantity. Typically this is the dissipation rate transport equation, but the turbulent frequency or enstrophy (i.e. mean square vorticity fluctuations) are sometimes used, too. In two-equation turbulence models the value of the eddy viscosity is directly affected by the turbulence quantities obtained from the transport equations. In possible higher order terms in the constitutive law the turbulence quantities enter directly through the turbulent time scale. As mentioned, the length scale transport equation remains necessary in Reynolds stress models, too, and is often considered the weak spot. Indeed many ad hoc adjustments are found in the literature in order to improve the results for a specific test case. For a generally applicable model this is to be avoided. Further research may still lead to improvements in this field.

Another aspect we consider sufficiently important to be included is the effect of rotation or streamline curvature on turbulence. This is readily incorporated in the definition of the eddy viscosity in two-equation turbulence models. In other words a lot of accuracy quality can be gained at very moderate (CPU time) expenses. In Reynolds stress models this is automatically accounted for.

The discussion becomes more subtle when it comes to second order terms in the constitutive law in two-equation models. Beyond any doubt certain aspects are lost when second order terms are omitted. Anisotropy in turbulent normal stresses cannot be described. A well-known example is the secondary vortex flow pattern in the corners of square channel flow. Another deficiency related to the lack of quality of the individual turbulent normal stresses due to omission of second order terms is the fact that application of a generalized gradient diffusion model for e.g. turbulent heat transfer is not straightforward. Second moment closure models do not suffer from these shortcomings. However, in the context of turbulent diffusion flames it is, in the author's opinion, not worthwile to include second order terms for the reasons mentioned, since the gain in quality of results is often small (unless of course one is particularly interested in e.g. turbulent normal stresses or other detailed flow features).

A more important issue may be the inevitable alignment of the turbulent stress tensor with the mean strain rate tensor when a first order expression is applied. Inclusion of second order terms removes this limitation. Reynolds stress models clearly do not rely on this alignment assumption, either. This may be the reason why unsteady simulations seem much more reliable when second moment closure models are applied than when a first order $k-\varepsilon$ model is applied[10]. With the latter typically hardly any unsteadiness is recovered, probably due to excessive values of eddy viscosity, in combination with alignment between the turbulent stress tensor and the mean strain rate tensor. Indeed this alignment implies that the production of turbulent kinetic energy (5) is as large as possible for certain individual tensor component values. This requires further investigation.

Inclusion of cubic terms in the constitutive law in two-equation turbulence models does not seem reasonable. These third order terms mainly have an effect in swirling flows and in principle may be valuable. In practice, however, the framework of steady simulations with two-equation turbulence models is not suited for flows with such a degree of complexity. As a result it is not feasible to add complexity within a frame that is not suited in the first place. For swirling flows, either with or without combustion, it is, in the author's opinion, much more useful to apply LES-type turbulence models or unsteady RANS calculations with a second moment closure model, as in [10].

4 Combustion Modelling

4.1 Conserved Scalar Approach

For the simulations with the pre-assumed PDF approach to model turbulence - chemistry interaction (see next section), combustion is modelled by means of a chemistry model based on a conserved scalar.

The normalized mixture fraction acts as the conserved scalar. It is defined as[38]:

$$\xi = \frac{\frac{2(Z_C - Z_{C,O})}{W_C} + \frac{Z_H - Z_{H,O}}{2W_H} - \frac{Z_O - Z_{O,O}}{W_O}}{\frac{2(Z_{C,F} - Z_{C,O})}{W_C} + \frac{Z_{H,F} - Z_{H,O}}{2W_H} - \frac{Z_{O,F} - Z_{O,O}}{W_O}} . \quad (9)$$

This quantity takes on the value $\xi = 0$ in the case of pure oxidizer and $\xi = 1$ for pure fuel.

The combustion model establishes the relationship between thermo-chemical quantities (species mass fractions, temperature, density) and the value of the mixture fraction. Only combustion at atmospheric pressure is considered and the ideal gas law is inherently applied to establish a relation between density and temperature (and composition).

Different chemistry models are possible. In the present chapter we restrict ourselves to application of full or constrained[39] chemical equilibrium and steady laminar flamelet modelling. In the latter some finite rate kinetics are indirectly accounted for in the construction of the laminar flamelet table. In its simplest form one table is constructed with a single strain rate. The value of this strain rate is assumed to be representative for the entire domain, which is a serious simplification of reality. A more advanced approach concerns the use of a library of flamelets with variable strain rate. Depending on the locally detected strain rate during the iterative procedure in the actual flame simulations, the corresponding flamelet table is chosen or interpolation is performed between tables from the library. We come back to this point during the discussion of the results.

Another way to incorporate finite rate kinetics, making use of 'progress variables', concerns the intrinsic low dimensional manifold (ILDM) technique[40]. Apart from mixture fraction, so-called 'progress variables' must be chosen. Typically CO_2 and H_2O species mass fractions are picked as independent variables. We do not go into any further detail here.

4.2 Finite Rate Kinetics

A major advantage of the transported PDF methodology as turbulence - chemistry interaction model (see next section) is the appearance of chemistry source terms in closed form. As a result, it is possible to include detailed chemistry and finite rate kinetics in a relatively straightforward manner.

Typically the PDF (or actually the MDF) transport equation is solved by following a large number of numerical particles, representing the PDF. A Monte Carlo solver is used due to the high dimensionality. Addition of species is in principle straightforward: it is merely an addition of properties to the particles followed. However, there are practical computing limitations to this procedure: the dimensionality, and thus the CPU time and memory requirements, increases as the number of species included increases. In practice, chemistry models with more than 20 species are rarely encountered in turbulent non-premixed flame simulations with the transported PDF approach.

In the literature several skeletal schemes are applied (e.g. [41, 42]). More recently so-called 'Augmented Reduced Mechanisms' (ARM) have also been applied (e.g. [9, 43, 44, 45]). ARM chemistry models consist of global reaction steps and are typically derived from the GRI2.11 mechanism[46] with the computer assisted reduction mechanism (CARM) code[47].

Note that, when only the mixture fraction is taken as property for the particles, the chemistry models of the previous section can still be applied (see later). A such finite rate kinetics can still be incorporated into the simulations through the use of ILDM or flamelet modelling in combination with the mixture fraction.

5 Turbulence - Chemistry Interaction

There is interaction between turbulence in the flow field and the combustion processes taking place. Influence from combustion on turbulence is usually 'accounted for' through the use of Favre averages (i.e. mass-weighted averages) instead of Reynolds averages (or 'ensemble averages'). It is then assumed that the turbulence models, developed for incompressible flows, remain valid without modification for variable density flows, provided that Favre averaged quantities are used. The words 'accounted for' are put in between brackets because the application of Favre averaging is merely a mathematical construction. There is in principle no a priori physical justification that indeed the turbulence models for incompressible flows can still be used. We do not dig into this discussion here, but clearly further research in this are might lead to valuable new insights.

The influence from turbulence on combustion is important to be incorporated. The use of mean values and fluctuations around those mean values, as is actually done in turbulence models, is not straightforward because of the strong non-linearity of the relationships in combustion processes. The solution to this averaging problem is found in the application of probability density functions, with a weighted integration. We restrict ourselves to non-premixed combustion again. The Favre averaged thermo-chemical quantities are determined from a weighted integration with the Favre PDF, with mixture fraction as integration variable:

$$\tilde{\phi} = \int_0^1 \phi(\zeta)\tilde{f}(\zeta)d\zeta \,. \tag{10}$$

The relation $\phi(\zeta)$ is determined by the chemistry model. In particular for density, where the Reynolds averaged value is required, the expression becomes:

$$(\overline{\rho})^{-1} = \int_0^1 (\rho(\zeta))^{-1}\tilde{f}(\zeta)d\zeta \,. \tag{11}$$

The two major PDF methods are now briefly addressed in the coming sections.

We do not consider conditional moment closure (CMC)[48] in this chapter, since our focus is on turbulence modelling. Still, CMC is becoming increasingly important as computer power increases.

5.1 Pre-assumed PDF

As explained in the previous section, chemistry is modelled through the establishment of relations between thermo-chemical quantities (species mass fractions, temperature, density) and mixture fraction. These relations are always highly non-linear, regardless of the chemistry model applied.

The classical way to incorporate the effect of turbulence on chemistry is through a 'pre-assumed' PDF approach: the class of functions the PDF belongs to, is pre-assumed and the exact shape of the local PDF is determined from the lowest two statistical moments (i.e. mean value and variance). Typically the PDF is pre-assumed to belong to the class of β-functions, because these can take on a wide variety of shapes, depending on the values for mean and variance.

The lowest two statistical moments are obtained from the transport equations for mean mixture fraction ξ and its variance g:

$$\frac{\partial(\rho\xi)}{\partial t}+\frac{\partial}{\partial x_j}(\rho\xi v_j)=\frac{\partial}{\partial x_j}[(\rho D+\frac{\mu_t}{\sigma_\xi})\frac{\partial\xi}{\partial x_j}] \tag{12}$$

$$\frac{\partial(\rho g)}{\partial t}+\frac{\partial}{\partial x_j}(\rho g v_j)=\frac{\partial}{\partial x_j}[(\rho D+\frac{\mu_t}{\sigma_g})\frac{\partial g}{\partial x_j}]+2\frac{\mu_t}{\sigma_\xi}\frac{\partial\xi}{\partial x_l}\frac{\partial\xi}{\partial x_l}-\rho\chi\,. \tag{13}$$

The scalar dissipation rate χ has to be modelled. It is typically directly related to the value of mixture fraction variance and the turbulent time scale:

$$\chi \;=\; c_g\rho g\frac{1}{\tau_t}\,. \tag{14}$$

The values for σ_ξ and σ_g range from 0.7 to 0.9 in the literature. The standard value for c_g is $c_g = 2$. Research is vividly ongoing in this area and new insights are developed. A recent special issue of the journal 'Flow, Turbulence and Combustion' (Vol. 72, 2004), entirely devoted to scalar dissipation, illustrates this. More advances modelling of the scalar dissipation rate in (13) may rely on a transport equation for χ. In this modelled transport equation higher order terms must be modelled again. We do not discuss this here any further.

The procedure is now as follows. From equations (12) and (13) the PDF $\tilde{f}$ is determined within the class of β-functions. Then integrations (10) and (11) are performed in order to determine average quantities. During the iterative solution of the global set of conservation equations the mean density is the most important quantity, because it is directly coupled back to the mean continuity equation and the momentum equations (and possibly the energy equation). As such it directly influences the mean flow field. In practice a table is constructed a priori with two independent entries (mean mixture fraction and variance). The Favre averaged quantities are then looked up and interpolated, so that the integration does not need to be repeated at every iteration step. An order of magnitude in CPU time is gained in this manner.

For completeness we mention that, when static enthalpy cannot be directly related to the mixture fraction, such as when radiation is important, a third independent entry appears in the table (e.g. static enthalpy). Here we do not deal with this matter. Radiation is not considered.

The importance of the turbulence model is again visible at different positions. First of all the eddy viscosity appears at several positions in transport equations (12) and (13): in the diffusion terms, but also in the production term of mixture fraction variance. Note that, when a second moment closure model is applied, more advanced diffusion models and production terms can be constructed. This has in general the advantage that better results are obtained for the mixture fraction variance than with two-equation turbulence models.

This may again be due to both an over-estimation of eddy viscosity on the one hand and the alignment of vectors on the other hand (similar to the discussion on the unsteadiness for the turbulence models: first order turbulence models always assume alignment of the turbulent stress tensor with the mean strain rate tensor and hardly any unsteadiness is observed).

A second point where the turbulence model exerts a strong influence is through the turbulent time scale in the model for the scalar dissipation rate. In principle a transport equation can be derived and modelled for this quantity, but there again the eddy viscosity enters and new terms have to be closed.

Last but not least the turbulence model also strongly influences the mean flow field, so that the mean convection terms in the transport equations for mean mixture fraction and mixture fraction variance are also determined by the turbulence model.

5.2 Transported PDF

Already back in 1985 the cornerstone article on transported PDF methodology to model interaction between turbulence and chemistry was published[49]. The general PDF transport equation, when velocity components are included as independent density variables, reads:

$$\begin{aligned}\rho(\overline{\psi})\frac{\partial f}{\partial t} &+ \rho(\overline{\psi})V_j\frac{\partial f}{\partial x_j} + (\rho(\overline{\psi})g_j - \frac{\partial < p >}{\partial x_j})\frac{\partial f}{\partial V_j} + \frac{\partial}{\partial \psi_\alpha}[\rho(\overline{\psi})S_\alpha(\overline{\psi})f] \\ &= \frac{\partial}{\partial V_j}[< -\frac{\partial \tau_{ij}}{\partial x_i} + \frac{\partial p'}{\partial x_j}|\overline{V},\overline{\psi} > f] + \frac{\partial}{\partial \psi_\alpha}[< \frac{\partial J_{i,k}}{\partial x_i}|\overline{V},\overline{\psi} > f] \ . \qquad (15)\end{aligned}$$

Note that $\overline{\psi}$ is a vector is sample space variables. The terms on the left hand side of eq. (15) appear in closed form: transport in physical space, transport in velocity space by gravity and the mean pressure gradient and transport in composition space by reaction. As such chemical source terms appear in closed form which is, as mentioned, a major strength of this method. The terms on the right hand side need to be modelled: transport in velocity space by the molecular viscous stresses and the fluctuating pressure gradient and transport in composition space by the molecular fluxes. We come back to this later.

If the velocity components are not included as independent density variables, in other words when a joint scalar PDF is used, transport equation equation (15) can be simplified. Integration over velocity space and expressing the transport equation in terms of the mass density function (MDF) $F(\psi) =< \rho > \tilde{f}_\phi(\psi)$ yields:

$$\frac{\partial F}{\partial t} + \frac{\partial}{\partial x_j}(V_j F) + \frac{\partial}{\partial \psi_\alpha}(S_\alpha(\overline{\psi})F) = \frac{\partial}{\partial \psi_\alpha}[< \frac{1}{\rho}\frac{\partial J_{i,k}}{\partial x_i}|\overline{\psi} > F] - \frac{\partial}{\partial x_j}[v"_j|\overline{\psi}] \ . \qquad (16)$$

Again only the terms on the right hand side need to be modelled: transport in composition space due to molecular mixing and transport in physical space due to velocity fluctuations.

When the velocity components are included as density variables (eq. (15)), the turbulence model is incorporated in the modelled PDF transport equation. Due to the high dimensionality of the problem, the PDF transport equation is often solved by means of a Lagrangian method, following a large number of particles. A Langevin type model can then be introduced to model the turbulent motion of the particles[14]. After integration a

Reynolds stress model can be related to the Langevin model[37], with the additional advantage that such a model is automatically realizable (i.e it always produces physically possible solutions). In practice, however, such a stand-alone method is prone to statistical bias error, which led to the development of the so-called 'hybrid' method[50] that is now mostly applied. In this hybrid model turbulence is modelled in a finite volume code and this information is incorporated in the PDF transport equation. Care must be taken that no consistency conditions are violated and correction algorithms have been developed to ensure this[51]. Note that the transport equation for a length scale determining quantity (e.g. turbulent dissipation rate) is still required: the one-point joint velocity-scalar PDF does not contain this information, so it must be externally provided. It is typical to use turbulent frequency as variable in this context[14].

When the velocity components are not included as PDF density variables (eq. (16)), it is clear that turbulence must still be modelled in the finite volume manner. In transported scalar PDF simulations it is relatively straightforward to maintain consistency: the mean velocity (deterministic particle motion) and turbulent diffusivity (random particle motion) are taken from the turbulence model. This can either be a two-equation model or a Reynolds stress model. Again the length scale transport equation remains necessary.

A major advantage of the transported PDF technology is, as mentioned, that the chemical source term appears in closed form. Still, due to the strong non-linearity the system of equations becomes very stiff, so that much computing time is devoted to the chemistry. A priori general tabulation, as is done in pre-assumed PDF calculations, is not possible due to the high dimensionality (one independent entry for each species in the chemistry mechanism). A large improvement in efficiency is obtained, though, by application of the ISAT ('In Situ Adaptive Tabulation') technique, as explained in [52]. A look-up table is constructed in practice during the iterative solution procedure for the global set of equations to be solved. Pope indeed recognized that only a small portion of the entire composition space is occupied in reality[53], so that once a sufficient number of points have been computed many other points can be found by interpolation between previously tabulated computed points, which is a much faster process. Still, a large portion of the total computing time is devoted to the chemistry part, so that, as previously mentioned, from a turbulence modelling point of view a major disadvantage of Reynolds stress models, compared to two-equation turbulence models, disappears.

The weak spot in the transported PDF technology is the micro-mixing model, governing the small scale mixing process. In the PDF transport equation this is the transport term in composition space. This is a very active research area. Classical mixing models are the IEM ('Interaction by Exchange with the Mean') model[54], the modified Curl"s CD ('Coalescence/Dispersion') model[55] and the EMST ('Euclidean Minimum Spanning Tree') model[56]. We do not go into the discussion of merits and shortcomings of each of these models. In general terms the more recent EMST model seems physically the most correct one.

It is worth highlighting the importance of the turbulence model in the transported PDF methodology. In the transported scalar PDF method (eq. (16)) the turbulence model significantly comes in at three places: first of all it strongly determines the mean flow field and as such the mean convection term in eq. (16) on the left hand side; secondly it determines the (turbulent) time scale in the micro-mixing model, so that it strongly influences the closure of this term on the right hand side of eq. (16); and thirdly it determines the turbulent transport in physical space (or the random motion of the particles, stated in another manner), i.e. the closure of the second term on the right hand side of eq. (16). When the velocity components are included as independent variables in the PDF, eq. (15) is solved and the turbulence model is already incorporated in the modelled PDF transport equation. Still, the turbulence model remains important for all the aspects mentioned.

Note that the domains on which the turbulence model exerts influence are the same as for the pre-assumed PDF simulations. This need not be surprising: integration of the transported PDF or MDF equation results in the transport equation for the statistical moments of the PDF. As such the transport equations for the mean mixture fraction and the mixture fraction variance in the pre-assumed PDF framework are recovered. The mixing rate in the micro-mixing model is then directly related to the scalar dissipation rate in the mixture fraction variance transport equation.

Finally it is noteworthy that length scale information must still be externally provided in the framework of one-point PDFs. This can be done by either incorporating the turbulent frequency as independent PDF variable (i.e. add this as a particle property) or by the solution of a turbulent length scale equation (or typically the dissipation rate transport equation). Again this equation then plays a very important role, as described earlier in this chapter.

6 Discussion of Results

Our purpose is not a quest for the ultimate models to use. Rather we try to give an impression as to what models have already been applied to different turbulent non-premixed flames, what the problems are and what success has already been booked. We will not present any detailed profiles, but rather give a more general description.

6.1 Piloted Jet Diffusion Flame ('Sandia Flame D')

As already mentioned the Sandia Flames have been very popular in the literature. In particular Flame D[3] has been simulated by various research groups.

Most of the results have been discussed at the TNF ('Turbulent Nonpremixed Flames') workshop series and can be consulted at the web site[9]. Here we focus on turbulence modelling related issues, although some statements about chemistry and turbulence - chemistry interaction are made, too.

A complete test case description is found at the TNF web site[9]. The burner configuration is shown in fig. 1. The central fuel jet consists of 75% air and 25% methane (by volume). It is surrounded by a co-flow air stream. These two flows mix and combustion takes place. The actual flame is stabilized by an annular pilot gas stream in between the fuel jet and the air stream. The pilot flame concerns a lean mixture of C_2H_2, H_2, air, CO_2 and N_2. It creates a high temperature region near the burner that stabilizes the flame. The burner configuration is placed inside a vertical wind tunnel but the flame is unconfined. The configuration has been operated at various Reynolds numbers of the central fuel jet. Based on the mean fuel velocity and the central diameter, the Reynolds number is $Re = V_m D_n / \nu = 22400$ for Flame D. This corresponds to a mean velocity of $49.6 m/s$. The mean pilot stream velocity is $11.4 m/s$ and the co-flow air has a free stream velocity of $0.9 m/s$. For Flames E and F the Reynolds number is $Re = 33600$ and $Re = 44800$ respectively and the pilot flame power is proportionally increased.

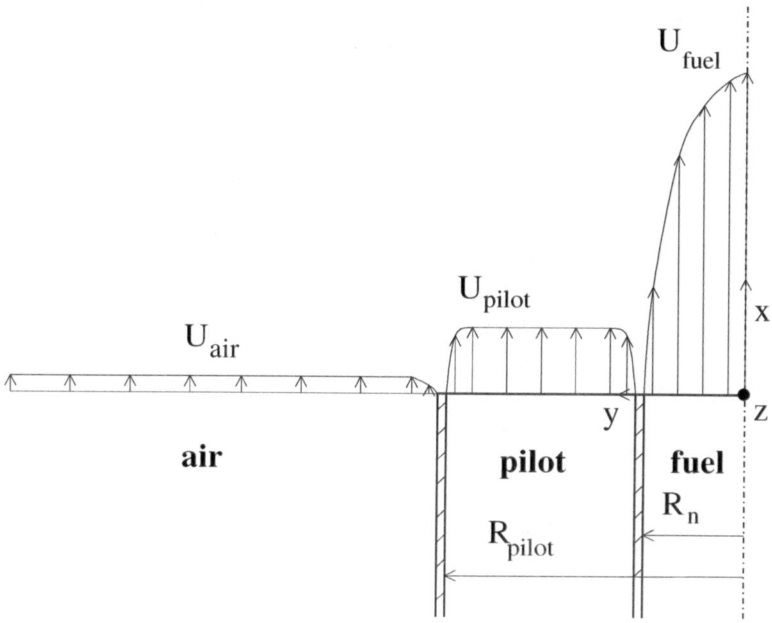

Figure 1: Burner configuration of the Sandia piloted jet non-premixed flames[3].

Figure 2 shows contours of mean temperatures for Flame D as obtained with the standard $k-\varepsilon$ model without model constant adjustment (left side) and with the non-linear $k-\varepsilon$ model of [16] (right side). The difference in flame structure due to the turbulence model is very clear. With the standard $k-\varepsilon$ model the flame is clearly much shorter. In [16] it is shown that the flame length and the radial flame structure is much better predicted with the non-linear two-equation model. The turbulent mixing pocess is very well reproduced.

The result of fig. 2 has been obtained in the framework of the pre-assumed β-PDF for turbulence - chemistry interaction. This is justified because there is relatively little local extinction. The constrained equilibrium chemistry model of [39] has been applied in order to improve CO species mass fraction concentrations, compared to full chemical equilibrium assumption.

Still, simulations have been reported where the transported PDF methodology has been applied in combination with detailed chemistry, so that finite rate kinetics effects are incorporated. In particular at the rich side of stoichiometric ($\xi_{st} \approx 0.35$), improvement of minor species (CO, OH and NO) was desired. Another argument in favour of the transported PDF approach for the Sandia flames is that local extinction becomes much more pronounced for Flames E and F, due to severe turbulence - chemistry interaction for the higher Reynolds numbers. Thus it is impossible to obtain reasonable results within the pre-assumed PDF framework for these two flames, since use is made of a chemistry model that assumes reaction as soon as the mixing is done.

Since the primary aim, apart from a more realistic description of turbulence - chemistry interaction and local extinction (and re-ignition) for flames E and F, was a better description

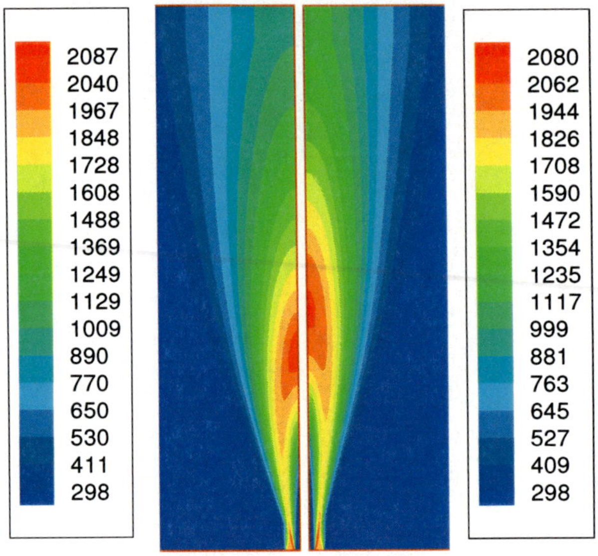

Figure 2: Difference in flame structure due to the turbulence model (left: standard $k-\varepsilon$ model; right: non-linear model[16]).

of the chemistry modelling, it is not surprising that the papers in the literature focus on this topic. Still some turbulence modelling aspects are discussed here, too. In [57, 58] Sandia flames D - F are simulated with a joint velocity - composition - turbulence frequency PDF method. This means that the velocity components and the turbulence frequency are included as independent PDF variables. Turbulence is thus modelled through a Langevin type model within the PDF transport equation. As already mentioned the flow is relatively simple, so that it is no real challenge for the turbulence model. In [57, 58] it was found that inclusion of C_2 species in the chemistry model is important for CO species mass fractions. In [57] an augmented reduced mechanism ('ARM2') is applied containing 19 species and 15 global reaction steps (derived from GRI-mech 1.2), whereas in [58] the ARM scheme of [44] is used, containing 16 species and 12 global reaction steps. It was found that inclusion of a radiation model is important for NO predictions. Apart from this NO prediction there are only small differences between ARM2 and the ARM mechanism of [44].

In [45] the same set of flames is studied with a joint scalar PDF method. Turbulence is modelled with the SSG Reynolds stress model [32]. Important to note is that the model constant $c_{\varepsilon 2}$ is lowered from $c_{\varepsilon 2} = 1.92$ to $c_{\varepsilon 2} = 1.8$ in order to correctly predict the fuel jet spreading rate. This illustrates that the turbulent dissipation rate transport equation remains a wek spot for second moment closure models and that no real gain in quality is obtained from solving the individual Reynolds stress components transport equations. In [45] it is confirmed that the inclusion of C_2 chemistry is important in order to obtain accurate results for the minor species.

For completeness it is mentioned that the Sandia Flame simulation results are rather sensitive to the inlet boundary conditions applied. This is mostly the case for LES results, where not only the level of mean velocities and turbulent kinetic energy must be imposed, but also the spectral distribution of the fluctuations. The major problem in RANS simulations is the determination of the turbulent length scale at the inlet or in other words the inlet boundary condition for the turbulent dissipation rate. In [59, 60] these problems are discussed extensively.

Finally we remark that the Sandia flames have a relatively strong pilot to stabilize the actual flames, so that in numerical transported PDF simulations it is relatively easy to obtain flame ignition when the experimental pilot flame power is appplied. Moreover, the flame is 'perfectly' axisymmetric, i.e. there are no three-dimensional effects due to the burner configuration. These features are different from the piloted jet diffusion flame of the next section.

6.2 Piloted Jet Diffusion Flame ('Delft Flame III')

This test case has been studied experimentally at Delft University[4, 5, 6]. It must be noted that the more recent measurements overrule the earlier measurements of [4], because they are more accurate.

The burner configuration is shown in fig. 3. The central fuel jet consists of Dutch natural gas. This jet emerges from the central pipe. The Reynolds number, based on mean velocity and central pipe diameter, is $Re = 9700$. Around the central jet a primary air annulus supplies the air for combustion. A secondary co-flow air stream surrounds the burner nozzle. Specific nozzle features are a constriction of the central pipe (from $d = 8mm$ to $d = 6mm$) and the conical shape of the primary air annulus. This implies at first sight more complicated inlet boundary conditions to impose in numerical simulations, but in [61] it is illustrated that the test case is rather insensitive to the detail of the inlet boundary conditions (in contrast to the previous test case). This makes the test case appealing to modellers, since one possible source of error is eliminated. The determination of the inlet boundary condition for the turbulent dissipation rate remains the most crucial part[59].

A more important practical feature for the simulations is that in the experimental nozzle the pilot streams emerge from twelve small separate holes, rather than from an annulus. As a consequence the test case is not perfectly axisymmetric and three-dimensional flow effects may come into play. On the other hand a large factor of CPU time is gained when the numerical simulations can be performed in a plane, relying on axisymmetry. For pre-assumed PDF simulations this poses no problem: the actual flame ignites anyway due to the infinitely fast chemistry, even in the absence of the pilot flames. In [61] it is indeed argued that the pilot flames can be omitted without disturbing the downstream flow field (and thus the actual flame). For transported PDF simulations it can be a problem. Without introduction of the pilot flame thermal power, the flame numerically extinguishes when finite rate chemical kinetics are incorporated. There are several options to incorporate the pilot flames: in [62] the pilot flames are omitted when the constrained equilibrium chemistry model is applied in combination with the mixture fraction approach for the particles in the transported PDF simulations, for the reason just mentioned (namely that the flame ignites anyway); when the ILDM[40] chemistry model is applied, the problem of numerical flame extinction is solved by imposing chemical equilibrium in the pilot flame region instead of application of ILDM; in [61] the pilot flames are modelled by addition of a volumetric heat source in the pilot flame region. Probably the best way to model the pilot flames is through a separate three-dimensional simulation, but this is ongoing research at the moment

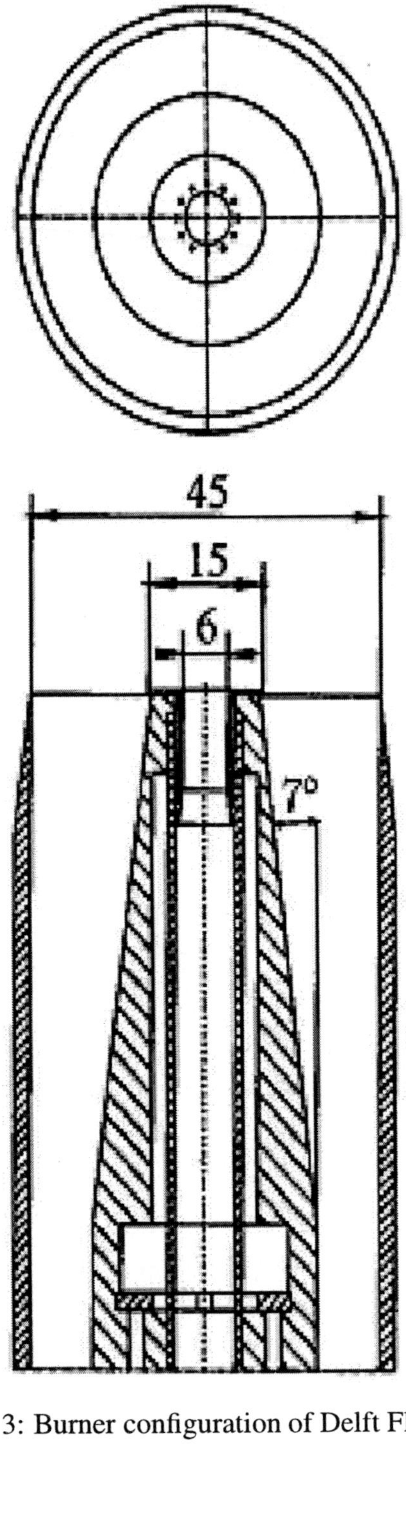

Figure 3: Burner configuration of Delft Flame III.

of writing.

Not many publications have appeared yet on this test case. To the best of the author's knowledge, Reynolds stress modelling has not yet been applied. In [62] the standard $k-\varepsilon$ turbulence model is applied, but with addition of the round jet correction of [63]. This is again necessary to correctly reproduce the turbulent mixing process. The transported joint velocity - scalar PDF approach is followed. In [61] the non-linear $k-\varepsilon$ model of [16], in a first order formulation, is applied. The quality of the flow field results is slightly better than in [62] and agreement with experimental data is reasonable. In [61] results are presented for both pre-assumed β-PDF simulations and transported scalar PDF calculations. It is illustrated that the chemistry model choice (chemical equilibrium vs. laminar flamelet model with a single constant strain rate) has a visible influence on the results, but the impact is much smaller than the turbulence model choice (non-linear $k-\varepsilon$ vs. standard $k-\varepsilon$). With the transported scalar PDF approach good agreement is obtained with experimental data.

6.3 Bluff Body Stabilised Diffusion Flame ('Sydney HM1')

This test case has been studied experimentally at Sydney University [7, 64]. The burner configuration is shown in fig. 4. It is described in detail in [7] and on the TNF web site[9]. The geometry is characteristic for some types of industrial burners. The central fuel jet consists of $50\% H_2$ and $50\% CH_4$ (by volume). It mixes with the co-flow air stream, resulting in a jet type non-premixed turbulent flame. The flame remains attached, because there is a recirculation zone behind the bluff-body. In this region, fuel and air are intensively mixed and hot combustion products are convected back towards the bluff body. Reaction takes place and creates a high-temperature region that stabilizes the actual flame.

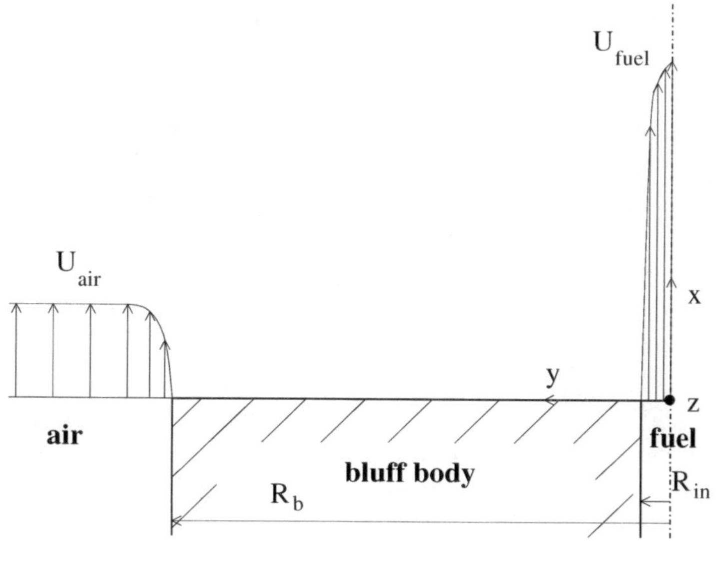

Figure 4: Bluff-body burner configuration of the Sydney non-premixed flames[7].

Experiments have been performed for various Reynolds numbers of the central fuel jet. For flame HM1 the Reynolds number, based on the mean fuel velocity and the central

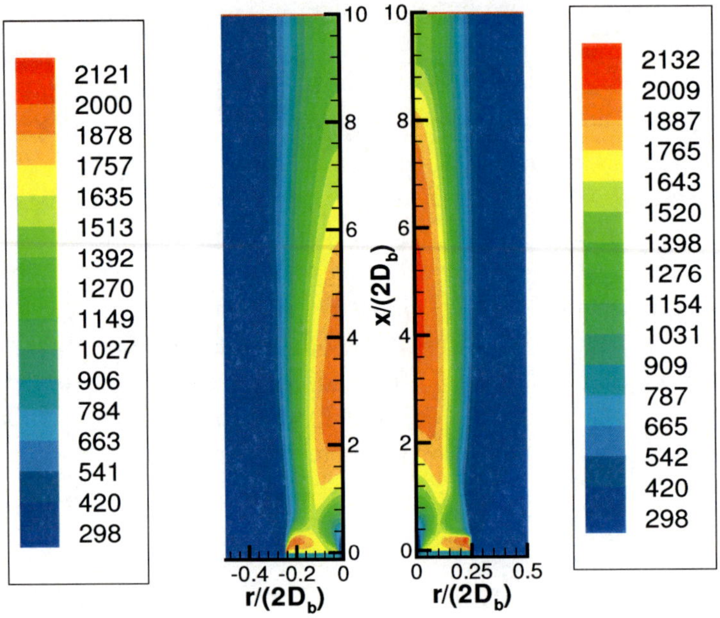

Figure 5: Difference in bluff-body burner flame structure due to the turbulence model (left: standard $k-\varepsilon$ model; right: non-linear model of [16]).

diameter, is $Re = V_m D/\nu = 15900$. This corresponds to a mean velocity of about $108m/s$. The co-flow air has a free stream velocity of $35m/s$. For flame HM3 the jet Reynolds number is $Re = 28700$ ($V_m = 195m/s$), while the co-flow air velocity remains unchanged. The configuration is placed in a vertical wind tunnel but the flame is unconfined.

The difference in flame structure (of Flame HM1) due to the turbulence model is illustrated by means of mean temperature contours in fig. 5. The left part shows the mean temperature pattern obtained with the standard $k-\varepsilon$ model. The pattern with the non-linear model of [16] is shown on the right side. As was the case for the Sandia piloted diffusion flame (fig. 2), the flame is more stretched with the latter model. Apart from problems downstream of the recirculation region (the 'neck' zone), which are common for all RANS results reported in the literature so far, agreement with experimental data is very good for the non-linear two-equation model[16], in contrast to the standard $k-\varepsilon$ model.

Two aspects are important for the quality in fig. 5. Firstly the streamline curvature effect on turbulence must be accounted for. This is particularly important for the recirculation region. Secondly, and this is more important, the dissipation rate transport equation must be more advanced than the standard transport equation. In [16] the coefficient $c_{\varepsilon 2}$ is made locally flow dependent, but the transport equation of [30] may be even a better alternative.

In [65] the performance of three Reynolds stress model has been evaluated: the basic LRR-IP model[2], the SSG model[32] and the JM model[33]. Differences in the pressure - rate-of-strain tensor modelling in the Reynolds stresses transport equations lead to visible differences in the mean flow field. This is due to the more complex flow field behind the bluff body. Still, the major result is that the global quality can only be improved when the model constant $c_{\varepsilon 1}$ is increased from $c_{\varepsilon 1} = 1.44$ to $c_{\varepsilon 1} = 1.6$. Once again this illustrates that the dissipation rate transport equation remains the most influential part of the model for this test case and that only a small gain in accuracy is obtained, compared to the non-linear $k-\varepsilon$ model of [16].

In [23] the performance of the standard $k-\varepsilon$ model, an explicit algebraic stress model and a non-linear first order $k-\varepsilon$ model has been compared. The findings confirm the above statements: differences are in general rather small, indicating that the constitutive law is not crucial for the test case under study. Still some improvement is observed when the eddy viscosity is made locally flow dependent, compared to the standard constant value for $c_\mu = 0.09$. The major impact, however, is again the decrease of $c_{\varepsilon 2}$ from $c_{\varepsilon 2} = 1.92$ to $c_{\varepsilon 2} = 1.83$. The problems mentioned downstream of the recirculation region are observed again in [23].

All the results mentioned so far have been obtained within the framework of the pre-assumed β-PDF modelling for turbulence - chemistry interaction. Chemistry has been modelled by either the constrained equilibrium[39] model[16, 65] or laminar flamelet modelling[23]. This choice hardly affects the results. When the focus is on chemistry modelling or turbulence - chemistry interaction it is again more fruitful to apply the transported PDF approach.

In [66] the joint velocity - composition - turbulence frequency PDF approach is taken. Turbulence is modelled through the transported PDF equation. It is reported that simulation results are very sensitive to a model constant ($C_{\omega 1}$) that is related to the turbulence frequency. This is completely in line with the sensitivity of RANS turbulence models to the dissipation rate transport equation. Note that the turbulent dissipation rate can be determined from the turbulence frequency (and the turbulent kinetic energy) and vice versa. It is also confirmed in [66] that the turbulence model has a much larger impact on the results than e.g. the micro-mixing modelling: the sensitivity to the model constant in the turbulence frequency is much stronger than the sensitivity to the model constant C_ϕ, which determines the scalar dissipation rate or the molecular mixing rate.

More recently results have been reported with the joint velocity - composition - turbulence frequency transported PDF approach for Sydney flames HM1, HM2 and HM3[67]. For flame HM1 very good agreement has been obtained, but larger deviations are observed for flames HM2 and HM3 (with the higher fuel jet mean velocity). This is attributed to the turbulence model, but it is a topic of ongoing research at the moment of writing.

7 Concluding Discussion

In numerical simulations of turbulent non-premixed flames the turbulence model is extremely important because fuel and oxidizer can only react as they mix. In this chapter an over-view has been given of RANS type turbulence models that have been applied in the field of turbulent non-premixed flames. Both two-equation and Reynolds stress models have been considered and some important aspects have been discussed.

An order of importance has been given with respect to the introduction of complexity in the model. Mostly the flow configuration is relatively simple in burner configuration (with swirling flows as an important exception). As a consequence a two-equation RANS turbulence model is often sufficient for many engineering purposes, in combination with a first order expression for the turbulent stresses. It is often (but not always) important to account for the effect of streamline curvature on turbulence. This can be done through the expression for the eddy viscosity, so that the constitutive law becomes non-linear, but remains of first order. The major reason why it usually does not pay off to use Reynolds stress models is that the detail of the individual turbulent stresses is of minor importance as long as the resulting force of the turbulent shear stress and the mean pressure are reproduced with a sufficient degree of accuracy.

There are some important exceptions, though. A first type of flows are flows with a high degree of large scale unsteadiness. Since the unsteadiness must then be resolved, LES type simulations are to be preferred above RANS simulations. Unsteadiness is indeed not well described with two-equation RANS models. Recent work has shown encouraging results for unsteady simulations with second moment closure models, so that there may be potential for Reynolds stress models in this area.

Swirling flows are a second exception. Two-equation models are not suited for such simulations in the author's opinion. As such it does not make much sense to try and repair such models with higher order constitutive laws for the Reynolds stresses. Second moment closure models have the potential to do a better job, but often unsteadiness (e.g. precessing vortex core) is important so that the arguments above are again valid, in favor of LES type simulations.

Finally a discussion has been devoted to turbulence modelling in the framework of transport PDF methodology. Since most of the computing time is devoted here to chemistry, one major disadvantage of second moment closure models (namely that they are much more time consuming than two-equation models) disappears. Moreover it is sometimes necessary to use Reynolds stress turbulence models for consistenvy reasons in the hybrid transported PDF approach. Still convergence can be a problem. The importance of the turbulence model has been highlighted in three aspects: the mean flow field (mean convection), the time scale in the micro-mixing models and the turbulent transport in physical space.

Acknowledgement

The author is Postdoctoral Fellow of the Fund of Scientific Research - Flanders (Belgium) (FWO-Vlaanderen). The author deeply acknowledges Prof. Dirk Roekaerts (Delft University of Technology, The Netherlands), dr. Bertrand Naud (CIEMAT, Spain) and Prof. Stephen B. Pope (Cornell University, USA) for their everlasting help and fruitful discussions and collaborations.

References

[1] W.P. Jones and B.E. Launder. The prediction of laminarization with a two-equation model of turbulence. *International Journal of Heat and Mass Transfer*, **15**(2):301–314, 1972.

[2] B.E. Launder, G.J. Reece, and W. Rodi. Progress in the development of a Reynolds-stress turbulence closure. *J. of Fluid Mechanics*, **68**:537–566, 1975.

[3] R.S. Barlow and J.H. Frank. Effects of turbulence on species mass fractions in methane/air jet flames. In The Combustion Institute, editor, *27th Symposium (International) on Combustion*, pages 1087–1095, Pittsburgh, 1998.

[4] T.W.J. Peeters, P.P.J. Stroomer, J.E. de Vries, and D.J.E.M. Roekaerts. Comparative experimental and numerical investigation of a piloted turbulent natural gas diffusion flame. *Proceedings of the Combustion Institute*, **25**:1241–1248, 1994.

[5] P.A. Nooren, M. Versluis, T.H. van der Meer, and R.S. Barlow. Raman-Rayleigh-LIF measurements of temperature and species concentrations in the delft piloted turbulent jet diffusion flame. *Applied Physics B - Lasers and Optics*, **71**(1):95–111, 2000.

[6] E.H. van Veen and Roekaerts D. On the accuracy of temperature measurements in turbulent jet diffusion flames by coherent anti-Stokes-Raman spectroscopy. *Combustion Science and Technology*, **175**(10):1893–1914, 2003.

[7] B.B. Dally, A.R. Masri, R.S. Barlow, and G.J. Fiechtner. Instantaneous and mean compositional structure of bluff-body stabilized nonpremixed flames. *Combustion and Flame*, **114**(1/2):119–148, 1998.

[8] B.B. Dally, D.F. Fletcher, and A.R. Masri. Flow and mixing fields of turbulent bluff-body jets and flames. *Combustion Theory and Modelling*, **2**:193–219, 1998.

[9] http://www.ca.sandia.gov/tnf/abstract.html.

[10] B. Wegner, A. Maltsev, C. Schneider, A. Sadiki, A. Dreizler, and J. Janicka. Assessment of unsteady rans in predicting swirl flow instability based on LES and experiments. *International Journal of Heat and Fluid Flow*, **25**(3):528–536, 2004.

[11] D.C. Wilcox. *Turbulence Modeling for CFD*. Griffin Printing, Glendale, California, 1993.

[12] F.R. Menter. Two-equation eddy-viscosity turbulence models for engineering applications. *AIAA J.*, **32**(8):1598–1605, 1994.

[13] P. Durbin. Near-wall turbulence closure without damping functions. *Theoretical and Computational Fluid Dynamics*, **3**(1):1–13, 1991.

[14] S.B. Pope. *Turbulent Flows*. Cambridge University Press, 2000.

[15] B. Merci, C. De Langhe, J. Vierendeels, and E. Dick. A quasi-realizable cubic low-Reynolds eddy-viscosity turbulence model with a new dissipation rate equation. *Flow, Turbulence and Combustion*, **66**(2):133–157, 2001.

[16] B. Merci, E. Dick, J. Vierendeels, D. Roekaerts, and T.W.J. Peeters. Application of a new cubic turbulence model to piloted and bluff-body diffusion flames. *Combustion and Flame*, **126**(1-2):1533–1556, 2001.

[17] S.B. Pope. A more general effective-viscosity hypothesis. *J. of Fluid Mechanics*, **72**:331–340, 1975. Part2.

[18] T.J. Craft, H. Iacovides, and J.H. Yoon. Progress in the use of non-linear two-equation models in the computation of convective heat-transfer in impinging and separated flows. *Flow, Turbulence and Combustion*, **63**:59–80, 2000.

[19] F.S. Lien and M.A. Leschziner. Multigrid acceleration for recirculating and turbulent flows computed with a non-orthogonal, collocated finite-volume scheme. *Comp. Meths. Appl. Mech. Eng.*, **118**:351–371, 1994.

[20] W. Rodi. A new algebraic relation for calculating the Reynolds stresses. *Z. angew. Math. Mech.*, **56**:219–221, 1976.

[21] T.B. Gatski and C.G. Speziale. On explicit algebraic stress models for complex turbulent flows. *J. of Fluid Mechanics*, **254**:59–78, 1993.

[22] S. Wallin and A.V. Johansson. An explicit algebraic Reynolds stress model for incompressible and compressible turbulent flows. *Journal of Fluid Mechanics*, **403**,

2000.

[23] J. Yan, F. Thiele, and M. Buffat. A turbulence model sensitivity study for CH_4/H_2 bluff-body stabilized flames. *Flow, Turbulence and Combustion*, **73**(1):1–24, 2004.

[24] C.G. Speziale, B.A. Younis, and S.A. Berger. Analysis and modelling of turbulent flow in an axially rotating pipe. *Journal of Fluid Mechanics*, **407**:1–26, 2000.

[25] F. Schmitt, B. Merci, E. Dick, and C. Hirsch. Direct study of the k-equation for a complex turbulent flow. *Journal of Turbulence*, **4**(paper 021), 2003.

[26] V. Yakhot and S.A. Orszag. Renormalization-group analysis of turbulence. *Physical Review Letters*, **57**(14):1722–1724, 1986.

[27] J.L. Lumley, Z. Yang, and A.M. Yaglom. A length-scale equation. *Flow, Turbulence and Combustion*, **63**(1–4):1–21, 2000.

[28] C.G. Speziale, B.A. Younis, R. Rubinstein, and Y. Zhou. On consistency conditions for rotating turbulent flows. *Phys. Fluids*, **10**(8):2108–2110, 1998.

[29] T.H. Shih, W.W. Liou, A. Shabbir, Z. Yang, and J. Zhu. A new $k-\varepsilon$ eddy viscosity model for high Reynolds number turbulent flows. *Computers Fluids*, **24**(3):227–238, 1995.

[30] B. Merci, E. Dick, and C. De Langhe. Application of an improved ε-equation to a piloted jet diffusion flame. *Combustion and Flame*, **131**(4):465–468, 2002.

[31] B. Merci and E. Dick. Predictive capabilities of an improved cubic $k-\varepsilon$ model for inert steady flows. *Flow, Turbulence and Combustion*, **68**(4):335–358, 2002.

[32] C.G. Speziale, S. Sarkar, and T.B. Gatski. Modelling the pressure-strain correlation of turbulence: An invariant dynamical systems approach. *Journal of Fluid Mechanics*, **227**:245–272, 1991.

[33] W.P. Jones and P. Musonge. Closure of the Reynolds stress and scalar flux equations. *Physics of Fluids*, **31**(12):3589–3604, 1988.

[34] J.C. Rotta. Statistische Theorie inhomogener Turbulenz. *Z. Phys.*, **129**:547–572, 1955.

[35] W.P. Jones. *Turbulence Modelling and Numerical Solution Methods*, pages 309–374. Turbulent Reacting Flows. Academic Press, 1994.

[36] S. Jakirlic, K. Hanjalic, and C. Trpoea. Modeling rotating and swirling turbulent flows: A perpetual challenge. *AIAA Journal*, **40**(10):1984–1996, 2002.

[37] S.B. Pope. On the relationship between stochastic lagrangian models of turbulence and second-moment closures. *Physics of Fluids*, **6**(2):973–985, 1994.

[38] R.W. Bilger, S.H. Starner, and R.J. Kee. On reduced mechanisms for methane/air combustion in nonpremixed flames. *Combustion and Flame*, **80**(2):135–149, 1990.

[39] R.W. Bilger and S.H. Starner. A simple model for carbon monoxide in laminar and turbulent hydrocarbon diffusion flames. *Combustion and Flame*, **51**:155–176, 1983.

[40] U. Maas and S.B. Pope. Implementation of simplified chemical kinetics based on intrinsic low-dimensional manifolds. In The Combustion Institute, editor, *Twenty-Fourth Symposium (International) on Combustion*, pages 103–112, 1992.

[41] S.M. Correa. Turbulence - chemistry interactions in the intermediate regime of premixed combustion. *Combustion and Flame*, **93**(1–2):41–60, 1993.

[42] M.D. Smooke and V. Giovangigli. *Premixed and nonpremixed test problem results.*, pages 29–47. Reduced kinetic mechanisms and asymptotic approximations for methane air flames - Lecture Notes in Physics, Vol. 384. Springer Verlag, 1991.

[43] H.P. Mallampalli, T.H. Fletcher, and J.-Y. Chen. Evaluation of CH_4/NO_x reduced mechanisms used for modeling lean premixed turbulent combustion of natural gas. *Journal of Engineering for Gas Turbines and Power - Transactions of the ASME*, **120**(4):703–712, 1998.

[44] C.J. Sung, C.K. Law, and J.-Y. Chen. An augmented reduced mechanism for methane oxidation with comprehensive global parametric evaluation. *Proceedings of the Combustion Institute*, **27**:295–304, 1998.

[45] R.P. Lindstedt, S.A. Louloudi, and E.M. Vaos. Joint scalar pdf modelling of pollutant formation in piloted turbulent jet diffusion flames with comprehensive chemistry. In Pittsburgh The Combustion Institute, editor, *Twenty-Eighth Symposium (International) on Combustion*, pages 23–39, 2000.

[46] http://www.me.berkely.edu/gri mech.

[47] J.-Y. Chen. Near-wall turbulence models for complex flows including separation. *Combustion Science and Technology*, **57**:89–94, 1988.

[48] A.Y. Klimenko and R.W. Bilger. Conditional moment closure for turbulent combustion. *Progress in Energy and Combustion Science*, **25**(6):595–687, 1999.

[49] S.B. Pope. PDF methods for turbulent reactive flows. *Prog. Energy Combust. Science*, **11**:119–192, 1985.

[50] P. Jenny, S.B. Pope, M. Muradoglu, and D.A. Caughey. A hybrid algorithm for the joint PDF equation of turbulent reactive flows. *Journal of Computational Physics*, **166**(2):218–252, 2001.

[51] M. Muradoglu, S.B. Pope, and D.A. Caughey. The hybrid method for the PDF equations of turbulent reactive flows: consistency conditions and correction algorithms. *Journal of Computational Physics*, **172**(2):841–878, 2001.

[52] S.B. Pope. Computationally efficient implementation of combustion chemistry using in situ adaptive tabulation. *Combustion Theory and Modelling*, **1**(1):41–63, 1997.

[53] S.B. Pope. Accessed compositions in turbulent reactive flows. *Flow, Turbulence and Combustion*, **72**(2–4):219–243, 2004.

[54] C. Dopazo and E.E. O'Brien. Probability density-function approach for a turbulent axisymmetric heated jet - centerline evolution. *Physics of fluids*, **17**(11):1968–1975, 1975.

[55] J. Janicka, W. Kolbe, and W. Kollmann. Closure of the transport-equation for the probability density-function of turbulent scalar fields. *Journal of Non-Equilibrium Thermodynamics*, **4**(1):47–66, 1979.

[56] S. Subramaniam and S.B. Pope. A mixing model for turbulent reactive flows based on Euclidean minimum spanning trees. *Combustion and Flame*, **115**(4):487–514, 1998.

[57] Q. Tang, J. Xu, and S.B. Pope. Probability density function calculations of local extinction an no production in piloted-jet turbulent methane/air flames. *Proceedings of the Combustion Institute*, **28**:131–139, 2000.

[58] J. Xu and S.B. Pope. PDF calculations of turbulent nonpremixed flames with local extinction. *Combustion and Flame*, **123**(3):281–308, 2000.

[59] B. Merci, E. Dick, J. Vierendeels, and C. De Langhe. Determination of ε at inlet boundaries. *Int. Journal of Numerical Methods for Heat and Fluid Flow*, **12**(1):65–80, 2002.

[60] B. Merci and E. Dick. Influence of computational aspects on simulations of a jet diffu-

sion flame. *Int. Journal of Numerical Methods for Heat and Fluid Flow*, **13**(7):1057–1078, 2003.

[61] B. Merci, D. Roekaerts, and B. Naud. Flow and mixing fields for transported scalar PDF simulations of a piloted jet diffusion flame ('delft flame III'). *Flow, Turbulence and Combustion*, **74**(3):239-272, 2005.

[62] P.A. Nooren, H.A. Wouters, T.W.J. Peeters, D. Roekaerts, U. Maas, and D. Schmidt. Monte carlo PDF modelling of a turbulent natural-gas diffusion flame. *Combust. Theory and Modelling*, **1**:79–96, 1997.

[63] S.B. Pope. An explanation of the turbulent round-jet/plane-jet anomaly. *AIAA Journal*, **16**(3):279–281, 1978.

[64] A.R. Masri. Combustion data base, the University of Sydney and the Combustion Research Facility, Sandia National Laboratories. Technical report, http://www.mech.eng.usyd.edu.au/research/energy/resources.html, 2000.

[65] G. Li, B. Naud, and D. Roekaerts. Numerical investigation of a bluff-body stabilised nonpremixed flame with differential Reynolds-stress models. *Flow, Turbulence and Combustion*, **70**(1–4):211–240, 2003.

[66] M. Muradoglu, K. Liu, and S.B. Pope. PDF modeling of a bluff-body stabilized turbulent flame. *Combustion and Flame*, **132**(1–2):115–137, 2003.

[67] K. Liu, S.B. Pope, and Caughey D.A. Calculations of bluff-body stabilized flames using a joint probability density function model with detailed chemistry. *Combustion and Flame*, **141**(1–2):89–117, 2005.

In: New Devel. in Combustion Research
Editor: William J. Carey, pp. 163-185

ISBN 1-59454-826-9

Chapter 3

Probabilistic Methods and Physical Factors in Turbulent Premixed Combustion

A. A. Burluka*
School of Mechanical Engineering,
The University of Leeds, Leeds, LS2 9JT, UK

Abstract

This work undertakes an analysis of parameters defining the rate of chemical reactions in turbulent premixed combustion with particular emphasis on the strong turbulence case characterised by a break-down of a simple relationship between the flame speed and turbulent rms velocity. It is argued that, in addition to commonly considered flow properties such as the laminar flame speed and rms turbulent velocity, both the details of chemical kinetics and the molecular transport coefficients are important in this case. Mathematical description of this combustion regime should be based on use of the probability density functions (pdf's), or similar probabilistic concepts, and possible frameworks for doing so are investigated. Two novel approaches are presented: one is based on consideration of a transport equation for two-point probability density function and the second is an extension of the reference scalar field (rsf) model with an improved representation of the small-scale mixing. While the former method is self-consistent and self-contained, nevertheless it is expressed in terms of a multi-dimensional function and is rather complicated. The latter method is based on a use of several reference fields possessing only one extra dimension and the small-scale mixing is modelled as a diffusion process in this extra dimension. The proposed model is capable of predicting the small-scale mixing in agreement with the DNS results for homogeneous turbulence. When applied to a reacting case it predicts very strong effects of molecular transport coefficients, consistently with the analysed experiments.

Keywords: Turbulent premixed combustion, molecular transport effects, small-scale mixing, probability density functions, reference scalar field (RSF).

*E-mail address: A.A.Burluka@leeds.ac.uk

1 Introduction

There exist two large groups of approaches to finding the average reaction rates in premixed turbulent combustion. The first group employs transport equations for several first moments of the fluctuating velocity and concentration fields, and the reaction rates are found from some *a'priori* assumptions about the structure of the regions where the chemical reactions take place. To this group belong the Eddy-Break-Up type models, e.g. [1], assuming that the reactions occur in infinitesimally thin flame fronts, laminar flamelet models, e.g. [2, 3], presuming that the reactions rates are the same as in (strained and curved) laminar flames, and various models based on a concept of the average flame surface, e.g. [4] or related to it average scalar dissipation, e.g. [5, 6]. The second group, sometimes referred to as "statistical" or "probabilistic" models, aims at calculation of statistical properties, such as probability density functions (pdf's), of reacting scalar, such as concentrations and temperature, fields and the reactions rates are calculated from their "exact" instantaneous rates (if those are known) taking into account the statistics of the scalars fluctuations. This type of models was pioneered in 1960 by a then second-year PhD student V.A. Frost [7] who suggested a transport equation for a progress variable one- and two-point pdfs. Following the work [8] Monté-Carlo method is commonly employed for solution of the pdf transport equation which is reformulated in terms of equations for properties of a discrete ensemble of statistical particles. Alternative concepts in this class include the Linear-Eddy model [9], reference scalar fields [10, 11] and recent attempts [12] to adapt the Conditional Moment Closure [13] for the premixed combustion. As compared to the model of the first class employing some presumed formula(ae) for the mean reaction rate(s), the models of the second class are invariably more complicated, heavier in required computational resources and the question then arises as to whether this additional complexity is justified by their larger generality and applicability.

Very recently, two papers of monumental proportions [14, 15] (featuring 350 and 571 referred papers, respectively) presented a comprehensive review of turbulent premixed flames properties and physical agents defining these properties. Also, these reviews analysed to which extent the models of the first group are capable of describing the experimental data on turbulent flame speed and thickness. A number of critical factors, e.g. molecular transport properties or unsteady character of a flame, have been highlighted in this analysis and the capability of a model to predict the influence of those factors was put to test. However, the models of the second group which can, at least in principle, incorporate exact expressions for instantaneous rates of elementary reactions have been left out of consideration in these reviews and one of the purposes of the present work is to consider them along the same guidelines as was adopted in [14, 15]. Namely, these reviews concentrated on definition of how the turbulent flame speed, or related to it the mass burning rate, and a flame thickness depend upon various mixture and flow properties.

This work concerns the implications which the conclusions of [14, 15] may have for the second group of models. Some experimental data are first considered in order to clarify what requirements the second group of models should satisfy in order to compete successfully against simpler models of the first group, in particular, the model advocated in [14] which is capable of predicting the burning rate for a fairly wide variation in relevant parameters. Following this review an attempt is made to construct a universal model and the two-point pdf description is invoked so that the information about turbulence scales is contained therein. It proved possible, departing from the ideas of Ievlev [16] closure, to construct an equation for joint velocity-scalars pdf. However, an engineering application of this approach seems too costly in computational resource and an alternative approach consisting in development of much simpler one-point description at the expense of loss of generality and such a description is proposed in terms of the reference scalar fields, following the ideas of [11].

2 Main Factors Determining the Rate of Turbulent Premixed Combustion

There exists a very large number of variables which may alter an average reaction rate in a turbulent premixed flame. Some of these variables, such as laminar flame speed u_n or molecular heat diffusivity κ are determined by the mixture composition and temperature, some, such as turbulence properties, are dependent on flow and flame geometry. Below presented a brief summary of how a particular variable can affect the average reaction rates; this summary repeats some of the findings in [14, 15], however, the emphasis here is put on phenomena and combustion regimes which may be relevant for the considered "probabilistic" models and which have not necessarily been covered in [14, 15].

2.1 Turbulence Parameters

Quite often in "combustion" literature , the turbulence is characterised in terms of its integral length-scale l_t and the rms velocity u' and the Reynolds number derived from them:

$$Re_t = \frac{u' l_t}{\nu} \tag{1}$$

where ν is the molecular viscosity. When turbulence is neither homogeneous neither isotropic which is presumably a case in any "practical" combustion chamber, e.g. an internal combustion spark-ignition (IC SI) engine, then various velocity components u_i have unequal rms values and each component has nine integral length scales defined from correlations $< u'_i(\vec{x},t)u'_j(\vec{x}+\vec{x}_k,t) >, j,k = 1,2,3$. While this is true in theory, in reality very little is known about inhomogeneity of turbulence and most often the turbulence description is given in terms of a single u' and either longitudinal or lateral length-scale l_t thus presuming local homogeneity and isotropy. Moreover, very often, spectral properties of turbulence are assumed to be given by the Kolmogorov's laws [17], thus assuming a steady velocity fluctuation spectrum. This may not necessarily be true in non-steady flames, e.g. in explosions, however, while there is an experimental evidence [18] that combustion modifies the turbulence spectrum, no other assessment of applicability of Kolmogorov's laws to combustion flows is known. As far as this can affect the burning rate, there exists an early experimental evidence [19] that the rate of combustion depends upon the spectral characteristics of the turbulence upstream of the flame *paribus cœteris*. This evidence is essentially the only available justification for the extra complexity of the so-called spectral combustion models [20, 21]. While it must not be discarded because of absence of published follow-up research, at the same time the influence of turbulence spectrum on burning rates would probably need a further verification with the modern experimental techniques. In the absence of measurements clarifying the turbulence structure in premixed flames the analysis is then limited solely to the effects of u', l_t and Re_t.

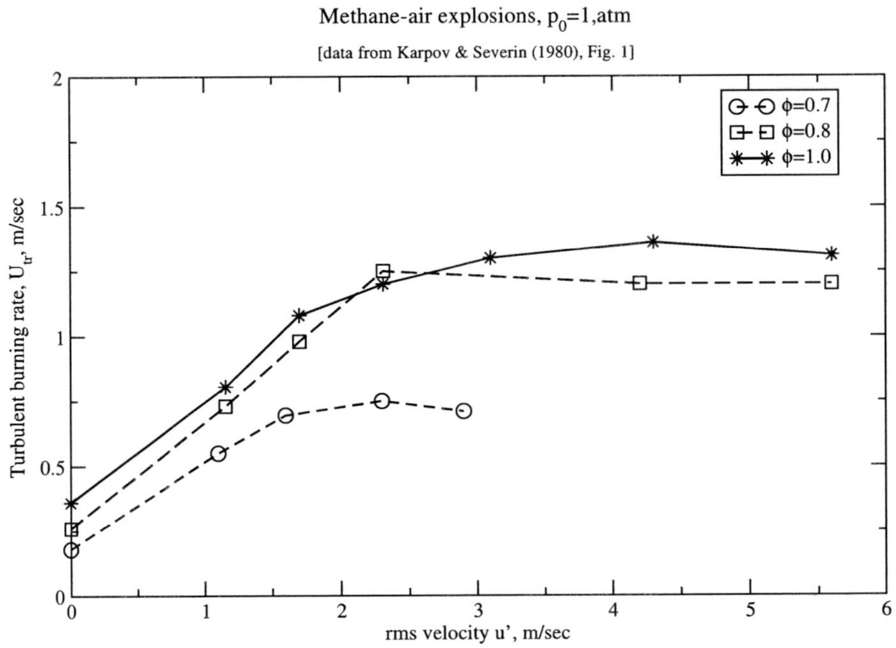

Figure 1: Mass burning rate in CH_4-air spherical explosions for different rms velocities averaged over flame radii $r_{fl} = 1.5 - 4.5\,cm$ at initial temperature of $T_0 = 20^0C$ and atmospheric pressure. The data are re-plotted from [24].

The review [14] concluded that a dependency of the burning rate U_t upon the integral length-scale l_t cannot be established without ambiguity for a number of reasons, controversy and paucity of available experimental data being arguably the main ones. At the same time, it shows a number of successful applications of Zimont-Lipatnikov model to a variety of flame configurations. Leaving aside the aspects related to unsteady flame development, that model employs the following expression for the turbulent burning velocity achieved by a planar flame at large times:

$$U_t = Const \cdot u' Da^{1/4} = Const \cdot u' \cdot \left(\frac{\tau_t}{\tau_{ch}}\right)^{1/4} = Const \cdot u'^{3/4} l_t^{1/4} u_n^{1/2} \kappa^{-1/4} \quad (2)$$

which shows rather weak sensitivity of U_t to the integral scale. Here τ_{ch} is the chemical time scale and κ is the molecular heat diffusivity.

Eq.(2) predicts that $U_t \sim u'^{3/4}$, however, the real dependency on u' is more complicated than that and this simple proportionality breaks down when u' is increased beyond a certain limit u'_m [22, 23]. When u' is increased beyond u'_m the corresponding increase of U_t is much smaller than given by Eq.(2) and, under certain conditions, it may even

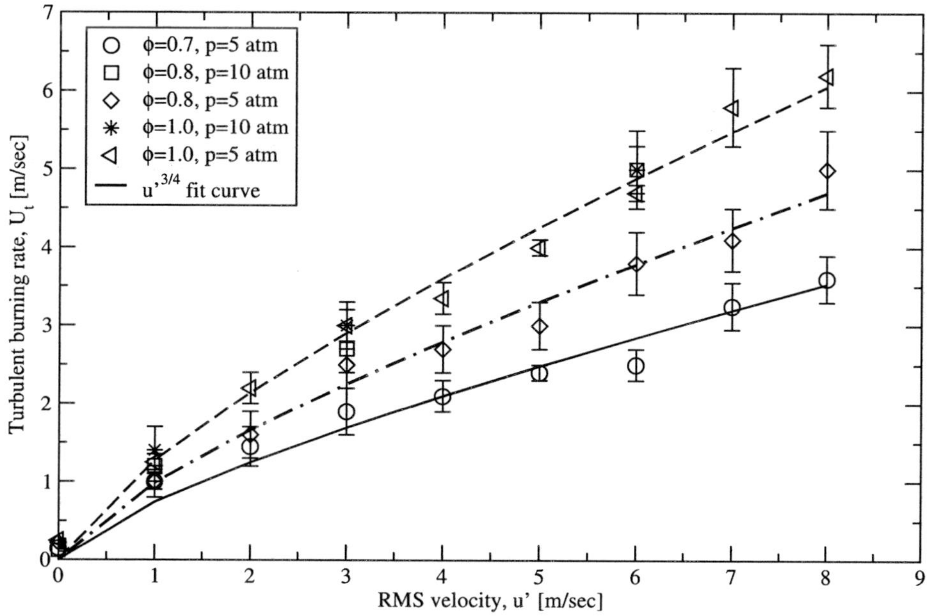

Figure 2: Mass burning rate in CH_4-air spherical explosions for different rms velocities measured at flame radius $r_{fl} = 6\,cm$ at initial temperature and pressures $T_0 = 358\,K$ and $p_0 = 5$ and $10\,bar$. Lines show fitted dependency of the form $U_{tr} = Const \cdot u'^{3/4}$ for explosions at $p_0 = 5\,bar$. The experimental points are taken from [27]

be negative. Finally, the combustion is extinguished when u' exceeds another threshold value u'_q [22, 23]. This situation is illustrated in Fig.(1) for methane-air explosions performed at an atmospheric pressure [22], so that for the mixtures with $\phi = 0.7,\ 0.8$ and 1.0 $u'_m \approx 1.8,\ 2.5$ and $3.0\,m/sec$, respectively, while $u'_q \approx 3.0, 5.7$ for the former two mixtures while no quenching was observed for the stoichiometric mixtures [24]. It must be stressed that the assessment of various models in [14, 15] has been performed for $u' \leq u'_m$ while it has been remarked that the stronger turbulence case with $u' \geq u'_m$ cannot, as a general rule, be adequately described with the models analysed therein without invoking additional assumptions.

It should be noted that a general rule for calculating u'_m has not yet been proposed, perhaps partly because all the changes in flame behaviour are gradual rather than sudden. One of the attempts to find a correlation between u'_q, Lewis number $Le = \frac{\kappa}{D_{def}}$ and Klimov-Williams criterion

$$K = Const \cdot \left(\frac{u'}{u_n}\right)^2 Re_t^{-1/2}$$

was undertaken in [25], however, it predicts that u'_q should decrease with the pressure rise and this trend is contradicted by all available experimental evidence, see e.g. [18], also see Fig. (2). In the absence of a well-supported by experiments theory predicting u'_m and u'_q, there exists no simple guidelines for validity domain of the simpler models such as developed in [14, 15] and this provides a justification for a further research with more complex models of the second class, sometimes referred to as "statistical" models, which, at least in principle, should not need extra hypothesis for the changing combustion regimes. Without any alteration of the model parameters a predictive model should describe burning rate saturation at $u' \approx u'_m$ followed by an eventual quenching at $u' \approx u'_q$. In other words, use of a more complicated model is justified provided any such model **must** firstly, describe adequately reaction rates at $u' \leq u'_m$ and reproduce the trends analysed in [14, 15] and, secondly, yield u'_m and u'_q as an outcome rather than be based on these values as input parameters.

2.2 Mixture Properties

Difficulty in finding both u'_m characterising slow increase in average reaction rate with increasing u' and u'_q is caused by the fact that these variables seem to depend on a large number of factors. This fact is illustrated in Figs. (2) and (3). The former presents the average mass burning rate measured from the pressure rise in a fan-stirred bomb for the same mixtures as in the Fig. (1) but for higher initial temperature and pressure. It can be seen, that derived from Eq. (2), a power-law dependency $U_{tr} = Const \cdot u'^{3/4}$ provides quite a good approximation to these data over the entire range of u' achievable in the experiment, clearly demonstrating applicability of the model advocated in the [14] for these conditions.

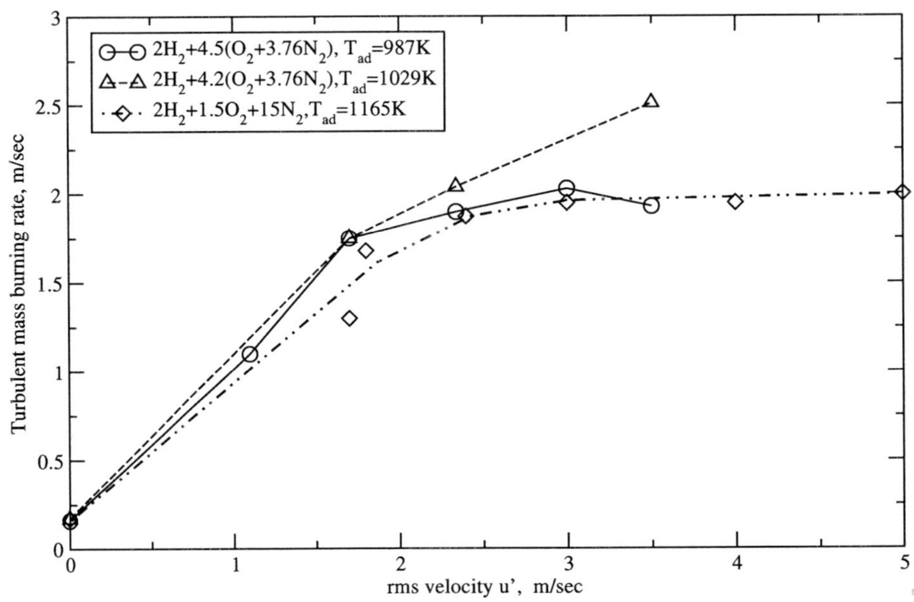

Figure 3: Mass burning rate for atmospheric explosions of diluted lean hydrogen-oxygen-nitrogen mixtures. The data are taken from [26]

The Fig. (3) presents the data obtained in diluted lean hydrogen-oxygen-nitrogen mixtures having the same laminar flame speed of $u_n = 17\,cm/sec$ [26] and very close values of the Lewis number but different adiabatic flame temperature. Because different degrees of dilution were used, while the rates of reactions involving the nitrogen as a third body may be different, however, it is most likely the observed difference in u'_m and u'_q is caused by variation in the adiabatic flame temperature. The laminar flame speed for methane-air flame at the equivalence ratio $\phi = 0.8$ and initial conditions of $T_0 = 358\,K$ and $p_0 = 5\,bar$ is $u_n \approx 16\,cm/sec$ [28]. Comparison of the curve in Fig. (2) with the curves in Fig. (3) shows clearly that the laminar flame speed cannot be the only parameter determining u'_m and u'_q.

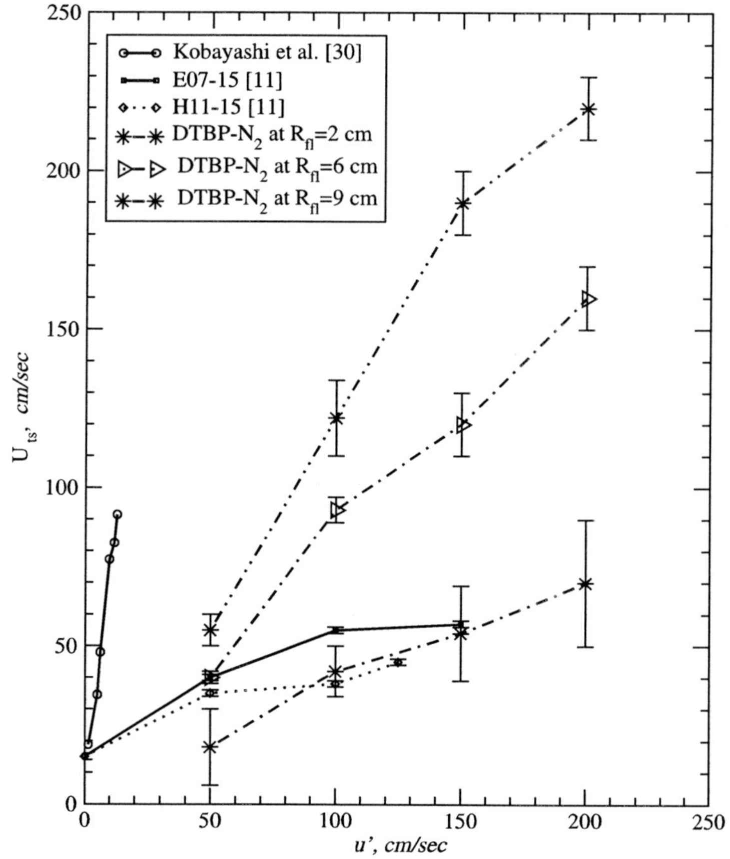

Figure 4: Comparison of the mass burning rate *vs.* rms turbulence in the DTBP decomposition flame with the literature values reported for the flames with the same $U - n$ and Le. (taken from [29]).

An attempt to explore to which extent the details of the flame chemical kinetics mechanism affect the burning rates was undertaken in [29]. In contrast to commonly employed hydro-carbon and hydrogen oxidation flames which are characterised with multi-step chain-branching chemical kinetics, this work reports a study of di-tert-butyl-peroxide (DTBP) thermal decomposition flame. This DTBP thermal decomposition flame in an inert gas atmosphere is described [29] with a single first-order reaction of which the rate constant is:

$$W = 2 \cdot 10^{15} \exp\left(-\frac{18282}{T}\right) \left[\frac{1}{sec}\right] \tag{3}$$

The thermal decomposition flame propagation was studied in the fan-stirred constant volume combustion vessel with a central spark ignition. Fig. (4) presents comparison of the flame mass burning rates measured for this simple kinetics flame with the literature [30, 31] values reported for the mixtures having very close values of u_n and Le. While the inspection of the Fig. (4) shows the effects of unsteady flame development, especially important in comparison between the freely propagating flames [29, 31] and Bunsen burner ones [30], it also indicates that the $U_t(u')$ curves for DTBP decomposition flames barely have any saturation effects, which means that for this mixture the value u'_m cannot be defined unambiguously. At the same time, for $u' \leq 1\, m/sec$ there is no noticeable difference in the burning rates for all freely propagating flames. This observation indicates that the chemical kinetics mechanism details should be accounted for in determination of the average reaction rates for "strong", i.e. $u' \geq u'_m$, turbulence, while their effects for "moderate" $u' \leq u'_m$ are most likely to be encapsulated in terms of u_n and Le, in full agreement with the analysis of [15].

2.3 Summary of Observations

The extensive analysis of experimental data in [14, 15] demonstrated that the average reaction rates in turbulent premixed combustion depend, in addition to the commonly considered turbulence rms velocity u', length-scale l_t and laminar flame speed u_n, upon the molecular transport coefficients, such as heat diffusivity κ and the Lewis number Le. The arguments presented above shows that when u' exceeds a certain value u'_m then, in addition to these factors, the adiabatic flame temperature and the details of chemical kinetics mechanism become important. Because u_n is itself a function of elementary reactions rates and the molecular transport properties, a model of turbulent premixed combustion should take these into account in order to be applicable across entire range of u' from weak turbulence to extinction. As is clearly shown by the available experimental data, the burning rate dependency upon the molecular transport coefficients does not seem likely to be expressed solely in terms of u_n and Le.

3 Models Incorporating "Exact" Reaction Rates

It should be emphasised that a majority of models is based on consideration of one-point one-time statistical properties which essentially means that the turbulence time- and length-scales should be provided from a turbulence sub-model distinct from model for combustion processes. Self-contained statistical model which does not require any external information about the turbulence length-scales is possible only if statistical properties of turbulent fields are considered at least in two points in space simultaneously. Equations for either characteristic functions (cf) or probability density functions (pdf) of velocity, density and reacting scalars at n points can be derived as follows [16, 32, 33].

Let $\hat{A}_n = (a^j, j = 1, \ldots n)$ be a set of values of variables describing the flow field: $(\underset{\rightarrow}{u}, \rho, Y_\alpha, \alpha = 1, \ldots l)$ at points $\underset{\rightarrow}{x}^k, k = 1, 2, \ldots n$. Let $P_n(\hat{A}_n)$ be a joint probability density function for the set $\hat{A}_n$. Generally speaking, the "balance", or transport equation for $P_n(\hat{A}_n)$ is a Kramers-Moyal equation [34]:

$$\frac{\partial P_n(\hat{A}_n)}{\partial t} + \hat{u}_i^k \frac{\partial P_n(\hat{A}_n)}{\partial x_i^k} = -\frac{\partial}{\partial \hat{a}_i^k} \left\langle F_i^k \right\rangle_{\hat{A}_n} P_n(\hat{A}_n) \tag{4}$$

where conditional rates of change $\langle F_i^k \rangle_{\hat{A}_n} = \left\langle \frac{da_i}{dt} \right\rangle_{\hat{A}_n}$ of variable a_i^k are expressed in terms of derivatives of this variable and, henceforth, require pdf $P_{n+1}(\hat{A}_{n+1})$ taken at $n+1$ points. To obtain these, the usual equations of motion (Navier-Stokes), continuity and scalar transport should be used:

$$\langle F_i^k \rangle_{\hat{A}_n} = \left\langle -\frac{1}{\rho^k} \cdot \frac{\partial p}{\partial x_i^k} + \frac{\partial}{\rho \partial x_j^k} \frac{\rho \nu \partial u_i^k}{\partial x_j^k} \right\rangle_{\hat{A}_n} \quad \text{for } \hat{a}_i^k = \hat{u}_i^k \tag{5}$$

$$\langle F_i^k \rangle_{\hat{A}_n} = \left\langle -\frac{\partial u_j^k}{\partial x_j^k} \right\rangle_{\hat{A}_n} \quad \text{for } \hat{a}_i^k = \hat{\rho}^k \tag{6}$$

$$\langle F_i^k \rangle_{\hat{A}_n} = \left\langle -\frac{\partial}{\rho \partial x_j^k} \frac{\rho D_j \partial Y_j^k}{\partial x_l^k} \right\rangle_{\hat{A}_n} + W\left(\underset{\rightarrow}{Y^k}\right) \quad \text{for } \hat{a}_i^k = \hat{Y}_j^k \tag{7}$$

Because $\langle F_i^k \rangle_{\hat{A}_n}$ depend upon the $n+1$-point pdf the Eqs. (5-7) require the so-called closure hypothesis expressing these conditional rates of change in terms of n-point pdf's and $\langle F_i^k \rangle_{\hat{A}_{n-1}}$. For the scalar values pdf such closure hypothesis for Eq. (7) are known as the micro- (or small-) scale mixing models and there exist a large number of research devoted to this topic, see the review in [35] . It should be noticed that none of the models reviewed in [35], neither more recent models employing the pdf concept [36, 37, 38, 39] incorporate the molecular transport coefficients except, perhaps, the fluid viscosity which may be present in the Reynolds number. There are two aspects causing a difficulty in introducing molecular diffusivities into Eq. (4). The first aspect is that the decay rate for scalar variance, or, in other words, the rate of scalar mixing is proportional to the inverse of the integral time scale τ_t^{-1} and according to all available evidence, both experimental, reviewed in [33] as well as obtained from DNS [40], depends rather weakly upon the molecular transport. The second aspect is related to a difficulty in modelling continuous diffusion flux in Monté-Carlo methods, commonly employed for solving Eq. (4) for $n = 1$. Below an attempt is made to construct a closure for Eqs. (5-7) consistent with the considerations presented in the Section 2.2 and directly incorporating molecular transport coefficients.

3.1 Closed Two-Point Joint Velocity-Scalars PDF Equation

In 1970, V. M. Ievlev [16, 32] extended the derivation of Eq.4 to the case of joint velocity-scalars pdf and proposed a general method of expressing the $\langle F_i^k \rangle_{\hat{A}_n}$ terms through n-point pdf's using exclusively the formal mathematical constraints such as normalisation and reduction properties. This methods allows to achieve a closed description of a turbulent flow for $n \geq 2$ without resorting to any *a'priori* ideas about structure of the turbulent vortices, spectra etc. However, the proposed method was sufficiently complicated and the only application of Ievlev's closure had so far been made for studying the mixing of one non-reacting scalar in a homogeneous and isotropic turbulence [41] where it was shown that the Ievlev's method predicts correct physical trends in the evolution of an initially bi-modal pdf.

The method [16] is based on an observation that the following properties of reduction, separation and coincidence which should be satisfied by any $P_n(\hat{A}_n)$ can be expressed as a following set of **integral** equations for $\left\langle F_i^k \right\rangle_{\hat{A}_n}$:

- Separation:

$$\int \left\langle F_i^j \right\rangle_{\hat{A}_n} P_n(\hat{A}_n) da^k = \left\langle F_i^j \right\rangle_{\hat{A}_n^{(k)}} P_{n-1}(\hat{A}_{n-1}^{(k)}) \quad j \neq k$$

- Coincidence and reduction:

$$\lim_{\underset{\rightarrow}{x}^k \rightarrow \infty, \underset{\rightarrow}{x}^j} \langle F_i^m \rangle_{\hat{A}_n} \rightarrow \langle F_i^m \rangle_{\hat{A}_n^{(k)}}$$

Henceforth, one may attempt construction of an approximate expression for $\left\langle F_i^k \right\rangle_{\hat{A}_n}$ as:

$$\left\langle F_i^k \right\rangle_{\hat{A}_n} = \sum_j g_i^{k,j} \left(\hat{A}_{n-1}^{(k)} \right) \left\langle F_i^k \right\rangle_{\hat{A}_{n-1}^{(j)}} + u_i^j \cdot \frac{\partial \xi_j^k}{\partial x_i^j} \tag{8}$$

where ξ is an arbitrary function and g_j depends upon the coordinates of all n points. Eq. (8) is the original Ievlev's formulation, and it does not provide a clear expression for g_j functions, hence, in its use one has to resort to numerical quadratures as was done in [41].

At the same time, a remarkable simplification for can be obtained if one considers an expression slightly more general than the original Ievlev approximation, Eq. (8):

$$\left\langle F_i^k \right\rangle_{\hat{A}_{1,2}} = g_i^{k,1} \left(\underset{\rightarrow}{x}^1, \underset{\rightarrow}{x}^2, \hat{A}_1 \right) \left\langle F_i^k \right\rangle_{\hat{A}_2} + g_i^{k,2} \left(\underset{\rightarrow}{x}^1, \underset{\rightarrow}{x}^2, \hat{A}_2 \right) \left\langle F_i^k \right\rangle_{\hat{A}_1} \tag{9}$$

It is easy to see that the generalisation is simply the postulated dependency of $g_i^{k,1}$ on $\hat{A}_1$. The next step is to integrate Eq. (9) with respect to $d\hat{A}_1$:

$$\left\langle F_i^k \right\rangle_{\hat{A}_2} = \left\langle g_i^{k,1} \right\rangle \left(\underset{\rightarrow}{x}^1, \underset{\rightarrow}{x}^2 \right) \left\langle F_i^k \right\rangle_{\hat{A}_2} + g_i^{k,2} \left(\underset{\rightarrow}{x}^1, \underset{\rightarrow}{x}^2, \hat{A}_2 \right) \left\langle F_i^k \right\rangle$$

then $d\hat{A}_2$:

$$\left\langle F_i^k \right\rangle_{\hat{A}_1} = g_i^{k,1} \left(\underset{\rightarrow}{x}^1, \underset{\rightarrow}{x}^2, \hat{A}_1 \right) \left\langle F_i^k \right\rangle + \left\langle g_i^{k,2} \right\rangle \left(\underset{\rightarrow}{x}^1, \underset{\rightarrow}{x}^2 \right) \left\langle F_i^k \right\rangle_{\hat{A}_1}$$

Second integration of either of the above equations leads to

$$\left\langle g_i^{k,1} \right\rangle \left(\underset{\rightarrow}{x}^1, \underset{\rightarrow}{x}^2 \right) + \left\langle g_i^{k,2} \right\rangle \left(\underset{\rightarrow}{x}^1, \underset{\rightarrow}{x}^2 \right) = 1$$

Expressing $g_i^{k,1}$ and $g_i^{k,2}$ from the above and substituting them back into Eq. (9) results in the sought closure model:

$$\left\langle F_i^k \right\rangle_{\hat{A}_{1,2}} = \frac{\langle F_i^k \rangle_{\hat{A}_1} \cdot \langle F_i^k \rangle_{\hat{A}_2}}{\langle F_i^k \rangle} \tag{10}$$

It is straightforward to see that this closure does indeed satisfy re-numbering, reduction, coincidence and separation properties.

Resulting equations are best illustrated on the example of the velocity pdf for an incompressible flow. In this case, the pressure gradient can be expressed from solution of a Poisson equation as, e.g. for an infinite unbounded flow:

$$-\frac{1}{\rho}\frac{\partial p(\underset{\rightarrow}{x})}{\partial x_i} = \frac{1}{4\pi}\int d^3\underset{\rightarrow}{x}' \cdot \frac{\partial^2 u_j u_k}{\partial x'_j \partial x'_k} \cdot \frac{x'_i - x_i}{\|\underset{\rightarrow}{x} - \underset{\rightarrow}{x}'\|^3}$$

the averaging of which results in a clearly non-local term:

$$\left\langle -\frac{1}{\rho}\frac{\partial p(\underset{\rightarrow}{x})}{\partial x_i} \right\rangle_{\hat{A}_m} = \frac{1}{4\pi}\int d^3\underset{\rightarrow}{x}' \cdot \frac{x'_i - x_i}{\|\underset{\rightarrow}{x} - \underset{\rightarrow}{x}'\|^3}\int d^3\underset{\rightarrow}{\hat{u}}'\, \hat{u}'_j \hat{u}'_k\, \frac{\partial^2 f\left(\underset{\rightarrow}{x}', \underset{\rightarrow}{x}^m, \hat{A}', \hat{A}_m\right)}{\partial x'_j \partial x'_k}$$

The small-scale mixing term is local and contains explicitly the molecular viscosity:

$$\left\langle \nu \frac{\partial^2 u_i^k}{\partial x_j^k \partial x_j^k} \right\rangle_{\hat{A}_m} = \nu \int d^3\underset{\rightarrow}{\hat{u}}'\, \hat{u}'_i \lim_{\underset{\rightarrow}{x}' \rightarrow \underset{\rightarrow}{x}^k} \frac{\partial^2 f\left(\underset{\rightarrow}{x}', \underset{\rightarrow}{x}^m, \hat{A}', \hat{A}_m\right)}{\partial x'_j \partial x'_j}$$

Using the reduction property for a two-point pdf f_2:

$$\lim_{\underset{\rightarrow}{x}^2 \rightarrow \underset{\rightarrow}{x}^1} f_2(\underset{\rightarrow}{u}^1, \underset{\rightarrow}{u}^2, \underset{\rightarrow}{x}^1, \underset{\rightarrow}{x}^2, t) = \delta(\underset{\rightarrow}{u}^1 - \underset{\rightarrow}{u}^2) f_1(\underset{\rightarrow}{u}^1, \underset{\rightarrow}{x}^1, t)$$

the small-scale mixing terms can be further simplified to the sum:

$$-\nu \sum_{k,j} \left(\int d\hat{u}_j^k \cdot \hat{u}_j^k \int dA^{(k)} \frac{\partial^2 P_2}{\partial x_l^k \partial x_l^k} \right)^{-1}$$

$$\left(\int d\hat{u}_j^k \cdot \hat{u}_j^k \frac{\partial^2 P_2}{\partial x_l^k \partial x_l^k} \right) \frac{\partial}{\partial \hat{u}_j^k} \left(\left(\int d\hat{u}_j^k \cdot \hat{u}_j^k \frac{\partial^2 P_1(\underset{\rightarrow}{\hat{u}}^k, \underset{\rightarrow}{x}^k, t)}{\partial x_l^k \partial x_l^k} \right) P_2 \right) \quad (11)$$

Similar terms proportional to the molecular diffusivities will arise in the micro-scale mixing description for the scalars within the same framework.

The resulting set of Eqs. (4,5,10) form a closed set of non-linear equations, solution of which is, as was shown in [16], the best possible approximation for the two-point statistical properties. However, it is formulated in terms of a 15 + 2×(number of scalars) dimensional function. For a single reacting scalar problem, its numerical solution with a grid discretisation would require approximately 10^{35} bytes so this would exceed any foreseeable computer capacity. Alternative to a grid discretisation, Monté-Carlo or discrete particles methods, do not seem suitable either because of non-linear terms involving various derivatives in Eq. (11). The work in progress consists in exploring whether a solution to Eqs. (4,5,10) can be found as a truncated series expansion in some suitable functions set.

While the Eqs. (4,5,10) present, in a sense, a complete description of a reacting turbulent flow, their complexity would, in all likeliness, preclude a routine use of them for practical purposes. Therefore, a model usable for applied calculations should be restricted to a one-point statistics and have the time- and length-scales as input variables; following the discussion in Section (2.2) it should account for the molecular transport properties and the elementary chemical reactions rates. Previously, the emphasis in application of the one-point pdf models to the premixed turbulent combustion was put on elaboration of the chemical source terms and refinement of turbulent transport models [42, 43]. Such approach is insufficient as can be deduced from the reported inability of such models to predict different $U_t(u')$ curves for methane- and propane-air mixtures [43] even for very limited variation of equivalence ratios and initial parameters. No published study based on pdf methods has succeeded in predicting either u'_m or u'_q even for a single mixture and deficiencies in the micro-scale mixing models rather than an approximative chemical kinetics mechanism is the most probable cause for this.

Among the models which contain an explicit account of molecular transport and chemical kinetics details, the Linear-Eddy Model [9] has attracted most interest. In this model the statistics of reacting scalars is found from an ensemble of one-dimensional scalar fields the temporal evolution of which is governed by a diffusion-reaction equation supplemented with inversion events mimicking the stirring action of turbulent velocity fluctuations. Indeed, an application of the LEM to premixed flames produced [44] some interesting results; however, this model requires a large number of realisations computed at an extremely fine resolution for large Re_t [50] and there are some conceptual difficulties with the extension of this model to inhomogeneous flows. Another model was proposed in [45] based on analysis of different possible distributions for mixing layer thickness; it still awaits its application to premixed combustion case. Below outlined is another alternative satisfying, at least potentially, the requirements outlined in Section 2.2.

3.2 Reference Scalar Field (RSF) Model

The basic idea of this model is to represent the scalar statistics in terms of a number of the so-called reference fields $Y_j\left(X; \underset{\rightarrow}{x}, t\right)$ possessing an extra probabilistic dimension $X \in [0, 1]$. In other words, in this model, the computation of one n-dimensional pdf is replaced with computation of n one-dimensional reference fields. Original formulation [10, 11] was limited to a single scalar case; X there was defined as the cumulative probability distribution of the progress variable: $X(\hat{Y}) = 1 - \int_0^{\hat{Y}} P(\hat{Y})d\hat{Y}$. In case of many reactive scalars a question arises whether the combustion process can be represented as a motion of point representing the state of the reactive mixture at a given point in the configuration space of $\underset{\rightarrow}{Y}$ along a one-dimesional trajectory, especially in a case of complex chemistry. This is a necessary condition for accurate representation of chemical reaction rates within the RSF model [46]. This question was considered in [47] where it was argued that while variations in reaction rates and the mixing time alter the shape of those trajectories, they remain one-dimensional and the RSF model tries to take advantage of this fact.

While the trajectories in the configuration space are assumed to be one-dimesional, connecting the fresh gas state $\underset{\rightarrow}{Y_0}$ with the burnt gas state $\underset{\rightarrow}{Y_b}$, generally speaking, a single parameter describing the completeness of combustion, such as the progress variable for a single scalar case, cannot be specified in a general case of $Le \neq 1$ and for different combustion regimes. Any attempt to use one of species concentration, or the temperature, as such a parameter is doomed to fail as it can be seen from measurements of e.g. [48, 49] where such parametrisation results in widely scattered set of points. The same problem of lack of simple ordering for several scalar configuration space was discussed in [37] as well. However, because the one-point statistical description cannot be self-contained and an approximation for the small-scale mixing is required anyway, it proves quite possible to extend the RSF governing equations for a multi-scalar case without making a recourse to an exact definition of X but imposing instead some conditions it should satisfy. For example, if the motion of the point representing the mixture state is parametrised in terms of some parameter ξ with pdf $P(\xi)$, e.g. ξ might be some kind of the residence time and there exist a single-valued vector function $\underset{\rightarrow}{Y}(\xi)$, then X may be naturally defined as either

$$X(\xi) = \int_{\xi_{min}}^{\hat{\xi}} P(\xi) d\xi \quad \text{or} \quad X(\xi) = 1 - \int_{\xi_{min}}^{\hat{\xi}} P(\xi) d\xi \quad \text{or} \tag{12}$$

and the reference fields are then defined as $Y(X) = Y(X(\hat{\xi})) = Y(\hat{\xi})$.

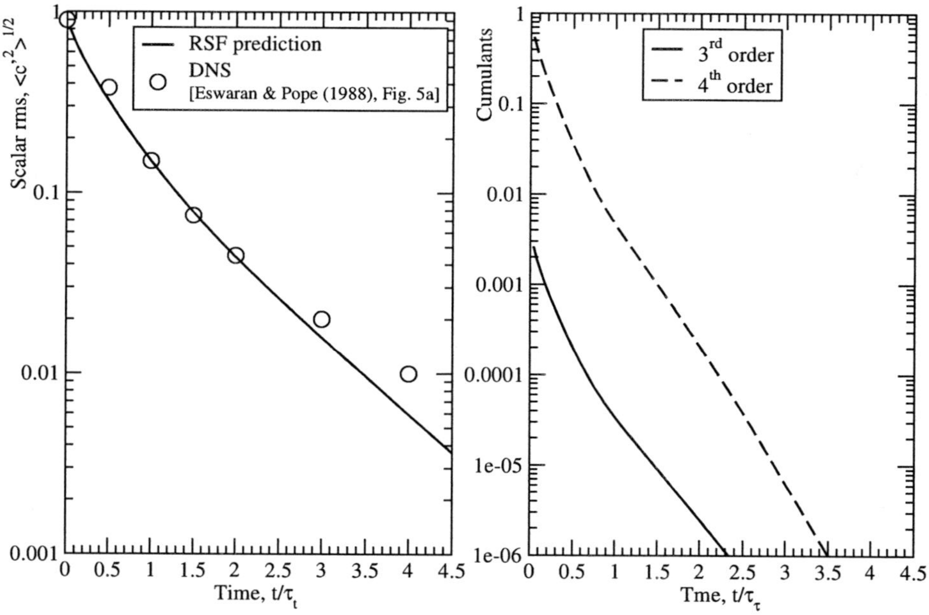

Figure 5: Scalar field moments *vs.* time for the initially bi-modal pdf for the conditions of DNS [40]. a) Comparison of the scalar rms with the DNS. b) The third and fourth cumulants.

The first requirement is that all single-point statistics of scalar fields could be found as:

$$\left\langle F\left(\underset{\rightarrow}{Y}\right)\right\rangle = \int_0^1 dX' \cdot F\left(\underset{\rightarrow}{Y}\left(X'; \underset{\rightarrow}{x}, t\right)\right) \tag{13}$$

The second requirement is that, exactly because X may have two definitions, see Eq. (12), the transport equations for $Y(X)$ should have coefficients symmetrical about the

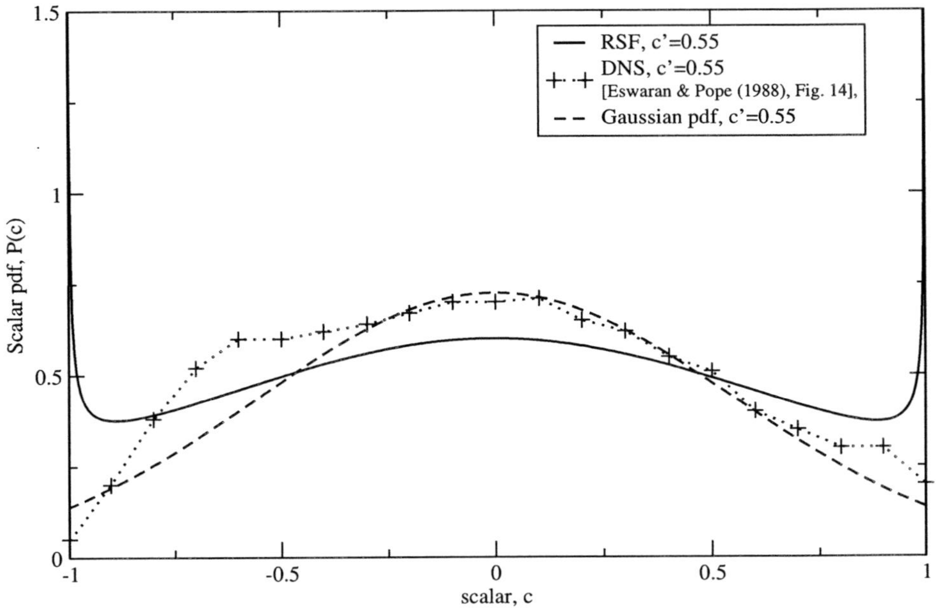

Figure 6: Scalar pdf for the initially bi-modal pdf for the conditions of DNS [40].

point $X = 1/2$. Similarly, employed small-scale mixing model for the rsf's should preserve the average scalar values in the absence of chemical reactions, produce correct rates of scalar variance decay and preserve the boundedness of scalars and possess a number of other properties enumerated in e.g. [37].

Original [11] formulation of the rsf's transport equations employed two separate terms describing the small-scale mixing, one of them taken following the IEM model [35]:

$$\frac{\partial Y_i\left(X; \underset{\rightarrow}{x}, t\right)}{\partial t} + \left\langle \underset{\rightarrow}{u} | \underset{\rightarrow}{Y}\left(\underset{\rightarrow}{x}, t\right) = \underset{\rightarrow}{Y}(X) \right\rangle \cdot \underset{\rightarrow}{\nabla} Y_i(X) =$$

$$W_i\left(\underset{\rightarrow}{Y}(X)\right) + \frac{D_i}{l_t^2} \cdot \frac{\partial^2 Y_i(X)}{\partial X^2} + \frac{\langle Y_i \rangle \left(\underset{\rightarrow}{x}, t\right) - Y_i(X)}{\tau_t} \tag{14}$$

where l_t and τ_t are integral length and time scales of turbulent scalar field. Boundary and initial conditions for Eq. (14) are discussed in [11]. Subsequent study [50] revealed that the deficiencies in the IEM mixing model [35] are inherited by the solutions of Eq. (14), moreover this equation predicts very little sensitivity of the results upon the molecular diffusivity D_i at $Re_t \gg 1$ contrary to the observations of the Sect. (2.2). To cure this deficiencies, a new small-scale mixing description is proposed here leading to the following RSF equation:

$$\frac{\partial Y_i\left(X; \underset{\rightarrow}{x}, t\right)}{\partial t} + \left(\underset{\rightarrow}{u}(X) \cdot \underset{\rightarrow}{\nabla}\right) Y_i(X) = W_i\left(\underset{\rightarrow}{Y}(X)\right) + \frac{\partial}{\partial X} D_i(X) \cdot \frac{\partial Y_i(X)}{\partial X} \tag{15}$$

where the small-scale mixing is described as a diffusion in X direction with variable diffusion coefficient $D_i(X)$ with the dimension of inverse time. The definition of the conditional

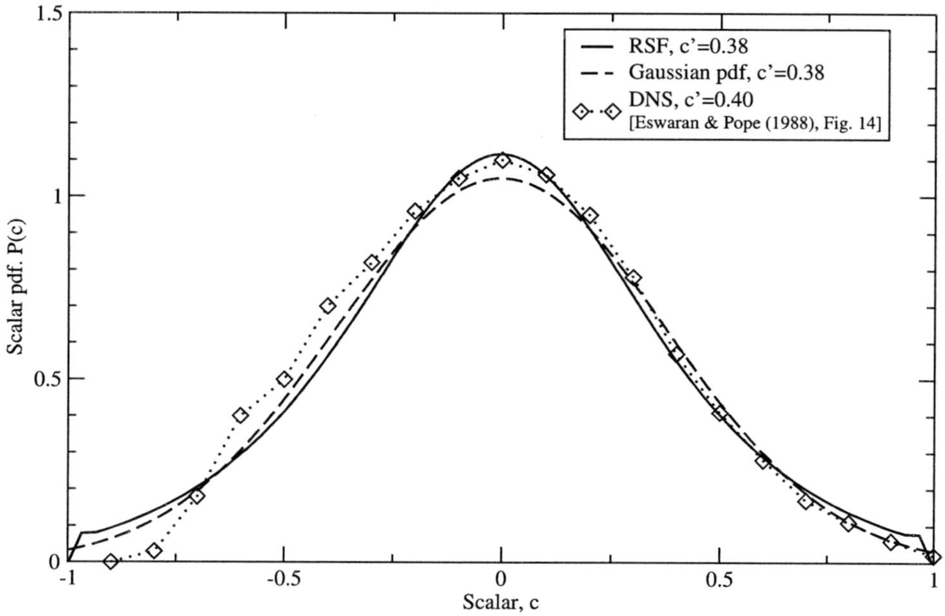

Figure 7: Scalar pdf for the initially bi-modal pdf for the conditions of DNS [40].

velocity $\underset{\rightarrow}{u}(X)$ as well as the derivation of its own transport equation is given in the subsequent section, but the only application of Eq. (15) that has so far been performed is a homogeneous combustion case.

The small-scale mixing in Eq. (15) is described as a diffusion process. Provided $D_i(X = 0) = D_i(X = 1) = 0$ and $D_i(X) > 0$ for $X \in (0, 1)$ and regardless of a particular shape of $D_i(X)$ satisfying these conditions, diffusion process representation preserves the mean scalar values and guarantees that the scalar values remain within the bounds they had at $t = 0$. At the same time, the correct rate of scalar variance decay in non-reacting homogeneous case is determined by the integral:

$$\frac{\partial}{\partial t}\int_0^1 (Y_i(X) - \langle Y_i \rangle)^2 \, dX = \frac{1}{2}\int_0^1 (Y_i(X) - \langle Y_i \rangle) \frac{\partial}{\partial X} D_i(X) \cdot \frac{\partial Y_i(X)}{\partial X} dX$$
$$= \frac{1}{2}\int_0^1 D_i(X) \cdot \left(\frac{\partial Y_i(X)}{\partial X}\right)^2 dX \qquad (16)$$

and one can see that the combination $\frac{1}{2} D_i(X) \cdot \left(\frac{\partial Y_i(X)}{\partial X}\right)^2$ is the conditional scalar dissipation and this is fully consistent with Eq. (13).

A specific shape of $D_i(X)$ should depend upon the molecular diffusivity $\mathcal{D}_i$ for the scalar Y_i but it the same time it should produce the rate of decay in Eq. (16) determined by the integral time scale τ_t with only a weak sensitivity upon $\mathcal{D}_i$. Further, considering that the Eq. (15) can be adimensionalised using the u', l_t and $\mathcal{D}_i$ as the scaling variables, the most general form for $D_i(X)$ should be $D_i(X, \underset{\rightarrow}{Y}(X), Re_t \cdot Sc_i)$ where $Sc_i = \frac{\mathcal{D}_i}{\nu}$ while the integral of the diffusivity in X-space must not depend on the molecular transport,

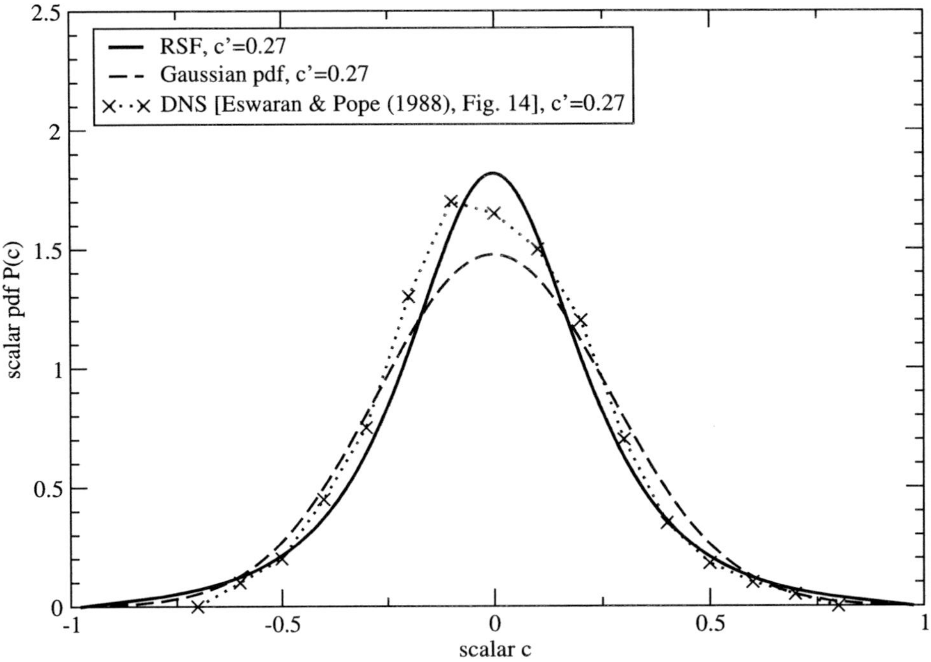

Figure 8: Scalar pdf for the initially bi-modal pdf for the conditions of DNS [40].

e.g.: $\int_0^1 D_i(X)dX = C_D\tau_t^{-1}$ where C_D is a constant. Independence of this constant of Re_t, molecular transport, or flow geometry would therefore be an indicator of how realistic is a particular approximation formula for $D_i(X, \underset{\rightarrow}{Y}(X), Re_t \cdot Sc_i)$. At the present, one has to resort to some guesswork to presume $D_i(X, \underset{\rightarrow}{Y}(X), Re_t \cdot Sc_i)$ and the to assess this presumed dependency following the guidelines outlined above. Below shown are the results obtained with the following "recipe" for $D_i(X)$:

$$D_i(X) = C_D\tau_t^{-1} \cdot (Re_t \cdot Sc_i) \cdot (X(1-X))^{\alpha(Re_t \cdot Sc_i)} \tag{17}$$
$$B(\alpha(\zeta)+1, \alpha(\zeta)+1) = \frac{\sqrt{\pi}\Gamma(\alpha+1)}{2^{2\alpha+1}\Gamma(\alpha+\frac{3}{2})} = \zeta$$

where $B(x, y)$ is the β-function and the last equation expresses the normalisation $\int_0^1 D_i(X)dX = C_D\tau_t^{-1}$. The exponent α is found as a solution to the last equation and it changes from 2 at $Re_t \cdot Sc_i = 30$, values typical for DNS research, to approximately 5 for more realistic case of $Re_t \cdot Sc_i = 2 \cdot 10^4$. Certainly, Eq. (17) is only a tentative expression, but, unlike any other small-scale mixing model developed so far [35], it should satisfy the physical requirements discussed above.

Assessment of Eq. (15) is still an on-going work, and the first results presented here concern its application to, firstly, a non-reactive single scalar mixing using the DNS data [40] as the reference and, secondly, finding whether this model is capable of predicting the influence of molecular transport coefficients at $Re_t \gg 1$ on the average reaction rate using a hypothetic homogeneous reactor with the chemical kinetics parameters describing the DTBP decomposition flame described in Sect. (2.2).

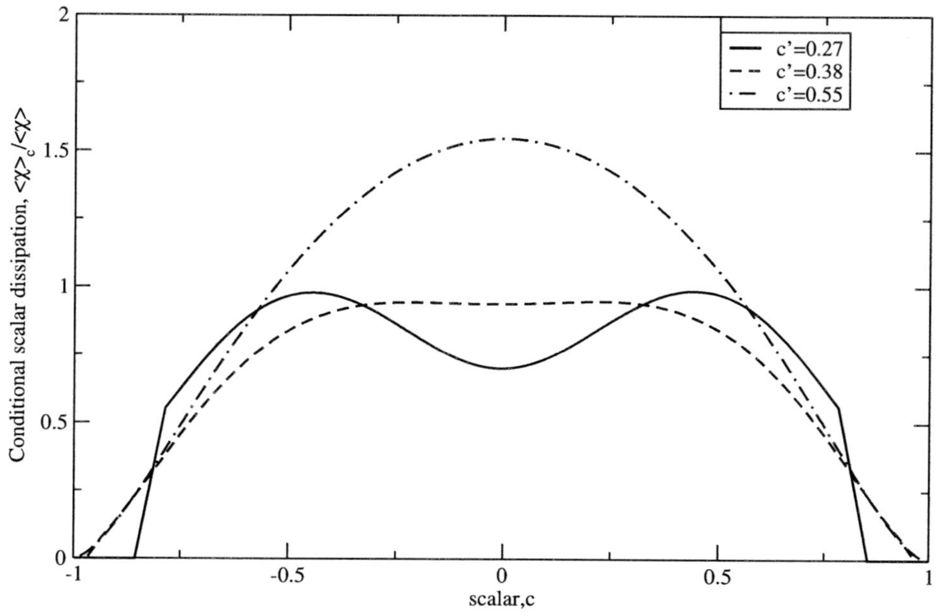

Figure 9: Conditional scalar dissipation for the initially bi-modal pdf for the conditions of DNS [40]. The curves shown correspond to the pdf shapes shown in Fig (8).

The work [40] presents the results of DNS in homogeneous and isotropic turbulence at $Re_t Sc \approx 28$ for the scalar $Y \in [-1, 1]$ with the zero mean and initial pdf close to the bi-modal shape. It was found that the particular value of $C_D = 0.18$ allows to reproduce well simultaneously both the rms scalar fluctuation decay, see Fig. (5a), and the corresponding pdf shapes, see Fig. (8). Moreover, predicted exponential decrease to zero of the third and fourth order cumulants of the scalar pdf, see Fig. (5b) is entirely consistent with the observation of [40, Fig. 21] of the pdf collapse to a universal shape parameterised solely in terms of the rms fluctuation. The profiles of the conditional scalar dissipation are shown in Fig. (9) and are also in a very good agreement with [40, Fig. 19]. However, influence of molecular transport cannot be reliably established in this test case because the it has not been varied in the DNS calculations.

Assessment of effects of molecular transport as predicted by the present form of the RSF model was therefore performed in a hypothetical case of a homogeneous constant pressure reactor fed with a premixture of $25\%vol.$ of DTBP vapour with nitrogen at the initial temperature $T_0 = 350\,K$. The combustion rate at the atmospheric pressure and is given by Eq. (3). This problem necessitates the use of two scalars which are the fuel mass fraction and the temperature. For the given initial conditions the mass diffusivity $\mathcal{D}_{DTBP} = 0.057\,cm^2/sec$, the thermal diffusivity $\kappa = 0.07\,cm^2/sec$ [29]. The turbulence integral length-scale was kept constant at $l_t = 2.0\,cm$, the rms velocity u' was varied. The principal value of interest was the maximum reaction rate W_{max} achieved in the transient regime when the rate of supply was superior to the blow-off limit. Obviously, the average rate of combustion equals the rate of supply below this limit and may therefore be made completely arbitrary.

There is a certain analogy between the conditions inside a turbulent flame brush and this hypothetical reactor. At the same time, the entrainment of a fresh mixture into the brush is governed by the turbulent diffusion [1] and is dependent on flow geometry and the local turbulence properties. In this simplified case, the rate of fresh mixture supply is an independent variable and one cannot establish a straightforward link between it and the turbulent burning rate discussed in Sect. (2.2). At the same time, one can expect that the influence of various factors on W_{max} and U_t is qualitatively similar which justifies the use of the former variable as an indicator of the average reaction rates in the turbulent flame brush.

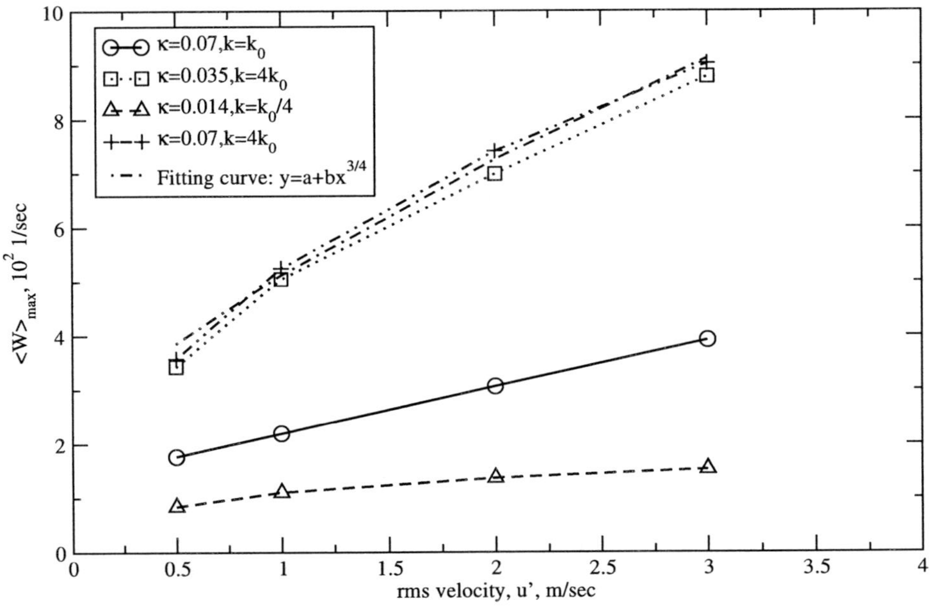

Figure 10: Maximum average reaction rate in a homogeneous reactor *vs.* rms turbulence velocity for different parameters.

Figure (10) presents the dependency of the maximum average reaction rate on u'. Variation in the Lewis number was achieved by changing κ simultaneously changing the pre-exponent in the Eq. (3) in such a way that the laminar flame speed u_n calculated for the first order reaction [51, Ch. 4, Eq.(3.29)] remains constant. The results so obtained do clearly demonstrate a very appreciable increase in $W_{max}(u')$ with reduction in κ, in agreement with the conclusions of [14, 15]. At the same time, varying κ only while keeping the constant value of the pre-exponential factor corresponds to maintaining the product $u_n \kappa$ constant while changing the Lewis number. It can be seen that such variation produces very little effect on W_{max}, somewhat differing from the scaling $u_n^2 \kappa$ following from Eq. (2). Finally, it may be remarked that for the case with the large Lewis number $Le = 2.48$ the increase in W_{max} with increase in u' is much slower, both absolutely and relatively, than for cases with smaller Le. All these trends support the analogy between W_{max} in a homogeneous reactor and the mass burning rate U_t within a turbulent flame brush.

3.3 Transport Equation for Conditional Velocity

Extension of the rsf model for inhomogeneous flows requires a definition of the conditionally averaged velocity $\langle u_k | Y(X) = \hat{c} \rangle = \langle u_k \rangle_{\hat{c}}$. Introduction of the eddy-diffusivity, e.g.:

$$\langle u_k \rangle_{\hat{c}} \cdot \frac{\partial P(\hat{c})}{\partial x_k} = \langle u_k \rangle \cdot \frac{\partial P(\hat{c})}{\partial x_k} - \frac{\partial}{\partial x_k} \frac{D_T}{Sc} \cdot \frac{\partial P(\hat{c})}{\partial x_k}$$

applied to the rsf has been shown to produce unphysical concentration [46] fields, and as an alternative, the calculation of the conditionally averaged velocity from its own transport equation is suggested. The departure point for this derivation is the transport equation for the joint scalar-velocity single point pdf [33]:

$$\frac{\partial P_{uc}\left(\underset{\rightarrow}{\hat{u}}, \hat{c}; \underset{\rightarrow}{x}, t\right)}{\partial t} + \left(\underset{\rightarrow}{\hat{u}} \cdot \underset{\rightarrow}{\nabla}\right) P_{uc} + \frac{\partial}{\partial \hat{c}} W(\hat{c}) P_{uc} = -\frac{\partial}{\partial \hat{c}} \left\langle \frac{1}{\rho} \underset{\rightarrow}{\nabla} D \underset{\rightarrow}{\nabla} c \right\rangle_{\hat{u}\hat{c}} P_{uc}$$
$$+ \sum_k \frac{\partial}{\partial \hat{u}_k} \left[\left\langle \frac{1}{\rho} \cdot \frac{\partial p}{\partial x_k} - \frac{1}{\rho} \underset{\rightarrow}{\nabla} D \underset{\rightarrow}{\nabla} u_k - F_k \right\rangle_{\hat{u}\hat{c}} P_{uc} \right] \quad (18)$$

This joint pdf can be expressed with the help of the Bayes theorem as:

$$P\left(\underset{\rightarrow}{\hat{u}}, \hat{c}; \underset{\rightarrow}{x}, t\right) = P\left(\underset{\rightarrow}{\hat{u}} | \hat{c}; \underset{\rightarrow}{x}, t\right) P\left(\hat{c}; \underset{\rightarrow}{x}, t\right)$$

where $P\left(\underset{\rightarrow}{\hat{u}} | \hat{c}; \underset{\rightarrow}{x}, t\right)$ is the conditional pdf of velocity at a given point $\underset{\rightarrow}{x}, t$. Applying this decompostition to the Eq. 18:

$$P_{\hat{c}} \cdot \left[\frac{\partial P_{u|\hat{c}}\left(\underset{\rightarrow}{\hat{u}} | \hat{c}; \underset{\rightarrow}{x}, t\right)}{\partial t} + \left(\underset{\rightarrow}{\hat{u}} \cdot \underset{\rightarrow}{\nabla}\right) P_{u|\hat{c}} + \frac{\partial}{\partial \hat{c}} W(\hat{c}) P_{u|\hat{c}} \right] +$$
$$P_{u|\hat{c}} \left[\frac{\partial P_{\hat{c}}\left(\hat{c}; \underset{\rightarrow}{x}, t\right)}{\partial t} + \left(\underset{\rightarrow}{\hat{u}} \cdot \underset{\rightarrow}{\nabla}\right) P_{\hat{c}} + \frac{\partial}{\partial \hat{c}} W(\hat{c}) P_{\hat{c}} \right] =$$
$$-P_{\hat{c}} \frac{\partial}{\partial \hat{c}} \left\langle \frac{1}{\rho} \underset{\rightarrow}{\nabla} \rho D \underset{\rightarrow}{\nabla} c \right\rangle_{\hat{u}\hat{c}} P_{u|\hat{c}} - P_{u|\hat{c}} \frac{\partial}{\partial \hat{c}} \left\langle \frac{1}{\rho} \underset{\rightarrow}{\nabla} \rho D \underset{\rightarrow}{\nabla} c \right\rangle_{\hat{u}\hat{c}} P_{\hat{c}}$$
$$+ P_{\hat{c}} \sum_k \frac{\partial}{\partial \hat{u}_k} \left[\left\langle \frac{1}{\rho} \cdot \frac{\partial p}{\partial x_k} - \frac{1}{\rho} \underset{\rightarrow}{\nabla} \rho \nu \underset{\rightarrow}{\nabla} u_k - F_k \right\rangle_{\hat{u}\hat{c}} P_{u|\hat{c}} \right]$$

and using the equation for the scalar pdf, obtained from the Eq.18 upon the integration with respect to the velocity variable:

$$\frac{\partial P_{\hat{c}}\left(\hat{c}; \underset{\rightarrow}{x}, t\right)}{\partial t} + \left(\underset{\rightarrow}{\hat{u}} \cdot \underset{\rightarrow}{\nabla}\right) P_{\hat{c}} + \frac{\partial}{\partial \hat{c}} W(\hat{c}) P_{\hat{c}} + \frac{\partial}{\partial \hat{c}} \left\langle \frac{1}{\rho} \underset{\rightarrow}{\nabla} \rho D \underset{\rightarrow}{\nabla} c \right\rangle_{\hat{c}} P_{\hat{c}} = 0 \quad (19)$$

the transport equation for the conditional pdf is obtained as:

$$P_{\hat{c}} \cdot \left[\frac{\partial P_{u|\hat{c}}\left(\underset{\rightarrow}{\hat{u}}|\hat{c}; \underset{\rightarrow}{x}, t\right)}{\partial t} + \left(\underset{\rightarrow}{\hat{u}} \cdot \underset{\rightarrow}{\nabla}\right) P_{u|\hat{c}} + \frac{\partial}{\partial \hat{c}} \left[W(\hat{c}) + \left\langle \frac{1}{\rho} \underset{\rightarrow}{\nabla} \rho D \underset{\rightarrow}{\nabla} c \right\rangle_{\hat{u}\hat{c}} \right] P_{u|\hat{c}} - \right.$$

$$\left. \sum_k \frac{\partial}{\partial \hat{u}_k} \left(\left\langle \frac{1}{\rho} \cdot \frac{\partial p}{\partial x_k} - \frac{1}{\rho} \underset{\rightarrow}{\nabla} \rho \nu \underset{\rightarrow}{\nabla} u_k - F_k \right\rangle_{\hat{u}\hat{c}} P_{u|\hat{c}} \right) \right] =$$

$$P_{u|\hat{c}} \left(\underset{\rightarrow}{\hat{u}} - \left\langle \underset{\rightarrow}{u} \right\rangle_{\hat{c}} \right) \cdot \underset{\rightarrow}{\nabla} P_{\hat{c}} + \frac{\partial}{\partial \hat{c}} \left[\left[\left\langle \frac{1}{\rho} \underset{\rightarrow}{\nabla} \rho D \underset{\rightarrow}{\nabla} c \right\rangle_{\hat{u}\hat{c}} - \left\langle \frac{1}{\rho} \underset{\rightarrow}{\nabla} \rho D \underset{\rightarrow}{\nabla} c \right\rangle_{\hat{c}} \right] P_{\hat{c}} \right] \quad (20)$$

Finally, the desired equation for the conditional velocity $\langle u_j \rangle_{\hat{c}}$ is obtained from the Eq. 20 upon myltiplying it by $\hat{u}_j$ and integrating it with respect to $d^3 \underset{\rightarrow}{\hat{u}}$ as:

$$P_{\hat{c}} \cdot \left[\frac{\partial \langle u_j \rangle_{\hat{c}}}{\partial t} + \frac{\partial \langle u_k u_j \rangle_{\hat{c}}}{\partial x_k} + \frac{\partial}{\partial \hat{c}} W(\hat{c}) \langle u_j \rangle_{\hat{c}} + \frac{\partial}{\partial \hat{c}} \left\langle \frac{u_j}{\rho} \underset{\rightarrow}{\nabla} \rho D \underset{\rightarrow}{\nabla} c \right\rangle_{\hat{c}} + \right.$$

$$\left. \left\langle \frac{1}{\rho} \cdot \frac{\partial p}{\partial x_j} \right\rangle_{\hat{c}} - \left\langle \frac{1}{\rho} \underset{\rightarrow}{\nabla} \rho \nu \underset{\rightarrow}{\nabla} u \right\rangle_{\hat{c}} \right] = \left(\langle u_k \rangle_{\hat{c}} \langle u_j \rangle_{\hat{c}} - \langle u_k u_j \rangle_{\hat{c}} \right) \cdot \frac{\partial P_{\hat{c}}}{\partial x_k} = \langle u'_j u'_k \rangle_{\hat{c}} \frac{\partial P_{\hat{c}}}{\partial x_k} \quad (21)$$

As any other "exact" equation, outside the functional formalism, in turbulence theory this equation contains the terms requiring some *ad hoc* approximations. The first three terms are closed, the fourth term, containing the correlation of the velocity and the scalar laplacian was analysed in [33] and the implication of that work is that this term can be presented as the product of the conditionally averaged values:

$$\left\langle \frac{u_j}{\rho} \underset{\rightarrow}{\nabla} \rho D \underset{\rightarrow}{\nabla} c \right\rangle_{\hat{c}} = \langle u_j \rangle_{\hat{c}} \cdot \left\langle \frac{1}{\rho} \underset{\rightarrow}{\nabla} \rho D \underset{\rightarrow}{\nabla} c \right\rangle_{\hat{c}} \quad (22)$$

The second co-factor is required in the so-called small-scale mixing closure for the Eq. 19, therefore this term in Eq. 21 does not require any additional closure hypothesis. Obviously, one has to use the same expressions for the similar terms in Eqs. 19 and 21.

The next term is the conditionally averaged acceleration caused by the pressure difference. Very little is known about pressure gradient fluctuations in a turbulent flow, and for combustion problems where the density difference between hot gas and cold products is large, a tentative closure for this term can be based on assumption that effects of pressure fluctuations are negligible compared with the density fluctuations:

$$\left\langle \frac{1}{\rho} \cdot \frac{\partial p}{\partial x_j} \right\rangle_{\hat{c}} = \frac{1}{\rho(\hat{c}, \langle p \rangle)} \cdot \frac{\partial \langle p \rangle}{\partial x_j} \quad (23)$$

The conditionally averaged Reynolds stress $\left\langle u'_j u'_k \right\rangle_{\hat{c}}$ arising in the second left-hand-side (convection) and the right-hand-side terms of the Eq. 21 may be modelled similarly to its unconditional conterpart with invoking a turbulent viscosity ν_t notion:

$$\langle u'_j u'_k \rangle_{\hat{c}} = -\frac{\nu_t}{2} \left(\frac{\partial \langle u'_k \rangle_{\hat{c}}}{\partial x_j} + \frac{\partial \left\langle u'_j \right\rangle_{\hat{c}}}{\partial x_k} \right) \quad (24)$$

This expression is arguably is the simplest possible starting point for application of the Eq. 19; the determination of $\nu_t \sim \langle u'^2 \rangle \tau_t$ require information about turbulence length- or time-scale which cannot, as a matter of principle, be provided from within a single-point statistical analysis, unlike the two-point description given by the Eqs. (4,5,10).

4 Conclusions

Analysis of [14, 15] demonstrate that account of molecular transport properties is essential for an accurate calculation of the rate of combustion. Those arguments are supplemented here by an observation that the details of chemical kinetics as well become important for stronger turbulence, in addition to all factors reviewed in [14, 15]. The present work explores possible frameworks of incorporating both molecular transport properties and details of chemical kinetics. While a completely self-contained and self-consistent description of a turbulent reacting flow is possible only at the expense of consideration two-point statistics, resulting complexity makes it an unlikely model for applied purposes. At the same time, descriptions limited to one-point statistics, as given by Eqs. 15,17 and 21, will necessarily involve some closure assumptions. It proved possible to find a formulation expressed in the Eqs. (15) and (17), in terms of the reference scalar fields possessing an extra probabilistic coordinate, which is capable of predicting very strong effects of molecular transport coefficients which are entirely in agreement with experimentally observed trends and also recover results of simpler model [14]. Application to inhomogeneous combustion flows, e.g. spherical explosions, will provide a further assessment of the proposed model.

Acknowledgments

Most grateful acknowledgment is due to Prof. C.G.W.Sheppard and Drs. R. Woolley, K. Liu and M.P.Ormsby for their kind permission to display the experimental data resulting from an on-going and yet unpublished work.

References

[1] Spalding D.B. *13^{t}h Symp. (Int.) on Comb.*, The Comb. Inst., Pittsburgh, p.649 (1971).

[2] Bradley D. and Lau A.K.C. *Pure & Appl. Chem.*, **62**, 803 (1990).

[3] Peters N. *21^{s}t Symp. (Int.) on Comb.*, The Comb. Inst., Pittsburgh, p.649 (1986).

[4] Candel S.M. and Poinsot Th.J. *Comb. Sci. and Tech.*, **70**, 1 (1990).

[5] Borghi R. and Mantel T. *Comb. Flame*, **96** 443 (1994).

[6] Mura A. and Borghi R. *Comb. Flame*, **133** 193 (2003).

[7] Frost V.A. in *Proc. of the 3^{r}d All-Union Symp. on Comb. Theory*, Izd-vo AN SSSR, Moscow, pp. 121-125 (1960).

[8] Pope S.B. *Comb. Sci. and Tech.*, **25**, 159 (1981).

[9] Kerstein A.R. *Comb. Sci. and Tech.*, **60**, 391 (1988).

[10] Burluka A.A. and Borghi R.*Archivum Combustionis*, **15**, 229 (1995).

[11] Burluka A.A., Gorokhovski M.A. and Borghi R. *Comb. Flame*, **109**, 173 (1997).

[12] Bilger R.W. *Comb. Flame* **138**, 188 (2004).

[13] Bilger R.W. and Klimenko A.Yu. *Progress Energy Comb. Sci.* **25**, 595 (1999).

[14] Lipatnikov A.N. and Chomiak J. *Progress Energy Comb. Sci.* **28**, 1 (2002).

[15] Lipatnikov A.N. and Chomiak J. *Progress Energy Comb. Sci.* **31**, 1 (2005).

[16] Ievlev V.M. *Izv. AN SSSR, Mechanica Zhidk. i Gaza* **No. 1**, 91 (1970). Also: Ievlev V.M. *Turbulent motion of high-temperature continuum media*, "Nauka", Moscow, 1975 (in Russian).

[17] Kolmogorov A.N. *Doklady AN SSSR* **30**, 299 (1941).

[18] Kobayashi H., Tamura T., Maruta K., Niioka T and Williams F.A. *XXVI^{t}h Symp. (Int.) on Comb.*, The Comb. Institute, 389 (1996).

[19] Gorbunov G.M., in *Flame stabilisation and combustion process development in a turbulent flow*, "Oborongiz", Moscow, 31 (1961) (work described and cited after: Schetinkov E.S. *Gaseous combustion physics*, Nauka, Moscow, 1965.)

[20] Gökalp I. *Rev. Gén. Therm. Fr.* **192**, 855 (1977).

[21] Ulitsky M., Ghenaï Ch., Gökalp I., Wang L.-P. and Collins L.R. *t Combust. Theory Modelling* **4**, 241 (2000).

[22] Sokolik A.S., Karpov V.P. and *Doklady*, (1965).

[23] Abdel-Gayed R.G. and Bradley D. *XVIth Symp. (Int.) on Comb.*, The Comb. Institute, 1725 (1976).

[24] Karpov V.P. and Severin E.S. *Fizika Goreniya i Vzryva* **16**, 45 (1980).

[25] Abdel-Gayed R.G. and Bradley D. *Comb. Flame*, **62**, 61 (1985).

[26] Betev A.S., Karpov V.P., Lipatnikov A.N. and Vardosanidze Z.P. *Arch. Combustionis* **15**, 187 (1995).

[27] Liu K., Ormsby M., Woolley R., Sheppard C.G.W. and Burluka A.A. *in preparation*.

[28] Gu X.J., Haq M.Z., Lawes M. and Woolley R. *Comb. Flame*, **121**, 41 (2000).

[29] Liu K., Ormsby M.P., Griffiths J.F. and Burluka A.A. *Flames of di-tert-butyl-peroxide decomposition as test case for turbulent combustion theories* (in preparation).

[30] H. Kobayashi, Y. Kawabata and K. Maruta, *Proc. Combust. Inst.*, **27** :941-948(1998).

[31] Nakahara M. and Kido H., *Mémoires of the Faculty of Engineering, Kyushu Univ.* **58**, 55 (1998).

[32] Ievlev V.M. *Dokl. Acad. Nauk SSSR* **208**, 1044 (1973).

[33] Kuznetsov V.R. and Sabel'nikov V.A. *Turbulence and Combustion*, Hemisphere Publ. Corp., New York, 1990.

[34] Gardiner C.W. *Handbook of Stochastic Methods*, 2nd Ed., Springer Verlag, Berlin, 1997.

[35] Dopazo C. in *Turbulent Reacting Flows*, Libby P.A. and Williams F.A., Eds., , Acad. Press, London, p. 375, (1994).

[36] Fox R.O *Phys. Fluids* **8** 2678 (1996).

[37] Subramaniam S. and Pope S.B. *Comb. Flame*, **115**, 487 (1998).

[38] Ren Zh. and Pope S.B. *Comb. Flame*, **136**, 208 (2004).

[39] Vicente W., Salinas M., Barrios E. and Dopazo C. *Comb. Sci. and Tech.*, **176**, 585 (2004).

[40] Eswaran and Pope S.B. *Phys Fluids* **31**, 506 (1988).

[41] Kuo Y.-Y. and O'Brien E.E. *Phys Fluids* **24**, 194 (1981).

[42] Anand M.S. and Pope S.B. *Comb. Flame*, **67**, 127 (1987).

[43] Hŭlek T. and Lindstedt R.P. *Comb. Flame*, **104**, 481 (1996).

[44] Smith T.M. and Menon S. *XXVIth Symp. (Int.) on Comb., The Comb. Institute*, **299** (1996).

[45] Jaberi F.A. and Givi P. *Comb. Sci. and Tech.*, **104**, 249 (1995).

[46] Burluka A.A. *Thèse de doctorat*, The Univ. of Rouen, Rouen, (1996).

[47] Borghi R. and Pourbaix E. *Physico-Chemical Hydrodyn.*, **2**, 65 (1981).

[48] Chen Y.-Ch. and Bilger R.W. *Comb. Flame*, **131**, 400 (2002).

[49] Chen Y.-Ch. and Bilger R.W. *Comb. Flame*, **138**, 155 (2004).

[50] Liu K. *PhD thesis*, The Univ. of Leeds, Leeds, (2002).

[51] Zel'dovich Ya. B., Barenblatt G.I., Librovich V.B. and Makhviladze G.M. *Mathematical theory of combustion and explosion*, Nauka, Moscow, 1980.

In: New Devel. in Combustion Research
Editor: William J. Carey, pp. 187-208
ISBN: 1-59454-826-9

Chapter 5

Study Diesel Spray Flame Structure and Emission Reduction Techniques Using Detailed Chemical Kinetics with CFD Models

Song-Charng Kong*

Engine Research Center, University of Wisconsin-Madison,
1500 Engineering Drive, Madison WI 53706, USA

Abstract

Detailed chemical kinetics was implemented into a state-of-the-art 3-D CFD code to simulate the complex diesel spray atomization, combustion and emission processes. Spray atomization is simulated by considering instability of the surface wave that results in fuel drops breakup from the fuel jet. The chemistry model uses a reaction mechanism to simulate the fuel oxidation and NOx formation processes. Soot emission is modeled by using a competing formation and oxidation rate formulation. The model is used to study the diesel lift-off flame structure under high ambient temperature and pressure conditions and results are compared with experimental imaging. Good levels of agreement are obtained in lift-off length, local equivalence ratio at the lift-off location, flame structure and soot distribution within the reaction zone. Results show that low-temperature cool flame heat release occurs prior to the lift-off location. As the cool flame mixture moves downstream of the spray plume, second-stage ignition occurs such that the flame is stabilized at the lift-off location. It is also found that the amount of soot emission is related to the local conditions at the lift-off location, which, in turn, is affected by ambient conditions. A decrease in ambient gas temperature results in a longer lift-off length (i.e., ignition delay) and less soot emission. Such a low-temperature combustion technology can be used to reduce soot and NOx emissions in diesel engine operations. The present model is also applied to simulate diesel engine combustion under conditions with highly premixed burn. Computed cylinder pressure, heat release rate, soot and NOx emissions are compared with experimental measurements with good agreement. Results show that NOx emission decreases monotonically as the start of injection (SOI) is retarded. However, soot emission increases as start of injection is retarded approaching top-dead-center and then decreases at further retarded timings, as also observed from engine-out

* E-mail address: kong@erc.wisc.edu, Tel: 608-265-6432, Fax: 608-262-6707

measurement. The model indicates that low soot emission at early SOI is due to better oxidation while low soot emission at late SOI is due to less formation. Results show that emission reduction strategies such as late injection and exhaust gas recirculation (EGR) can be used such that combustion occurs at a relatively low temperature environment.

Introduction

The internal combustion (IC) engine is a complex chemical system. To predict the behavior of such a system, one must be able to model a wide variety of physical and chemical processes, including the mechanical device itself, fluid dynamics of gas exchange and fuel-air mixing, and chemical processes associated with fuel combustion. Among the these processes, the energy released by chemical reactions is what drives the system and the computational combustion models can play a critical role in understanding the behavior of such a complex system. To simulate the combustion process in a reciprocating piston engine, chemical reactions need to be solved in the context of a three-dimensional, multiphase, transient, turbulent flow problem with moving boundaries (i.e., pistons).

Diesel engine manufacturers are facing stringent emission regulations. A better understanding of the diesel spray combustion process is crucial to help design low emission diesel engines. Motivated by the need to better understand the particulate matters (soot) and nitrogen oxides (NOx) formation processes in diesel sprays, researchers have made direct optical measurements in both engines [1—5] and high-temperature, high-pressure combustion chambers [6—8]. Experimental data have been used to construct a conceptual diesel spray combustion image [1] and show that the details of the flame structure are crucial to the soot formation process during the mixing-controlled combustion phase [7,8]. The lifted flame consists of a diffusion flame at the periphery of the fuel jet (where NOx is formed) and a rich reaction zone located downstream of the lift-off length in the central region of the fuel jet (where soot is formed).

Experimental investigations have also helped numerical model development and there have been numerous combustion simulations using Computational Fluid Dynamics (CFD) models [9—12]. A predictive numerical model can also help understand the diesel spray combustion process and provide insights to the details of flame structure. Development and applications of engine CFD models have become increasingly important and effective in analyzing the complex diesel combustion process.

A single-step approach was first used to simulate the diesel combustion energy release rate by using a turbulent eddy dissipation timescale [9]. A model was developed that used separate submodels to simulate autoignition, main combustion, and NOx and soot formation processes [10]. A PDF-timescale model was further introduced to account for the chemistry-controlled premixed burn and mixing-controlled diffusion combustion [11]. The above models only consider the major combustion species that are sufficient to accurately predict the equilibrium thermodynamic temperature. Examples of other combustion models include the progress variable approach [12], the intrinsic low dimensional manifold (ILDM) method [13] and the coherent flamelet model [14].

In order to introduce more complex chemistry, the Representative Interactive Flamelet (RIF) model was developed that views a turbulent flame as an ensemble of flamelet structures representing the instantaneous flame surface that is corrugated by the turbulent flow field [15,16]. The model can account for turbulence-chemistry interactions and allows consideration of detailed chemistry. The flamelet model concept was further improved to include more features for various applications [17—20]. Effects of vaporization on turbulence-chemistry interaction have also been incorporated [21].

Complex chemistry can also be used for combustion simulations by integrating detailed chemistry solver into the engine CFD code directly [22—24]. The model simulates the overall ignition, combustion and emission processes by using a reaction mechanism. This approach is particularly suited for simulating the novel low-temperature diesel combustion process that encompasses different combustion regimes in an engine cycle. To enhance the computational efficiency, a data retrieval methodology has also been developed to reduce the computer time in obtaining chemistry solutions [25].

This study will describe a numerical model that uses direct integration of detailed chemistry with the engine CFD code. The model uses detailed chemical kinetics to simulate the diesel lift-off flame, and its combustion and emission formation. The model is validated using experimental combustion and emission data from a combustion vessel and from a heavy-duty diesel engine under various operating conditions.

Model Formulation

Engine CFD Code

The CFD code is a version of KIVA-3V [26] with improvements in various physical and chemistry models developed at the Engine Research Center, University of Wisconsin—Madison. The major model improvements include the spray atomization, drop-wall impingement, wall heat transfer, piston-ring crevice flow, combustion, soot and NOx emissions [27—29]. The RNG k-ε turbulence model was used for in-cylinder flow simulations.

The KIVA-3V code is a computer program for the numerical calculation of transient, multi-dimensional, chemically reactive fluid flows with sprays. The interactions between drops and flow turbulence are modeled by coupling the mass, momentum, and energy transfer between the liquid and gas phases. Details of the governing equations and numerical schemes can be found in the original literature [26]. The code is widely used in the engine industry and research community and has provided valuable insights into the in-cylinder combustion and emission formation processes. Major model improvements including fuel atomization, combustion chemistry and emission formation mechanisms will be described in the following.

Fuel Spray Atomization

The atomization of the diesel fuel jet is simulated by a breakup model that monitors the growth of unstable surface waves [27,28]. The breakup model assumes that aerodynamic forces at a liquid-gas interface and the resulting surface waves are responsible for the atomization. Nozzle cavitation and flow turbulence effects are introduced through an initial disturbance level that is accounted for in model constants.

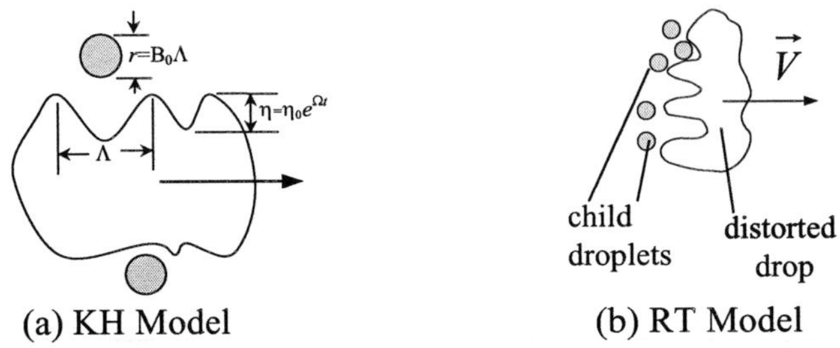

Figure 1 Schematic of Kelvin-Helmholtz and Rayleight-Taylor breakup models.

In the dense spray region near the nozzle exit, the model considers the breakup of the injected liquid to be due to the relative velocity between the gas and liquid phases. The growth of Kelvin-Helmholtz (KH) instabilities induces the shearing-off of droplets from the liquid surface, as shown in Fig. 1(a). In the KH model, the stability analysis results in a dispersion relation as follows.

$$\frac{\Lambda}{r} = 9.02\frac{\left(1+0.45Z^{0.5}\right)\left(1+0.4T^{0.7}\right)}{\left(1+0.87We_2^{1.67}\right)^{0.6}} \tag{1}$$

$$\Omega\left(\frac{\rho_1 a^3}{\sigma}\right)^{0.5} = \frac{0.34+0.38We_2^{1.5}}{\left(1+Z\right)\left(1+1.4T^{0.6}\right)} \tag{2}$$

In the above equations, Z is the Ohnesorge number, We the Weber number, ρ the liquid density, T the Taylor parameter, r the parent drop radius, subscript 1 the liquid phase and 2 the gas phase. The breakup timescale and resulting drop size are calculated from the wavelength (Λ) and frequency (Ω) of the fastest growing surface wave. The result of the KH breakup mechanism is the continuous shearing-off of droplets from the liquid surface. The rate-of-change of the drop radius and the resulting "child" droplet size are related to the frequency and wavelength of the fastest growing surface wave,

$$\frac{dr}{dt} = \frac{r'-r}{\tau} \tag{3}$$

$$\tau = B_1\frac{3.788r}{\Omega\Lambda} \tag{4}$$

where r is the drop radius, r' the equilibrium drop radius, τ the breakup timescale, and B_1 the model constant that can be related to nozzle geometry and initial jet disturbance.

On the other hand, at high relative velocities, drops are flattened and fragmented by relative large-wavelength waves whose wavelengths and growth rates are consistent with estimates from Rayleigh-Taylor (RT) instability theory,

$$\Lambda = 2\pi\sqrt{\frac{3\sigma}{a\rho}} \tag{5}$$

$$\Omega = \sqrt{\frac{2a}{3}\left[\frac{a\rho}{3\sigma}\right]^{1/4}} \tag{6}$$

where σ is the liquid surface tension, ρ the liquid density, and a the acceleration. Thus, the RT model simulates the sudden breakup of a highly distorted drop as also shown in Fig. 1(b). The breakup timescale and resulting drop size can be related to the acceleration/deceleration of the moving drops. Both KH and RT breakup mechanisms are implemented into the fuel spray atomization model. The KH model is used to simulate the drop breakup within the breakup length. Beyond the breakup length, a competing KH and RT breakup mechanism is used. The drop breaks up when either one of the breakup criteria is reached [27,28].

Integration of Engine CFD and Combustion Chemistry

Since detailed reaction mechanisms are used to simulate the diesel fuel chemistry, the CHEMKIN chemistry solver [30] was integrated into the KIVA-3V code for solving the chemistry during multi-dimensional engine simulations. The chemistry and flow solutions were then coupled. It should be noted that the convection and diffusion transport between different computational cells are modeled by the RNG k-ε turbulence model.

The interpreter of CHEMKIN is first executed to generate a binary linking file that contains all the reaction and species information as an input to KIVA. An interface program was developed such that CHEMKIN is used as the chemistry subroutine in KIVA. Basically the reaction mechanism is solved for every computational cell at each timestep. Species and thermodynamic conditions are passed to the CHEMKIN solver to obtain the full chemistry solutions.

The reactive mixture in each grid cell is treated as a closed system in which the rate of change of each individual species is

$$\frac{dY_k}{dt} = v\dot{\omega}_k W_k \tag{7}$$

where v is the specific volume, Y_k is the mass fraction , $\dot{\omega}_k$ is the production rate and W_k is the molecular weight of species k. By using the ideal gas mixture assumption, the energy equation can be obtained under constant volume conditions as

$$c_v \frac{dT}{dt} + v\sum_{k=1}^{K} e_k \dot{\omega}_k W_k = 0 \tag{8}$$

where c_v is the mean specific heat of the mixture, T is temperature and e_k is the internal energy of species k. This formulation is consistent with the KIVA formulation that solves the internal energy equation, and the volume of the computational cell is not updated until the final rezoning phase. During the chemistry solution, the CFD timestep is taken as the integration time in the chemistry solver DVODE to obtain new mixture conditions.

Despite the fact that both the species production rate and energy equations are solved by the DVODE solver, only the new species concentration is passed to the interface program which, in turn, calculates the change in specific internal energy of each computational cell. The new gas phase temperature obtained from the chemistry solver is not immediately updated in the CFD code. This strategy is also consistent with the KIVA formulation that retains the change in specific internal energy and updates the gas temperature at the end of each computational timestep.

Experience shows that the DVODE utility is relatively stable since it uses a fully implicit scheme and performs internal iterations within the CFD timestep. The maximum number of iterations within DVODE can be increased and/or the CFD timestep can be decreased if necessary. An alternative semi-implicit chemistry solver has also been developed to enhance the computational efficiency [31].

Fuel Oxidation Chemistry

A skeletal reaction mechanism for n-heptane [32] was developed to simulate diesel fuel chemistry due to their similar ignition characteristics and cetane number. This mechanism was obtained from a larger mechanism using an interactive reduction scheme that utilizes SENKIN, XSENKPLOT and genetic algorithm optimization. The mechanism reduction steps are summarized as follows.

Step 1: Identify the most essential species and reactions using XSENKPLOT
Step 2: Eliminate the unimportant species and reactions to form the new mechanism
Step 3: Validate the new mechanism
Step 4: Add or remove reactions/species to improve the new mechanism
Step 5: Assimilate two or more reactions into a single reaction to obtain the final mechanism
Step 6: Use micro-genetic algorithm optimization techniques to optimize rate constants

The resulting mechanism has 30 species and 65 reactions [32]. The mechanism retains the main features of the detailed mechanism and includes reactions of polycyclic aromatic hydrocarbons. The mechanism was validated using constant-volume ignition delay data in a shock tube and also from engine combustion experiments.

Reduced NO Reaction Mechanism

A NOx formation mechanism was obtained by reducing the Gas Research Institute (GRI) NO mechanism [33], which contains additional 22 species and 101 reactions pertaining to the formation of nitric oxides, in addition to the fuel oxidation mechanism. The GRI NO mechanism was first integrated with the fuel oxidation mechanism to be used in the SENKIN simulations. Both constant volume ignition delay and zero-dimensional HCCI engine combustion simulations were performed using SENKIN. The XSENKPLOT utility was then used to help construct the reduced NO mechanism. The resulting NO mechanism contains only four additional species (N, NO, NO_2, N_2O) and nine reactions. Note that the sum of NO and NO_2 is used to compared with the engine-out NOx emissions measurements in this study.

Phenomenological Soot Model

Soot emissions are predicted using a phenomenological soot model [34] that was incorporated into the KIVA/CHEMKIN code. Two competing processes are considered in this model, namely soot formation and oxidation. The rate of change of soot mass $\dot{M}_s$ within a computational cell is determined from the soot formation rate $\dot{M}_{sf}$ and soot oxidation rate $\dot{M}_{so}$.

$$\frac{dM_s}{dt} = \frac{dM_{sf}}{dt} - \frac{dM_{so}}{dt} \qquad (9)$$

The formation rate uses an Arrhenius expression and the oxidation rate is based on a carbon oxidation model, described as

$$\frac{dM_{sf}}{dt} = A_{sf} M_{C2H2} P^n \exp(-\frac{E_{sf}}{RT}) \qquad (10)$$

$$\frac{dM_{so}}{dt} = \frac{6Mw_c}{\rho_s D_s} M_s R_{Total} \,. \qquad (11)$$

The formation rate reaction uses acetylene (C_2H_2) as the inception species for soot formation, i.e., M_{C2H2} in Eq. (10). This is because acetylene is the most relevant species pertaining to soot formation in hydrocarbon fuels. The pre-exponential constant A_{sf} was adjusted accordingly for the present implementation. On the other hand, the soot oxidation rate is determined by the Nagle-Strickland-Constable model that considers carbon oxidation by two reaction pathways whose rates depend on surface chemistry of two different reactive sites, as in the original model [34].

Results and Discussions

Spray Combustion in a Constant-Volume Chamber

Experiments conducted in an optically accessible, constant-volume combustion chamber at the Combustion Research Facilities, Sandia National Laboratory were used for model validations [7,8]. The vessel has a cubical combustion chamber, 108 mm on a side. The high-temperature and high-pressure environments are created by burning a specified premixed mixture before the start of fuel injection. Optical diagnostics of diesel spray combustion are performed under ambient conditions similar to those in typical diesel engines at the time of injection. A wide range of pressure, temperature and density conditions were considered in the Sandia experiments and some of the cases were used to validate the present numerical model. The baseline experimental conditions for model validations are listed in Table 1.

Table 1 Experimental conditions for model validations [7,8]

Fuel	#2 Diesel
Injection system	Common-rail
Injection profile	Top-hat
Injector orifice diameter	50μm, 100μm, 180μm
Orifice pressure drop	138 MPa
Discharge Coefficient	0.80, 0.80, 0.77
Fuel temperature	436K
Ambient temperature	850~1300K
Ambient density	7.3, 14.8, 30.0 kg/m^3
O_2 concentration	21%

The computations used a 0.5-degree sector mesh with 2 mm grid size in both radial and axial directions. The computational domain was 12.6 cm in diameter and 10 cm in height such that the total volume is the same as that of the combustion chamber in experiments. Uniform chamber temperature, pressure and species concentration based on experimental data were assumed initially without considering combustion radicals.

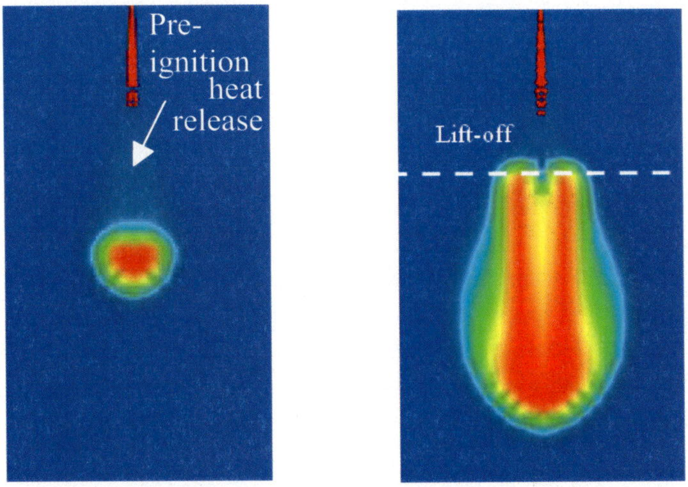

(a) 1.4 ms ASI (b) 2.7 ms ASI

Figure 2 Sample images of the predicted fuel spray and gas temperature distributions for d_{nozz}=100 μm, T_{amb}=900 K, P_{amb}=138 MPa, ρ_{amb}=14.8 kg/m^3. Color scale: 900 to 2600 K.

A typical image of predicted fuel spray and gas temperature distributions of a free diesel lift-off flame is shown in Fig. 2. It can be seen that the liquid fuel undergoes atomization, vaporization, and mixing with entrained air before the lift-off location and then enters the reaction zones. Chemical reactions take place once the fuel is injected and mixes with air, and lead to autoignition at the tip of the spray plume, as seen in Fig. 2(a). Note that in Fig. 2(a), the light colors seen between the liquid spray tip and the ignition location indicates a continuous temperature rise as a result of pre-ignition chemical reactions. The flame is then stabilized at the lift-off location, as seen in Fig. 2(b). The lift-off length is thus determined by both fluid mechanics and chemical kinetics that take place during the fuel/air mixing process prior to the lift-off location.

It is believed that chemical reactions prior to the lift-off location play an important role such that the flame is stabilized due to successive ignition events of the incoming fuel-air mixture. For example, it has been demonstrated that after the flame is established, if the chemical reactions before the lift-off location are suddenly deactivated, the flame is blown downstream and extinguished. It is noted that the high gas velocity of the injected diesel fuel jet would require an unreasonably high turbulent flame speed to balance the incoming reactive mixture in order to stabilize a free standing diesel flame. Nonetheless, more research is needed to study the physics of diesel flame lift-off under engine conditions.

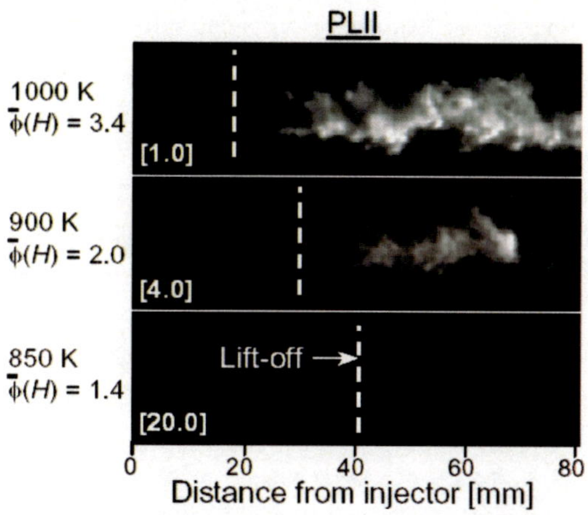

(a) Experimental lift-off length and PLII images

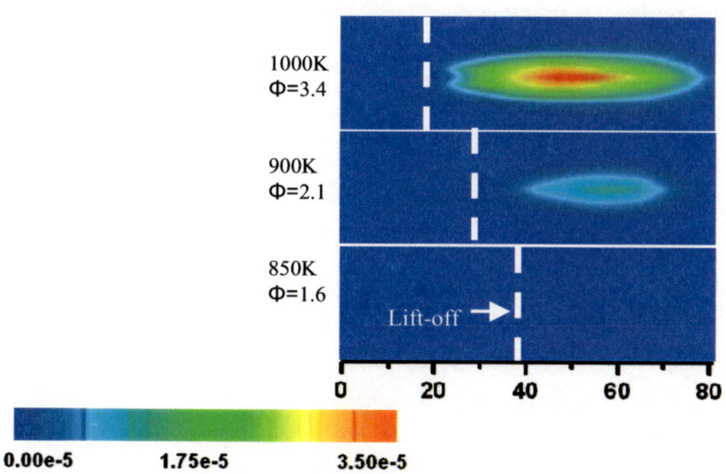

(b) Predicted lift-off length and soot mass fraction

Figure 3 Comparisons between PLII images and predicted soot mass fractions at the central plane of the fuel jet at 3.2 ms ASI. The equivalence ratios were estimated at the lift-off length. Relative PLII camera gain is given in brackets. d_{nozz}=100 μm, P_{inj}=138 MPa, ρ_{amb}=14.8 kg/m^3.

Planar laser-induced incandescence (PLII) images of soot on the cut-plane along the spray axis were compared with model predictions, as shown in Fig. 3. The injector orifice is located at the far left center of each image, with fuel being injected to the right. The flame lift-off location was determined from the OH chemiluminescence images where the OH chemiluminescence intensity is approximately 50% of that just downstream of the initial rapid rise in the OH chemiluminescence. The cross-sectional average equivalence ratio at the flame lift-off length was estimated and is given on the left of the PLII images.

The present simulated soot mass fraction distributions are given in Fig. 3(b) to compare with the PLII images. The predicted lift-off length was determined from the contour of OH species using a similar method as in the experiments. The predicted equivalence ratio at the lift-off length is also given on the left of the images. The color scale of the predicted soot mass fraction is also shown in Fig. 3.

Both experiments and simulations show that as the ambient gas temperature decreases, lift-off length increases, soot concentration decreases, and equivalence ratio at the lift-off length also decreases. In the simulations, the two sooting cases (e.g., 1000 K and 900 K) are found to have soot mass fraction of the order of 1.0E-5 while the predicted maximum soot mass fraction is only about 1.0E-8 for the 850 K case. Other comparisons between model results and experimental images suggest that a soot mass fraction of 1.0E-5 can be used as the criterion to specify sooting and non-sooting conditions in the simulations. This criterion is used later in this paper to determine the sooting limit of injectors with different orifice diameters.

Figure 4 shows comparisons of measured time-averaged KL factors and predicted soot mass fraction along the spray axis at 3.2 ms after start of injection. The KL factor is an indication of optical thickness derived from laser-extinction soot measurements [8]. The KL factor is proportional to the mass of soot along the line of sight of the extinction measurement, so it is compared with the predicted soot mass that is integrated along the same line of sight. Due to the different nature of the KL factor and predicted soot mass, both measured KL factors and predicted soot mass are normalized to allow qualitative comparisons, as shown in Fig. 4.

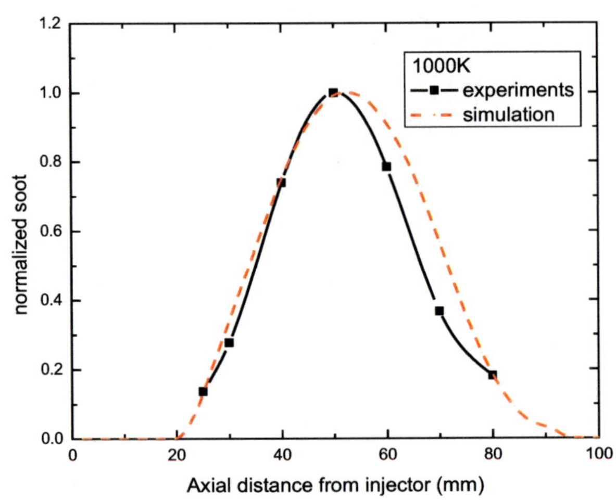

Figure 4 Comparisons of measured time-averaged KL factors and predicted soot mass along the central axis of the fuel jet for the same conditions as in Fig. 3. Both measured and predicted data were normalized to allow qualitative comparison. Results were acquired at 3.2 ms ASI for d_{nozz}=100 μm, T_{amb}=1000 K, P_{inj}=138 MPa, ρ_{amb}=14.8 kg/m^3.

Comparisons of the normalized curves show good agreement in the general trend of the soot distribution along the jet central axis. The agreement in the shape of the curves indicates that the transient features of soot formation and oxidation processes are captured by the present model.

The measured radial soot distribution 50 mm downstream of the injector was also compared with simulations in Figure 5. As before, both measured KL factors and predicted soot mass were normalized to allow qualitative comparisons. Figure 5 also indicates that the simulation results match the experiments very well even for such a small length scale (note that the length scale is 20 mm in the radial direction while it is 100 mm for the axial direction).

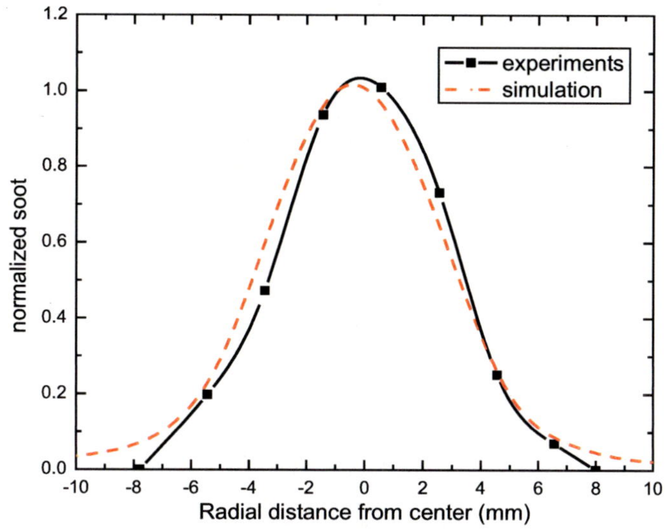

Figure 5 Comparisons of measured time-averaged KL factors and predicted soot mass as a function of radial distance from the jet centerline at an axial distance of 50 mm from the orifice (vertical solid line in Fig. 3). Both measured and predicted data were normalized to allow qualitative comparison. Results were acquired at 3.2 ms ASI for the same conditions as in Fig. 3.

Detailed diesel flame structure is explored by using the numerical models as shown in Figs. 6—8 for the same conditions as in Fig. 2. Figure 6 shows the distributions of instantaneous heat release rate and gas temperature at 2.7 ms after start of injection. Vigorous heat release is observed at the lift-off location due to the successive autoignition of incoming fresh fuel-air mixture. The model predicts negligible heat release inside the spray plume due to oxygen starvation despite the gas temperature is relatively high. The fuel-rich mixture in the center of the spray plume diffuses outward and react with surrounding air to form a diffusion flame, as seen in Fig. 6. The high temperature regions coincide with the location of the diffusion flame as can be seen by comparing both sides of the contours in Fig. 6. The contours of CO and CO_2 in Fig. 7 further confirm that there exists a region of high concentration of intermediate species (e.g., CO) in the center.

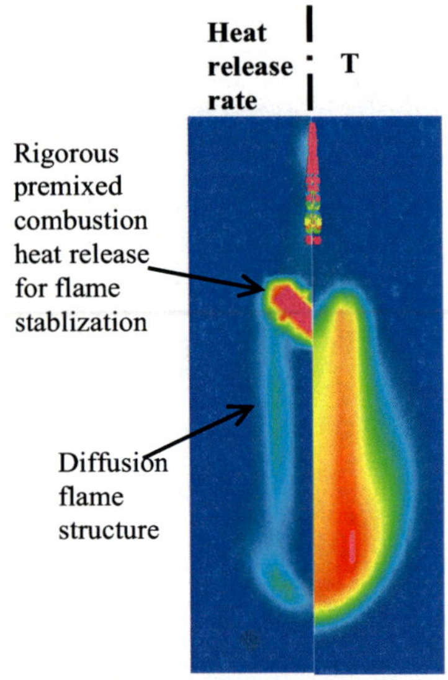

Figure 6 Contours of instantaneous heat release rate and combustion temperature at 2.7 ms after start of injection.

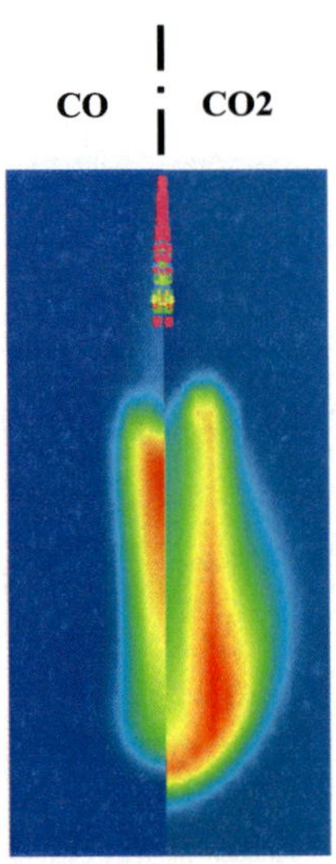

Figure 7 Contours of CO and CO2 at 2.7 ms after start of injection. Results show that high concentration of CO and other intermediate combustion products reside in the center the spray plume.

The distributions of acetylene and soot particles are shown in Fig. 8, which indicates a large amount of soot precursors, as represented by acetylene in the present phenomenological soot model, form at the lift-off location where fuel rich mixture autoignites. The soot precursors then form soot particles further downstream of the spray plume. It implies that the total soot emissions can be related to the soot precursors at the lift-off location, which, in turn, are related to the local mixture conditions. Figure 9 shows the flame lift-off length, equivalence ratio and ignition delay at corresponding lift-off location for various ambient temperature conditions. The ignition delay is based on the local equivalence ratio under corresponding ambient temperature. Under the same injection pressure and amount of fuel, as the ambient temperature decreases, the lift-off length increases such that it allows more time for fuel air mixing. Thus, a lower local equivalence ration can be achieved to result in less soot formation. Such low soot emission characteristics under low ambient temperature environments can be used to optimize the diesel engine combustion, as will be discussed in the next section.

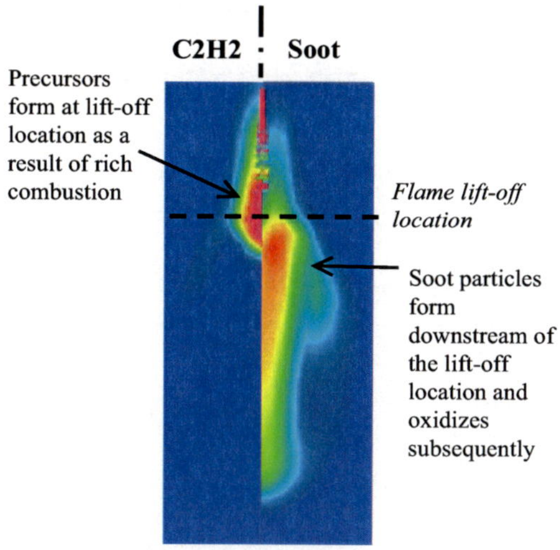

Figure 8 Contours of acetylene and soot particles at 2.7 ms after start of injection.

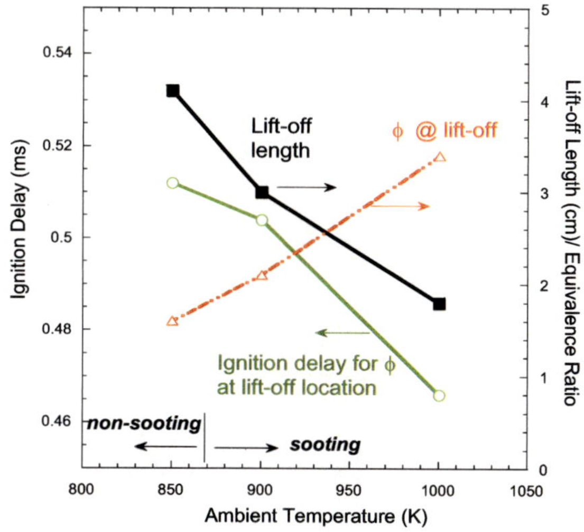

Figure 9 Flame lift-off length, equivalence ratios and ignition delay at corresponding lift-off locations for various ambient temperatures.

Diesel Engine Combustion and Emissions

The present model was also applied to simulate the combustion process in a Caterpillar heavy-duty diesel engine [35]. The engine is a single-cylinder engine whose specifications are listed in Table 1. The in-cylinder pressure, gaseous and soot emissions were recorded for model validations. A high-pressure injector was used for fuel injection with the maximum injection pressure up to 190 MPa [19]. The engine operating conditions were optimized to achieve low NO_x and particulate emissions and fuel consumption. The parameters that were varied included start-of-injection timings and EGR. The experimental conditions used for model validation are also listed in Table 1, and include three EGR levels.

Table 2 Caterpillar 3410E engine specifications [35]

Bore × Stroke	137.2 mm × 165.1 mm
Compression Ratio	16.1:1
Displacement	2.44 L
Connecting Rod Length	261.6 mm
Squish Height	1.57 mm
Combustion Chamber Geometry	In-Piston Mexican Hat with Sharp Edged Crater
Piston	Articulated
Charge Mixture Motion	Quiescent
Maximum Injection Pressure	190 MPa
Number of Nozzle Holes	6
Nozzle Hole Diameter	0.214 mm
Included Spray Angle	145°
Injection Rate Shape	Rising
Experimental conditions for model validation	
Case Group	SOI (ATDC)
A (8% EGR)	-20, -15, -10, -5, 0, +5
B (27% EGR)	-20, -15, -10, -5, 0, +5
C (40% EGR)	-20, -15, -10, -5, 0, +5

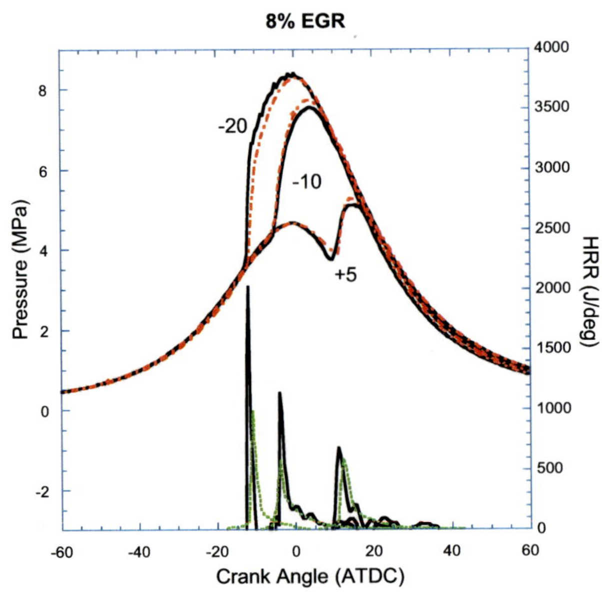

Figure 10 Comparisons of measured (solid line) and predicted (dotted) cylinder pressure and heat release rate data for 8% EGR cases (SOI=–20, –10 and +5 ATDC).

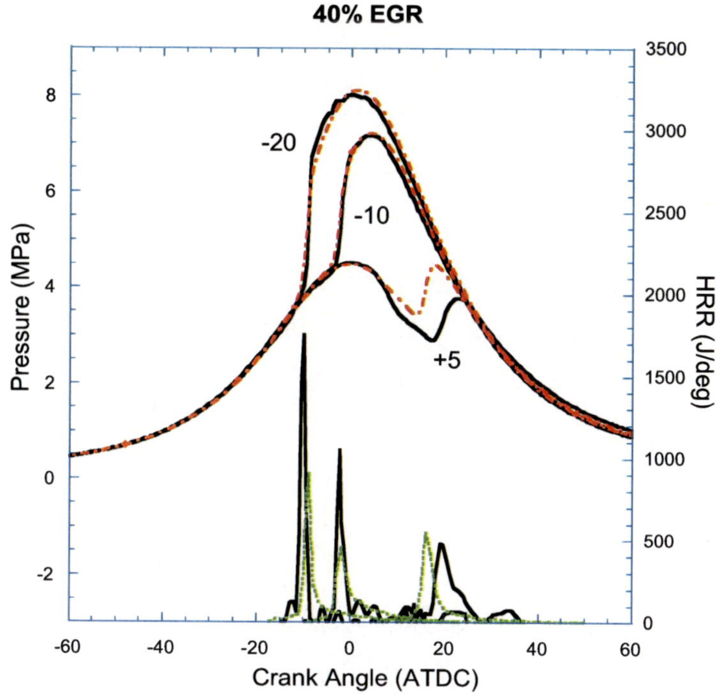

Figure 11 Comparisons of measured (solid line) and predicted (dotted) cylinder pressure and heat release rate data for 40% EGR cases (SOI=–20, –10 and +5 ATDC).

Comparisons of cylinder pressure and heat release rate data are shown in Figures 10 and 11 for various start-of-injection timings and EGR levels. The model is seen to perform well over a wide range of engine conditions. The heat release rate data does not exhibit the distinct premixed and diffusion burn characteristics of conventional diesel engines. The highly premixed burned features are well captured by the model.

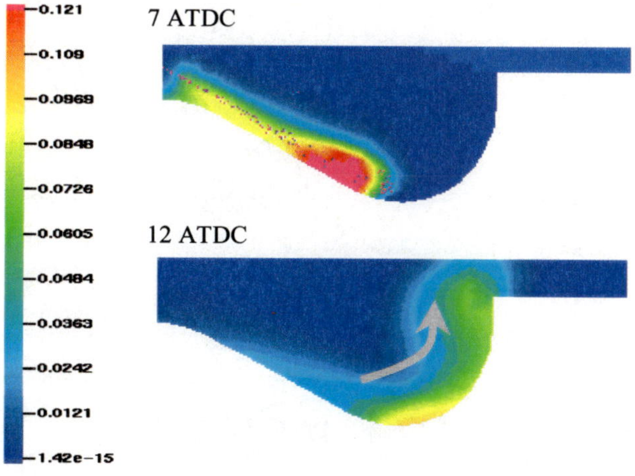

Figure 12 Distribution of liquid drops and fuel vapor prior to ignition (40% EGR, SOI=+5 ATDC). Arrow indicates the flow pattern generated by the fuel injection and reverse squish flow.

The fuel spray and vapor distributions in the cylinder are shown in Fig. 12 for 40% EGR, SOI=+5 ATDC. It can be seen that nearly all the liquid fuel has vaporized prior to the onset of combustion. The fuel spray impinges on the piston surface and generates a counter-clockwise flow motion, as indicated in the figure. The fuel-air mixture is then pushed to the squish region where ignition takes place, as can be seen from the temperature contours in Fig. 13. Combustion takes place mostly in the squish region and at the outer edge of the bowl.

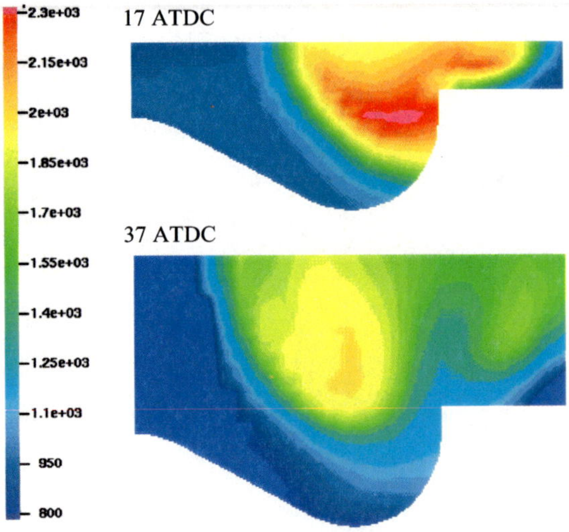

Figure 13 Temperature contours predicted by the present model (40% EGR, SOI=+5 ATDC).

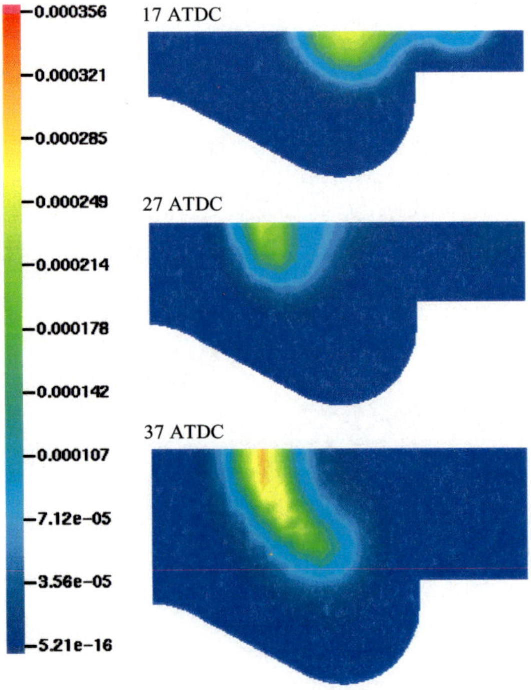

Figure 14 Soot mass fraction contours predicted by the present model (40% EGR, SOI=+5 ATDC).

Soot contours predicted by the model are shown in Fig. 14. Soot is initially formed near the piston bowl edge under the cylinder head and then is moved following the counter-clockwise flow motion slightly to the left. As the piston further expands, the soot cloud is stretched and oxidized by the surrounding air. It is found that better soot oxidation can be achieved if the soot cloud is pushed to the squish region in order to better utilize the fresh air in the squish region, instead of residing near the bottom the piston bowl.

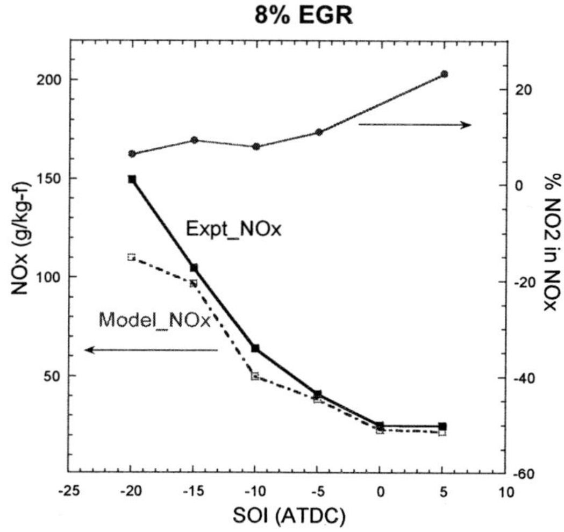

Figure 15 Comparisons of measured (solid line) and predicted (dotted) NOx emissions data for 8% EGR cases. The percentage of predicted NO_2 in NOx is also shown.

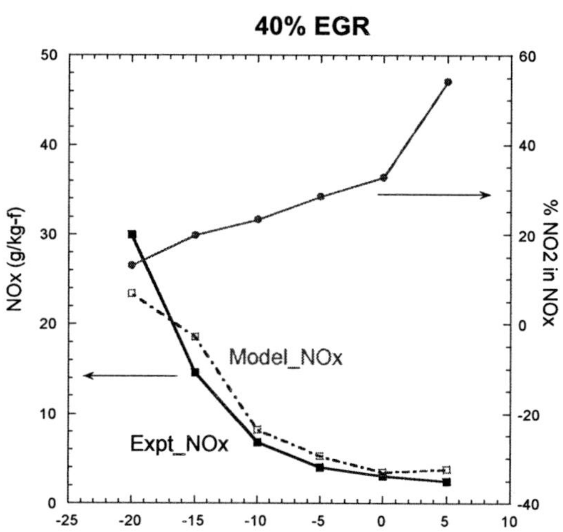

Figure 16 Comparisons of measured (solid line) and predicted (dotted) NOx emissions data for 40% EGR cases. The percentage of predicted NO_2 in NOx is also shown.

Comparisons of measured and predicted NOx emissions data are shown in Figs. 15 and 16. It can be seen the trend of NOx reduction with respect to the injection timing retardation is well predicted. On the other hand, it is found that the percentage of NO_2 increases as the injection timing is retarded. The results indicate that NO_2 becomes more significant when the combustion takes place at lower temperature environments. Figure 17 further shows the average and peak in-cylinder gas temperature. It is seen that the maximum of the average gas temperature is below 1500 K. Thus, it should be noted that the conventional Zel'dovich NO mechanism may under-predict NOx production for the HCCI engine that features low temperature combustion.

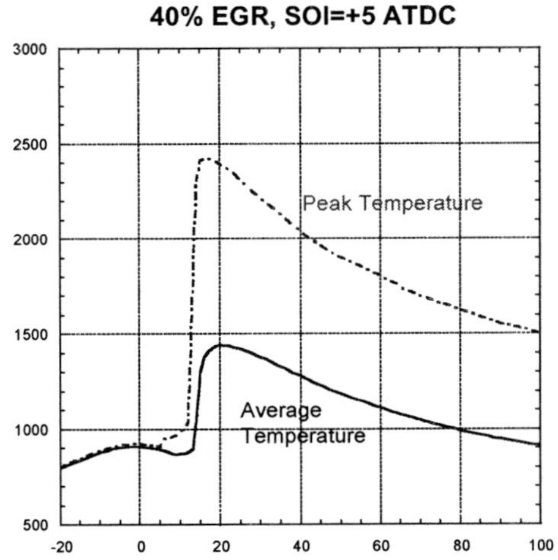

Figure 17 In-cylinder average temperature and peak temperature for 40% EGR, SOI=+5 ATDC case.

The predicted engine-out soot emissions were also compared with the measurements, as shown in Fig. 18. It can be seen that the overall trend of soot emissions was captured very well with respect to the start-of-injection timing. It is of interest to note that engine-out soot emissions reach a peak value when fuel is injected near top-dead-center. The present model also predicts correctly the soot reduction seen at further retarded injection timing (e.g., SOI=+5 ATDC) for all different EGR levels.

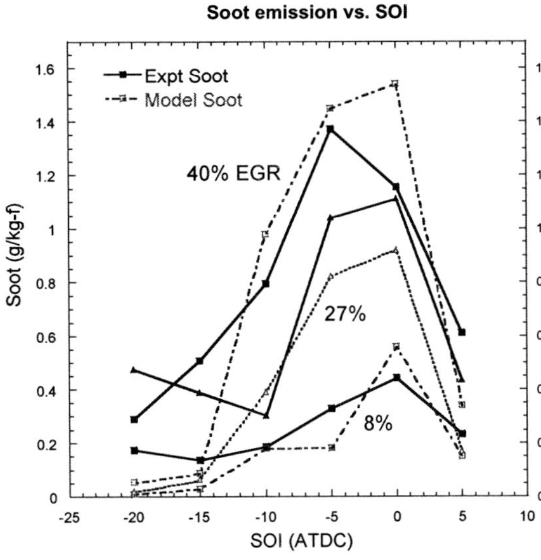

Figure 18 Measured and predicted engine-out soot emissions for cases listed in Table 1.

The decrease in soot emissions at the further retarded injection timings (e.g., SOI=+5 ATDC) is not consistent with the conventional understanding of soot-NOx trend-off trend. The numerical model can explain the soot emission reduction seen as the injection is further retarded past TDC. Figure 19 shows the total in-cylinder soot mass evolutions for three different injection timings for 40 % EGR cases, i.e., SOI=–10, 0, +5 ATDC. The model results indicate that the lower exhaust soot emissions for SOI=–10 ATDC is due to a better oxidation process as compared to that of SOI=0 ATDC. On the other hand, a lower soot emission for SOI=5 ATDC is because less soot is formed in the cylinder as a result of the low temperature combustion (as also can be seen from the low cylinder pressure in Figs. 10 and 11 for SOI=5 ATDC). The above low temperature combustion characteristics can be utilized for the low-emission diesel PCCI operation.

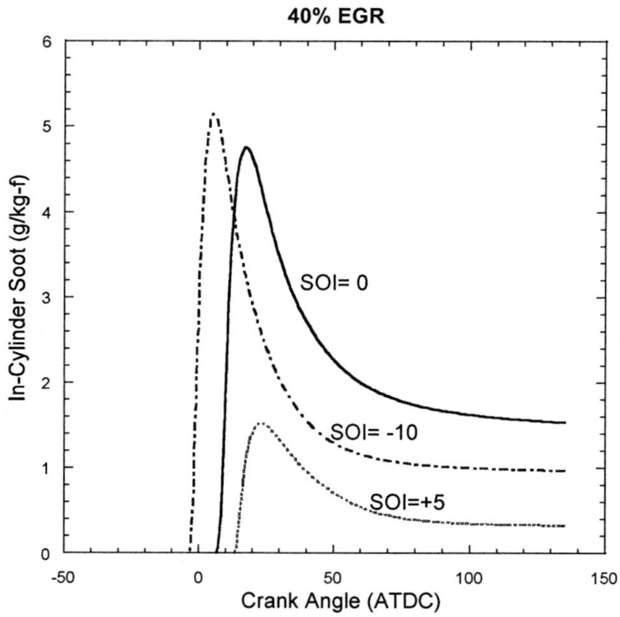

Figure 19 In-cylinder soot mass histories for 40% EGR cases at three different injection timings. Values at exhaust valve opening (130 ATDC) are shown in Fig. 11.

Conclusion

The complex spray combustion process in diesel engines is studied using the integrated CFD model with detailed chemical kinetics. The model can simulate the transient fuel injection, spray dynamics, combustion and emission processes simultaneously without any decoupling of fluid dynamics and chemistry. A skeletal n-heptane mechanism was used to simulate the diesel fuel oxidation chemistry with good levels of agreement in combustion characteristics.

The model predicts well the lift-off length, local equivalence ratio and flame structure for diesel spray combustion in a constant-volume combustion chamber. The model shows that vigorous heat release occurs at the lift-off location as a result of autoignition of incoming fuel-air mixture. The successive autoignition is thought to be the stablization mechanism of such a standing diesel spray flame.

Significant amounts of soot precursors are formed at the lift-off location and soot particles are formed downstream of the spray plume. As the ambient temperature is decreased, the lift-off length increases, local equivalence ratio at the lift-off location decreases and soot emissions decreases as a result of better mixing prior to the lift-off location. Such low soot emission characteristics can be utilized to improve the diesel engine emissions.

The present model is also applied to simulate diesel engine combustion with highly premixed burn conditions. Engine in-cylinder pressure, heat release rate data, soot and NOx emissions are well predicted. Results show that NOx emission decreases monotonically as the start of injection (SOI) is retarded. However, soot emission increases as start of injection is retarded approaching top-dead-center and then decreases at further retarded timings, as also observed from engine-out measurements. The model indicates that low soot emission at early SOI is due to better oxidation while low soot emission at late SOI is due to less formation. Emission reduction strategies can be used to achieve low-emission diesel engine operating by allowing combustion to occur at a relatively low temperature environment.

Acknowledgments

The author acknowledges the financial support by the DOE/Sandia National Labs and Caterpillar, Inc. The author also thanks Drs. L. M. Pickett and D. L. Siebers of Sandia National Labs for providing experimental PLII images for model validations.

References

[1] Dec, J.E., 1997, "A Conceptual Model of DI Diesel Combustion Based on Laser Sheet Imaging," SAE 970873.

[2] Dec, J.E. and Canaan, R.E., 1998, "PLIF Imaging of NO Formation in a DI Diesel Engine," SAE 980147.

[3] Zhao, H. and Ladommatos, N., 1998, "Optical Diagnostics for Soot and Temperature Measurement in Diesel Engines," *Prog. Energy Combust. Sci.* **24**, pp. 221—255.

[4] Dec, J.E. and Tree, D.R., 2001, "Diffusion Flame/Wall Interactions in a Heavy-Duty Diesel Engine," *SAE* 2001-01-1295.

[5] Musculus, M.P., Dec, J.E. and Tree, D.R., 2002, "Effects of Fuel Parameters and Diffusion Flame Lift-Off on Soot Formation in a Heavy-Duty DI Diesel Engine," SAE 2002-01-0889.

[6] Bruneaux, G., Verhoeven, D. and Baritaud, T., 1999, "High-pressure diesel spray and combustion visualization in a transparent model diesel engine," SAE 1999-01-3648.

[7] Pickett, L.M. and Siebers, D.L., 2004, "Non-Sooting, Low Flame Temperature Mixing-Controlled DI Diesel Combustion," SAE 2004-01-1399.

[8] Pickett, L.M. and Siebers, D.L., 2004, "Soot in Diesel Fuel Jets: Effects of Ambient Temperature, Ambient Density, and Injection Pressure," *Combust. Flame*, **138**, pp.114—135.

[9] Magnussen, B.F. and Hjertager, B.H., 1976, "On Mathematical Modeling of Turbulent Combustion with Special Emphasis on Soot Formation and Combustion," *Proc. Combut. Inst.*, **16**, pp. 719—729.

[10] Kong, S.C., Han, Z.W. and Reitz, R.D., 1995, "The Development and Application of a Diesel Ignition and Combustion Model for Multidimensional Engine Simulations," SAE Paper 950278, *Journal of Engines*, Vol. 104, pp. 502-518.

[11] Rao, S. and Rutland, C.J., 2002, "A Flamelet Time Scale Model for Non-Premixed Combustion Including Unsteady Effects," *Comb. Flame*, **133**, pp. 189—191.

[12] Steiner, R., Bauer, C., Kruger, C., Otto, F. and Maas, U., 2004, "3D Simulation of DI-Diesel Combustion Applying a Progress Variable Approach Accounting for Complex Chemistry," SAE 2004-01-0106.

[13] Correa, C., Niemann, H., Schramm, B. and Warnatz, J., 2000, "Use of ILDM Reduced Chemistry in Direct Injection Diesel Engines," THIESEL 2000 Conference, pp.163—170.

[14] Beard, P., Colin, O. and Miche, M., 2003, "Improved Modeling of DI Diesel Engines Using Subgrid Descriptions of Spray and Combustion," SAE 2003-01-0008.

[15] Pitsch, H., Barths, H. and Peters, N., 1996, "Three Dimensional Modeling of NOx and Soot Formation in DI-Diesel Engines Using Detailed Chemistry Based on the Interactive Flamelet Approach," SAE 962057.

[16] Peters, N., 1984, "Laminar Diffusion Flamelet Models in a Non-Premixed Turbulent Combustion," *Prog. Energy Combust. Sci.*, **10**, pp.319—339.

[17] Barths, H., Antoni, C. and Peters, N., 1998, "Three-Dimensional Simulation of Pollutant Formation in a DI Diesel Engine Using Multiple Interactive Flamelets," SAE 982459.

[18] Hasse, C., Bikas, G. and Peters, N., 2000, "Modeling DI-Diesel Combustion Using the Eulerian Particle Flamelet Model (EPFM)," SAE 2000-01-2934.

[19] Hasse, C. and Peters, N., 2004 "A Two Mixture Fraction Flamelet Model Applied to Split Injections in a DI Diesel Engine," *Proc. Combust. Inst.*, **30**.

[20] Hergart, C., Barths, H. and Peters, N., 1999, "Modeling the Combustion in a Small-Bore Diesel Engine Using a Method Based on Representative Interactive Flamelets," SAE 1999-01-3550.

[21] Kim, H., Heo, N., Kim, Y., Lee, J.-H. and Lee, J.K., 2004, "Numerical Study of Combustion Processes and Pollutant Formation in HSDI Diesel Engines," SAE 2004-01-0126.

[22] Kong, S.C. and Reitz, R.D. "Use of Detailed Chemical Kinetics to Study HCCI Engine Combustion with Consideration of Turbulent Mixing Effects," *J. Engng Gas Turbines Power*, Vol. 124, pp.702-707, 2002.

[23] Kong, S.-C. and Reitz, R.D., 2002, "Application of Detailed Chemistry and CFD for Predicting Direct Injection HCCI Engine Combustion and Emissions," *Proc. Combust. Inst.*, **29**, pp. 663—669.

[24] Kong, S.C. and Reitz, R.D. "Numerical Study of Premixed HCCI Engine Combustion and Its Sensitivity to Computational Mesh and Model Uncertainties," *Combust Theory Modeling*, Vol. 7, pp.417-433, 2003.

[25] Zhang, Y.Z., Kung, E.H. and Haworth, D.C., 2004, "A PDF Method for Multidimensional Modeling of HCCI Engine Combustion: Effects of

Turbulence/Chemistry Interactions on Ignition Timing and Emissions," *Proc. Combust. Inst.*, **30**.

[26] Amsden, A.A., 1997, "KIVA-3V: A Block-Structured KIVA Program for Engines with Vertical or Canted Valves," LA-13313-MS.

[27] Patterson, M.A. and Reitz, R.D., 1998, "Modeling the Effects of Fuel Spray Characteristics on Diesel Engine Combustion and Emissions," SAE 980131.

[28] Kong, S.C., Senecal, P.K. and Reitz, R.D, 1999, "Developments in Spray Modeling in Diesel and Direct-Injection Gasoline Engines," *Oil & Gas Science and Technology – Rev. IFP*, Vol. 54, No. 2, pp. 197-204.

[29] Han, Z.W. and Reitz, R.D., 1997, "A Temperature Wall Function Formulation for Variable-Density Turbulent Flows with Application to Engine Convective Heat Transfer Modeling," *Int. Journal of Heat and Mass Transfer*, Vol. 40, No.3, pp.613—625, 1997.

[30] Kee, R. J., Rupley, F. M. and Miller, J. A., 1989, "CHEMKIN-II: A FORTRAN Chemical Kinetics Package for the Analyses of Gas Phase Chemical Kinetics," *Sandia Report, SAND* **89**-8009.

[31] Liang, L., Jung, C., Kong, S.C. and Reitz, R.D., 2005, "Development of a Semi-Implicit Solver for Detailed Chemsitry Simulations," ASME ICES2005-1005.

[32] Patel, A., Kong, S.-C. and Reitz, R.D., 2004, "Development and Validation of a Reduced Reaction Mechanism for HCCI Engine Simulations," SAE 2004-01-0558.

[33] Smith, G.P., Golden, D.M., Frenklach, M., Moriarty, N.W., Eiteneer, B., Goldenberg, M., Bowman, C.T., Hanson, R.K., Song, S., Gardiner, W.C., Lissianski, V.V. and Qin, Z., 2000, http://www.me.berkeley.edu/gri_mech/.

[34] Han, Z., Uludogan, A., Hampson, G.J. and Reitz, R.D., 1996, "Mechanism of Soot and NOx Emission Reduction Using Multiple-Injection in a Diesel Engine," SAE 960633.

[35] Klingbeil, A. E., 2002, "Particulate and NO_X Reduction in a Heavy-Duty Diesel Engine Using High Levels of Exhaust Gas Recirculation and Very Early and Very Late Injection", MS Thesis, University of Wisconsin-Madison.

In: New Devel. in Combustion Research
Editor: William J. Carey, pp. 209-226

ISBN: 1-59454-826-9

Chapter 6

MODELING TURBULENT PREMIXED CH_4 FLAMES WITH FLOW RECIRCULATION USING REACTION PROGRESS VARIABLES

Jim. B.W. Kok * ***and Cornelis. J. A. de Hoon***
University of Twente, P.O. Box 217, 7500 AE Enschede, The Netherlands

Abstract

A model is presented to predict turbulent combustion of premixed CH_4 flames at gas turbine conditions. Mixing, combustion and heat loss are described with five scalar variables. These are the mixture fraction, enthalpy variable and three reaction progress variables for combustion of CH_4, H_2 and CO. The effect of turbulence on combustion is taken into account by weighting with an assumed shape probability density function. In a gas turbine the flame is stabilized in an internal flow recirculation area. In this configuration not only the mean but also the variance of the methane progress variable has to be taken into account. The model is tested on a premixed lean CH_4 flame stabilized by a coflow of burnt stoichiometric CH_4 flow. Essential in this configuration is the turbulent mixing of areas burnt and unburnt. The calculated velocity field and CH_4 concentration field have been compared with experiments done by P. Moreau and A. Boutier on this geometry. The results show good agreement concerning the velocity field. The reaction progress of CH_4 to H_2 and CO has to be assessed for good accuracy. Good results can be obtained when the derived thermochemical database is considered in detail for both source terms and calculated species concentrations.

Nomenclature

C	composed concentration
D	diffusion coefficient
f	mixture fraction
g	variance of mixture fraction
i	enthalpy loss variable

Greek symbols

ε	dissipation rate
μ	viscosity
ρ	density
σ	Prandtl number

* E-mail address: j.b.w.kok@utwente.nl

k	turbulent kinetic energy
l_t	turbulent macro length scale
p	pressure
P	probability density function
r	H_2 reaction progress
s	CO reaction progress
S	source term
t	CH_4 reaction progress
t''	variance of CH_4 reaction progress
T	temperature
$\underline{u}$	velocity vector
u'	turbulent fluctuating velocity
V	weighting function for s
W	weighting function for
x	generalized coordinate
X	mole number
Y	mass fraction
Z	weighting function for t
φ	equivalence ratio

Subscripts

ff	second derivative related to *f*
i	first derivative related to *i*
	species number

Superscripts

pe	partial equilibrium
~	Favre averaging
-	PDF averaging

1 Introduction

In this chapter the modeling of turbulent premixed combustion of CH_4 in a flame with internal flow recirculation is discussed. Important physical processes that need to be described are turbulence, chemical reactions, mixing of multiple differently premixed inlet flows and heat transfer. For turbulence modeling k-ε closure is used here, which is standard available in a commercial CFD code, but other models can be used as well. The work here focuses on the modeling of chemical reaction of methane air mixtures under turbulent conditions. The resulting model was implemented in a commercially available CFD code.

A progress variable model with 2 combustion scalars (r,s), mixture fraction (f) and enthalpy (i) (hence denoted FIRS) for the non-adiabatic turbulent combustion of syngas has been presented by [1], [2]. Subsequently the model was extended to premixed combustion of CH_4 using a 19 reaction step mechanism of CH_4 combustion with an extra combustion scalar t and denoted with FIRST (see [3]).

The work presented here is the result of the further development of the FIRST model, taking into account variance of methane combustion. The latter is important in flames with flow recirculation. In section 2 the reaction progress variables are defined. Section 3 deals with the mixing processes and associated variables in a partially premixed flame. The turbulent transport equations for the reaction progress variables are discussed in section 4. These transport equations have source terms for chemical reaction. These are analyzed in section 5. With the model simulations are performed of a set of a experiments performed by P. Moreau and A. Boutier, see [4]. In section 6 the set-up of these experiments and of the numerical simulations is described. The comparison of the improved FIRST model with the Moreau data is shown in section 7. Finally in section 8 conclusions are formulated.

2 Progress of Chemical Reaction

The combustion process in the turbulent premixed flame is described by three reaction progress variables, indicating the conversion of methane to hydrogen and carbon monoxide and the reaction progress in the hydrogen and carbon monoxide related chemical reactions. Earlier in [5], [6] a one progress variable model was introduced for the adiabatic combustion of turbulent hydrogen diffusion flames with a 6 species and 8 chemical reactions mechanism. Later in [7] and [8] a two reaction progress variable model was used for the adiabatic combustion of CO/H_2 mixtures. Recently a two reaction progress variable model for the non-adiabatic turbulent combustion of CO/H_2 mixtures with 10 species, 21 reactions was derived systematically in [1] and [2]. In this section the model is generalized to the combustion of methane, taking into account variance of methane reaction progress. A detailed chemical reaction system with 8 elementary methane reactions (specified in appendix I), is reduced to a system depending on one progress variable. For this variable a composed concentration is chosen, such that the highly non-linear source terms of the two-body reactions in the reaction mechanism, having high activation energies, cancel each other. This means that only the source terms of reactions without activation energy remain in the net source term of the combined mass fraction [3,7]. This technique (used earlier in [9] to reduce the stiffness of laminar flame equations) is demonstrated below for the reduction of the methane chemical reaction mechanism.

Consider a set of M elementary reactions involving N species and L elements. The chemical reactions with subscript k between the species with subscript i and with chemical symbol χ_i can be represented in a general form as ([10]):

$$\sum_{i=1}^{N} \nu_{ik}\chi_i \Leftrightarrow \sum_{i=1}^{N} \nu'_{ik}\chi_i \qquad k=1,\ldots,M \tag{1}$$

The net rate of production of species i due to all M chemical reactions is then given by the i_{th} component of the species rate of production vector:

$$\frac{dC_i}{dt} = \omega_i = \sum_{k=1}^{M} v_{ik}\varpi_k \tag{2}$$

with components of the reaction tensor defined by $v_{ik} = \nu'_{ik} - \nu_{ik}$.

A composed variable will be determined for the methane mechanism (specified in the appendix) as follows. The procedure is similar to the procedure for the hydrogen mechanism in [3]. The reactive species that are involved with methane conversion exclusively are CH_4, CH_3, CH_2O, CHO. The number of species is 4 and there is one element (C) additional to the H_2 reaction processes. Hence, the dimension of the reactive species space is 4-1=3. If this dimension is reduced to 1 (by taking one progress variable for methane), in total 3 source terms can be eliminated from the composed variable source term. Choosing for the independent species in the reactive subspace CH_4, CH_3, CH_2O, the composed concentration variable C_t is given by the expression:

$$C_t = a_1 C_{CH_4} + a_2 C_{CH_3} + a_3 C_{CH_2O} \tag{3}$$

The coefficients a_1 to a_3 are to be determined in such a way, that the chemical source term of the composed concentration does not depend on fast reactions with a large activation energy. In fact this way all locally fast reactive chemical time scales are moved to an implicit level of calculation. The chemical source term of C_t is (specified for the chemical reaction mechanism in the appendix) according to the expression above:

$$\frac{dC_t}{dt} = \omega_1(-a_1 + a_2) + \omega_2(-a_1 + a_2) + \omega_3(-a_2 + a_3) + \omega_4(-a_3) + \omega_5(-a_3) \ldots\ldots \tag{4}$$

In view of their large activation energies, the coefficients of the independent reactions 1 and 2 are eliminated and the source of fast reaction 3:

Reaction 1: $-a_1+a_2 \quad =0$ (5)

Reaction 2: $-a_1+a_2 \quad =0$ (6)

Reaction 3: $-a_2+a_3 \quad =0$ (7)

Scaling C_t such that a_1=1.00, the solution of this system of equations for the 2 independent linear equations with 2 remaining unknowns is:

a_1= 1.00 a_2= 1.00 a_3= 1.00 (8)

When substituted in equation (3) the composed concentration becomes then:

$$C_t = C_{CH_4} + C_{CH_3} + C_{CH_2O} \tag{9}$$

In a procedure similar to the CH_4 mechanism a composed concentration is chosen for the hydrogen variable, such that the highly non-linear source terms of the two-body reactions in the reaction mechanism cancel each other (see [1]-[3] and [5]-[8]). In the CH_4 mechanism, H_2 is both a product and a reactant. Hence in the chemical source term of the hydrogen combined concentration a production and a consumption term will be present. This will be discussed further in connection with the thermochemical database in section 4. The proposed combined concentration for the hydrogen progress, including the species H, O, OH and HO_2 becomes:

$$C_r = C_{H_2} + C_H + \frac{1}{2} C_O + \frac{1}{2} C_{OH} - \frac{1}{2} C_{HO_2} \tag{10}$$

The species in the carbon monoxide related chemical reactions are determined by the CO concentration:

$$C_s = C_{CO} \qquad (11)$$

The concentrations of the 13 species, composed of the elements C, H and O, are now mapped on and determined by the 3 composed concentrations. The chemical reaction source terms of the composed species have low activation energies and are not very sensitive to turbulent fluctuations. This allows an approach with an assumed shape Probability Density Function weighting.

3 Mixing of Fuel and Air

A partially premixed flame is characterized by a separate inflow of several streams that each are premixed with a different ratio between fuel and air. Depending on the burner construction, chemistry and flow conditions, these streams mix in the flame. In the present case only two streams are considered and the conventional mixture fraction formulation can be used. The mixing of fuel and air is then described with two scalar transport equations: one for the mean mixture fraction f and one for its 2nd moment (the variance $g = \langle f''^2 \rangle$). The mixture fraction f equals 1 in a pure fuel stream and 0 in a pure air stream, according to the definition below:

$$f = \frac{Y_{fuel} - Y_{fuel.2}}{Y_{fuel.1} - Y_{fuel2}} \qquad (12)$$

The turbulent transport terms in the equations for the first two moments are closed with the gradient hypothesis. This results in the following turbulent transport equation for the mean of the mixture fraction:

$$\nabla(\bar{\rho}\underline{\tilde{u}}\,\tilde{f}\,) - \nabla(D\nabla\tilde{f}\,) = 0\,. \qquad (13)$$

The diffusion coefficient D is the sum of molecular diffusion coefficient D_m and the turbulent diffusion coefficient D_t. The turbulent diffusion coefficient is defined as the ratio of the kinematic turbulent viscosity and the turbulent Schmidt number. For the variance of f: $g = \langle f''^2 \rangle$ the following turbulent transport equation can be derived:

$$\nabla(\bar{\rho}\underline{\tilde{u}}\, g) - \nabla(D\nabla g\,) = \tfrac{1}{2} C_{g_1}\bar{\rho}\chi - C_{g_2}\bar{\rho}\alpha \qquad (14)$$

with rate of scalar dissipation $\chi = 2D_m \left(\frac{\partial f}{\partial x_i}\right)^2$ and mixing rate $\alpha = \frac{\tilde{\varepsilon}}{k} g$. In equations 13 and 14, $\underline{\tilde{u}}$ is the Favre averaged velocity vector. In eq. 14 it can be observed that the rate of growth of g is determined by the balance between the rate of scalar dissipation and the mixing rate.

4 Mean Turbulent Transport Equations for the Reaction Progress Variables

The composed concentrations are independent variables, but in a turbulent flow they are statistically correlated with each other. To reduce the statistical correlation of the composed concentrations, all composed concentrations are made dimensionless with the completely burnt adiabatic and the unburned adiabatic situation. These adiabatic reference concentrations depend on the mixture fraction only. The combined mass fraction of the methane variable becomes then:

$$t = \frac{Y_{C_t} - Y_{C_t\,unburnt}}{Y_{C_t\,burnt} - Y_{C_t\,unburnt}} = \frac{Y_{C_t} - Y_{C_t\,unburnt}}{Z(f)} \qquad (15)$$

Similar dimensionless expressions are formulated for the other variables:

$$r = \frac{Y_{C_r} - Y_{C_r}^{unburnt}}{Y_{C_r}^{burnt} - Y_{C_r}^{unburnt}} = \frac{Y_{C_r} - Y_{C_r}^{uburnt}}{W(f)}, \qquad (16)$$

and

$$s = \frac{Y_{C_s} - Y_{C_s}^{pb}}{Y_{C_s}^{u} - Y_{C_s}^{pb}} = \frac{Y_{C_s} - Y_{C_s}^{pb}}{V(f)}. \qquad (17)$$

The main reason for introducing scaled variables such as *t*, is that in this way the statistical correlation is reduced. The statistical correlation between the scaled variables *r, s* and *t* will in general be much lower than between the non-scaled variables.

In the methane variable t, the normalization function *Z(f)* depends on mixture fraction. Substituting this definition in a conventional convection-diffusion equation for the combined mass fraction, with gradient closure for the turbulent flux term and applying Favre decomposition and averaging yields:

$$\nabla(\overline{\rho}\underline{\tilde{u}}\tilde{t}) - \nabla(D\nabla\tilde{t}) = \overline{S}_t + \left(\frac{\overline{Z_{ff}}}{Z}\right)\frac{1}{2}\,\tilde{t}\,C_{g_2}\tilde{\alpha}\overline{\rho}. \qquad (18)$$

In this equation the 1st rhs term is the chemical source term and the 2nd the term due to mixing between two differently premixed inlet streams. In case the inlet streams are of identical composition, the 2nd rhs term vanishes.

For the variance of the *t*-variable the following equation can be derived:

$$\nabla(\overline{\rho}\underline{\tilde{u}}\overline{t''^2}) - \nabla(D\nabla\overline{t''^2}) = 2\frac{\mu_t}{\sigma_t}(\Delta\tilde{t}\,)^2 - C_{t_2}\overline{\rho}\frac{\tilde{\varepsilon}}{\tilde{k}}\overline{t''^2} + 2\overline{t'' S_t}\,. \tag{19}$$

It must be observed that in flames with a recirculation area, like the Moreau flame [4] studied here, the variance of t is very important, as flows with $t = 0$ and 1 are mixed in the combustion zone. This relevance of the variance of t is relevant also to all swirl stabilized flames, where the flame is anchored on the recirculation area. The variance of the mixture fraction, on the other hand, is in the Moreau flame considered to be of less importance, as the Moreau flame has 2 premixed inlet flows that do not differ much in composition prior to combustion.

The Favre mean transport equation for the hydrogen progress variable r is derived as:

$$\nabla(\overline{\rho}\underline{\tilde{u}}\tilde{r}) - \nabla(D\nabla\tilde{r}) = \overline{S}_r^{+} + \overline{S}_r^{-} + \left(\frac{\overline{W_{ff}}}{W}\right)\frac{1}{2}C_{g2}\overline{\rho}\frac{\tilde{\varepsilon}}{\tilde{k}}g\tilde{r} \tag{20}$$

Here it can be seen, that the chemical source term of r can be split in a production (+) and a consumption (-) term.

Mean of carbon monoxide conversion:

The combustion of carbon monoxide plays a minor part in the overall combustion of methane, as CO can be assumed not to deviate significantly from equilibrium. If the non-equilibrium of CO is however taken into account the combustion of CO is governed by:

$$\nabla(\overline{\rho}\,\underline{\tilde{u}}\,\tilde{s} - (D\nabla\tilde{s})) = (\tilde{s}-1)\left(\frac{\overline{V_{ff}}}{V}\right)\frac{1}{2}\,C_{g_2}\overline{\rho}\,\tilde{\alpha}\,g + \overline{S_S}\,. \tag{21}$$

5 Chemical Reaction Source Terms

All instantaneous concentrations and source terms are stored as a function of *f,r,s,t* in a thermochemical data base. All concentrations and the source terms of all reactions are known when *f,r,s,t* are specified. Use is made of elemental balances, enthalpy balance, steady state of the intermediary species CH_3, CH_2O, HCO, HO_2 and partial equilibrium assumptions for reactions 9-12. Although not used here, identical results were obtained in earlier research with the more mathematical ILDM method (see [11]) for a hydrogen flame. The reduction leads to a system of nonlinear algebraic equations that is solved on a mesh of *f,r,s,t*. To speed up flame calculations, the concentrations and source terms are stored in advance on a 4 dimensional tabulation mesh in a database.

The source terms due to chemical reaction, as calculated by the mapping method described above, are of interest to further analysis. Specifically the chemical source terms in the mean of t- and r-equation are important as they initiate and drive the combustion process. In fig. 1 the non weighted (i.e. laminar) chemical source term at the rhs of the *t*-equation (9) is shown as a function of t and r for a premixed CH_4/air mixture at 600 K inlet temperature. This shows that the source term has a maximum at approximately $t=0.7$ and $r=0.7$. It becomes small at small *t* and vanishes at *t=1*. The value of the source term at small *t* is very small, but important for ignition processes and is at the moment the subject of research. The trajectory of the chemical system in a premixed natural gas flame can now easily be envisaged. Upstream of the flame the reaction progress variable will be near zero and increase slowly due to a small but positive source term (due to ignition processes). After ignition the progress variables t and r will be around 0.7 and the source term reaches a maximum value. The progress variables will increase fast to equilibrium with a value of 1. At that point the source term will have vanished.

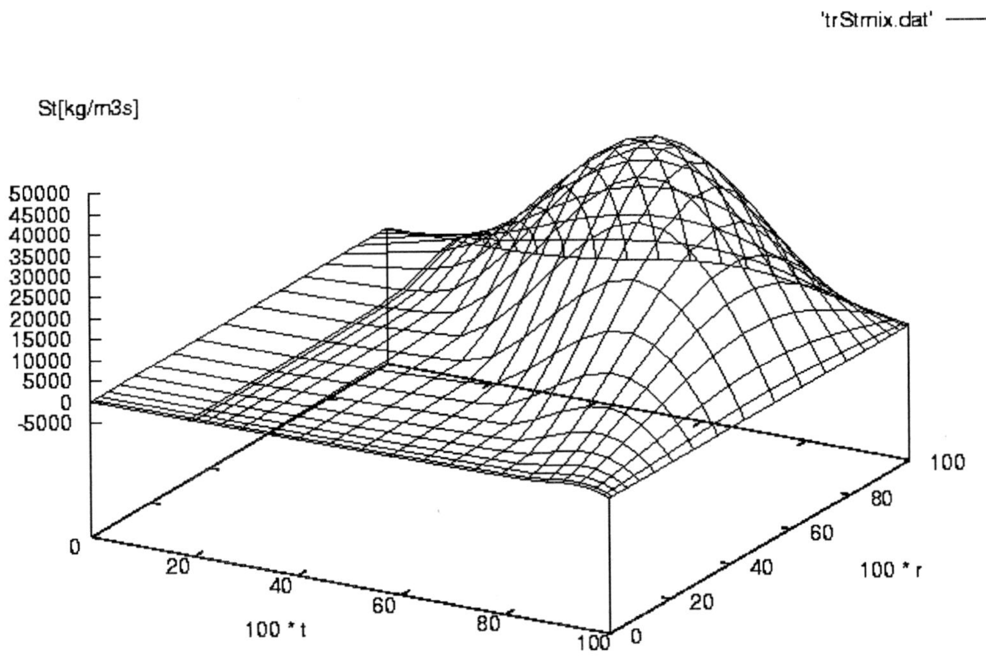

Figure 1: Chemical source term for the t variable as a function of t and r.

Some comments can be made regarding the amplitude of the S_t term. Compared to the chemical source term S_t on basis of a global reaction of CH_4 to H_2 and CO presented in [12] the presently calculated S_t is of very similar shape but approximately a factor 10 larger. This inconsistency in modeling result could be caused by the fact that the mechanism used for the CH_4 conversion to H_2 and CO is not detailed enough and a larger mechanism should be used. The most recent research (see [13]) indicates however that the inaccuracy in the amplitude of the source term is caused by the small number of progress variables used. This is inevitable in a turbulent flow simulation.

As remarked above, the shape of the source term as a function of mixture fraction and progress variables is however correct. Therefore in order to optimize the accuracy of the combustion model, in the simulation results shown in the next sections the S_t term is optimized to give good flame evolution data. This is like adjusting the Damkohler number switch in the Eddy Break Up model. It appeared that good results were obtained by scaling the chemical source term by a multiplication factor 0.1, which is consistent with the data in [12]. It will be shown below that it leads to results with a correct flame evolution. The source terms for the variance of t and for the mean of r were not adjusted. All local mass fractions were calculated from the set of algebraic equations.

The r progress variable is related to hydrogen concentration evolution and hence to an intermediate species in a premixed natural gas flame. This means that the intermediate species related to the scalar variable is produced from naught and consumed again. In that situation the chemical source term of r can be split in a production (+) and a consumption (-) term. The chemical source term of r is depicted as a function of *r* and *t* for premixed CH_4 at 600 K inlet temperature in figs. 2 and 3 for the production and consumption effects respectively. The system is brought out of equilibrium by the progress of the t variable during ignition and flame propagation. The production source term (fig. 2) is in this case negative and increasing. The consumption (fig. 3) is observed to increase fast when the progress variables t and r deviate from equilibrium, thereby forcing the hydrogen system to a burnt situation. When both t and r have reached unity all source terms vanish again.

In order to take turbulent fluctuations into account, the instantaneous concentrations and source terms are weighted over a Probability Density Function. The Favre averaged mean of the source terms, mass fractions and temperature is calculated by convolution with a joint probability density function P:

$$\widetilde{\phi} = \frac{1}{\overline{\rho}} \int_0^1 \rho \phi \ P(f,r,s,t) \ df \ dr \ ds \ dt. \tag{22}$$

As the independent variables are normalised, they can be assumed to be statistically independent and the joint PDF can be factorised in the individual PDF's:

$$\widetilde{\phi} = \frac{1}{\overline{\rho}} \int_0^1 \rho \phi \ P_f(f) \ P_r(r) \ P_s(s) \ P_t(t) \ df \ dr \ ds \ dt. \tag{23}$$

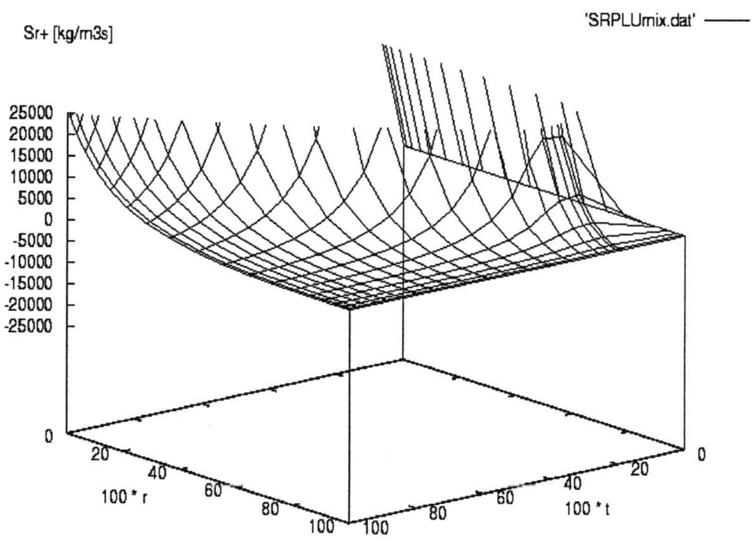

Figure 2: Positive chemical source term for r (shown until Sr=25000),

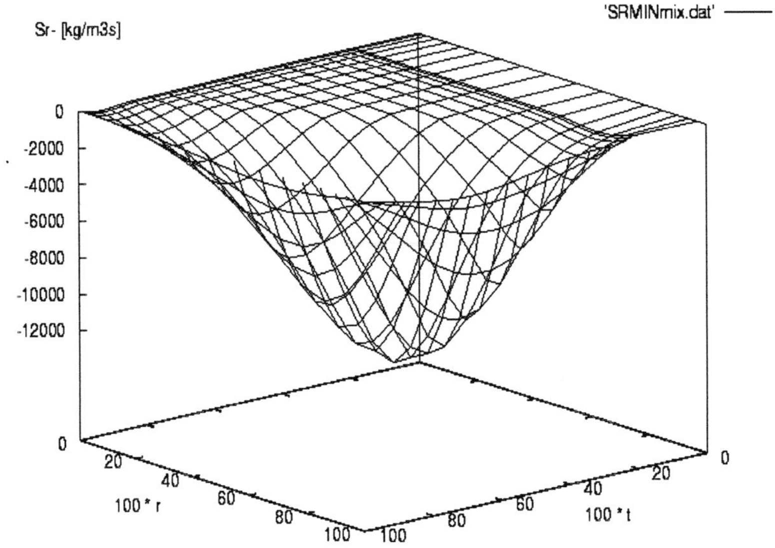

Figure 3: Negative chemical source term for r, as a function of t and r.

The individual Probability Density functions are chosen with a presumed shape. For the mixture fraction f and the methane reaction progress t variable a Beta function is chosen, because these variables are affected by turbulent fluctuations. Its local value is calculated from the mean and variance as described by their turbulent transport equations. For the other variables a Delta function is taken.

The transport equations and database are coupled through user scalars to a commercial CFD code (CFX 4.3), that solves the convective/diffusive equations together with the flow field and turbulence equations. Here a conventional k-ε model was used for the latter.

6 Set-Up of Experiments and Numerical Simulations

The Moreau Experiments

In order to study the development of turbulent combustion in a premixed CH_4/air flow, a test setup was designed by P. Moreau and A. Boutier ([4]) in which combustion is initiated and stabilized by a parallel flow of combusted high temperature gases. In this highly turbulent flow, laser velocimetry was used to measure velocities and its fluctuations. The experimental setup of the Moreau experiments is illustrated in fig 4. The test section in this figure is the combustion chamber where velocity, concentration and temperature have been measured at several cross sections. The combustion chamber is rectangular with a length of *1300 mm* and a square cross section of *100x100 mm*^{2} and has two inlets. The first inlet (*y=0-20 mm*) is a flow with hot burned gases, which is formed during the combustion of CH_4 and air in the auxiliary burner (see fig. 4). The second inlet (*y=20-100 mm*) is a fresh mixture of CH_4 and air. The inlet conditions are given in [4]. To prevent the non-cooled wall from overheating, the measurements were performed as a sequence of different tests, where each test had a duration of about 30s. In the measurements no wall temperatures are provided, therefore in the simulations a constant wall temperature of 600 K is assumed.

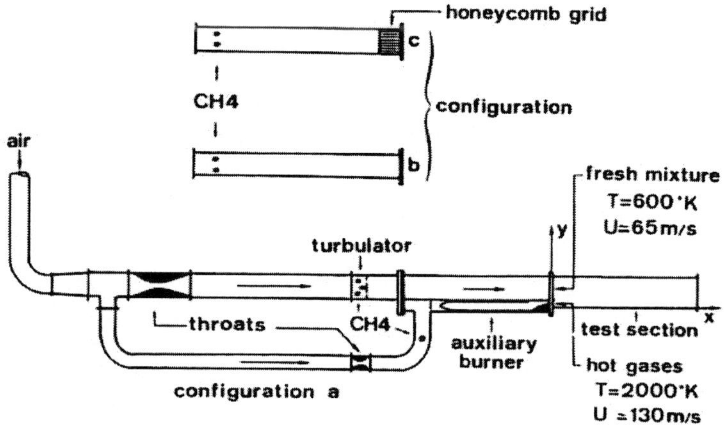

Figure 4: Experimental Set-up of the Moreau experiments.

One should notice the measured 2000 K inlet temperature in the hot inlet. This temperature is not the adiabatic temperature of a stoichiometric CH_4/air mixture. The adiabatic temperature is about 2300 K, if the pre-combustion temperature is 600 K. This low measured temperature can be caused by heat loss in the auxiliary burner or by non-equilibrium of the combusted gas. Another possibility is that the temperature measurement at this very high level was very (probably) very inaccurate. In the literature no explanation is given for this fact. It should be noted that the difference of 300 K is large and will have its influence.

In simulations of this Moreau c.s. experiment, the set up of the combusted inlet flow with non-adiabatic temperature is quite complicated. One possibility is to have a long inlet channel designed to mimic the heat loss leading to a reduction of the temperature from adiabatic to 2000 K. but this is quite troublesome. Therefore in the calculations presented here, the possible heat loss of the hot inlet is neglected and an inlet temperature of 2300 K is taken.

The Simulation Set-Up

Simulations of the Moreau experiments have been done with the FIRST combustion model, using the commercial CFD package CFX 4.3. The flow solver and the k-ε turbulence model of this package are used to calculate the flow field. The combustion model is implemented by adding user scalar transport equations, describing the turbulent transport equations for the mean and variance of the combustion scalars (eqs 18-21). The combustion scalar source terms and the density are returned to the solver by the reaction progress variables' thermochemical database at every iteration. The thermochemical database stores the species concentrations, temperature, density and scalar source terms, and is calculated preliminary to the simulations. The chemical source terms for t and r used in the simulations were described in the previous section.

A computational mesh consisting of *1500x95* (LxH) grid points is used in the simulations of the Moreau case. With a resolution in x and y direction of: $\Delta x = 0.0008$m, $\Delta y = 0.001$m.

In the experimental setup as operated by Moreau C.S. homogeneous turbulence was created using configuration c indicated in fig. 4. For the simulations the associated turbulent kinetic energy k is taken from the measurements given in [4] and the kinetic energy dissipation rate in the inlets was estimated using equation (24) (see [14]):

$$\varepsilon_{inl} = \frac{k_{inl}^{1.5}}{0.3\,D} \tag{24}$$

The *k-ε* turbulence model constants used in the simulations are given in table 1.

Table 1: Turbulent modeling constants used in CFX.

Turbulence model		
	$C\varepsilon_1 = 1.44$	$\sigma_k = 1.00$
$C_\mu = 0.09$	$C\varepsilon_2 = 1.92$	$\sigma_k = 1.30$
	$C\varepsilon_3 = 0.00$	$\sigma_{pr} = 0.90$

With the FIRST model, in first instance, it is chosen to simulate the Moreau experiments with the inlet temperature of the hot flow 2300 K and the *s* variable in equilibrium (*s=0*). The latter is done for simplicity and because the database had to be extrapolated over a large range of *s*. In the calculations this assumption this will not have a large effect, because especially at higher temperatures the deviation of s from equilibrium will be small. The combustion is assumed to be adiabatic. The combustion scalar transport equations that have been solved are:

- Mean of mixture fraction equation.
- Mean of t progress variable equation.
- Mean of r progress variable equation.
- Variance of the t progress variable equation.

For the Moreau experiments the mixture fraction *f* is defined such that *f=1* at the hot inlet and *f=0* at the cold inlet. Because the two inlets do not differ much in equivalence ratio, the effect of the variance of *f* can be neglected. It is assumed that the variance of the *t* variable is the most important, because a burned flow (*t=1*) is right next to a unburned flow (*t=0*). Due to this fact the gradients in *t* are large and turbulence has a significant influence due to the shear layer that exists between the two flows. The *r* variable is set to 1 in both inlets, because they do not contain H_2.

7 Results of the Simulation

The results of the combustion simulations are given in figures 5 to 8. Figure 5 shows the progress of the consumption of CH_4 in the t progress variable field. It can be seen that the *t* variable changes rapidly from 0 at the freshly premixed inlet to 1 at the ignition front traveling from the hot already combusted inlet flow. Due to the rapid CH4 conversion to CO, H_2 and intermediate species the concentration of these species increases very quickly in the reaction zone. When H_2 is produced by the CH_4 combustion, this induces the *r* variable to decrease from 1, as the hydrogen chemical system is forced from equilibrium. This is counteracted by chemical reaction and reflected by the r source term increasing to a positive value. This way H_2 is predicted to be converted to the reaction product H_2O. The hydrogen chemical reaction system goes back to equilibrium and the r progress variable returns to 1.

This process can be seen in the calculated mean of r variable field in figure 6. It is apparent that hydrogen production and consumption is preceded by methane conversion and that the time scale of methane conversion is much shorter than the hydrogen reaction time scales. The calculated temperature field is shown in fig. 7. In line with the conversion processes of CH_4 and H_2 the temperature rises accordingly in the combustion field. Near the exit of the combustor the combustion process is completed and the temperature profile becomes flat at a level of approximately 2200 K.

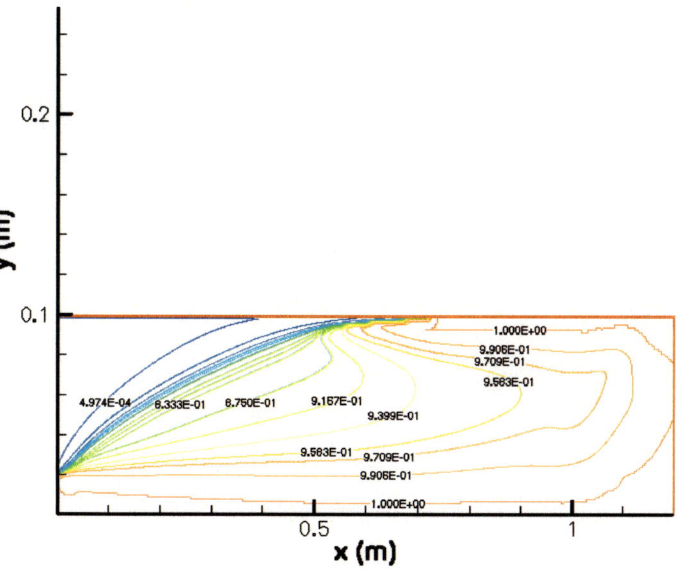

Figure 5: Contours of the t-variable field in the combustor.

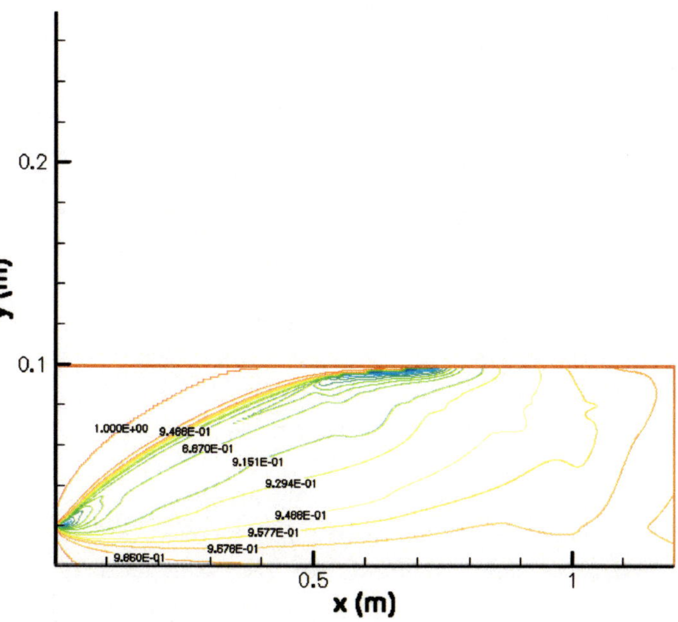

Figure 6: Contours of the r-variable field in the combustor.

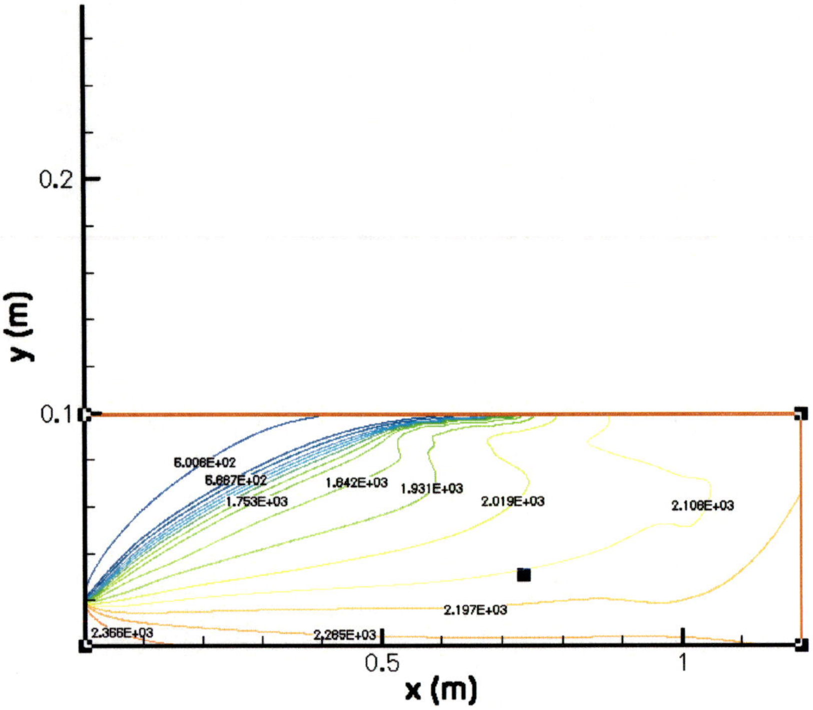

Figure 7: Contours of the temperature field in the combustor.

In figures 8 and 9 the comparison with the available measurement data is presented. Figure 8 shows at 4 axial locations radial profiles of the measured (dotted line) and predicted (drawn line) axial velocity. At low values of axial location the velocity in the burnt inlet flow ($0<y<0.01$ m) is 150 m/s and much lower (60 m/s) in the unburnt flow at $y>0.01$ m. At increasing axial distance the unburnt gas is mixed with the burnt gas at the shear layer, ignites and is combusted. Due to this, the axial velocity in the burnt gas increases even more, to 230 m/s and the axial velocity at larger radial coordinate increases gradually to about 150-200 m/s. It can be seen that the u-velocity profiles are quite well predicted. There is some overestimation of the velocity at $0<y<0.01$ m due to the set 2300 K inlet temperature in the lower part of the combustion chamber, where measured was 2000 K at the inlet. These velocity profiles indicate that the density evolution, and hence the progress of chemical reaction is well predicted.

This is not confirmed by the comparison of the measured (dotted line) and predicted (drawn line) CH_4 concentration radial profiles as shown in figure 9. Both in the experiments and the simulation at x=122 mm the CH_4 concentration is about 12% at large radial distance y and remains constant (unburnt gas) until y=0.04 m. At smaller y, when the hot combusted inlet flow is approached, the measurements show a slow decrease of the CH_4 concentration and hence slow combustion. In the prediction of the reaction progress variable model the rate of reaction is predicted much faster and hence a fast decrease of the CH_4 concentration with decreasing radial coordinate. This situation is also seen at larger axial distances, but with the combustion front moved to larger radial distances as well. The predicted combustion front is much steeper than the measured front and indicates that the combustion rate of CH_4 combustion is predicted to be much faster when compared to the experiments. This is however in contradiction with the good comparison with the velocity data, which is impossible without good prediction of density, temperature and major species conversion. As the accuracy of the measurement of methane concentration in this experiment is quite uncertain, this discrepancy has to be attributed to measurement inaccuracies. More accurate measurements with known error limits are welcomed to confirm this.

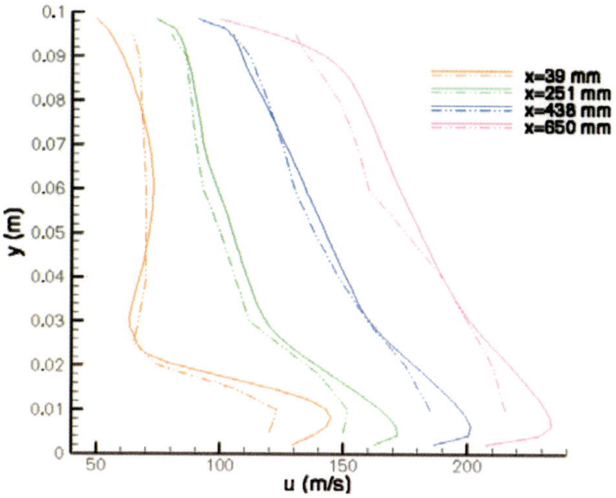

Figure 8: Measured and predicted radial profiles of axial velocity.

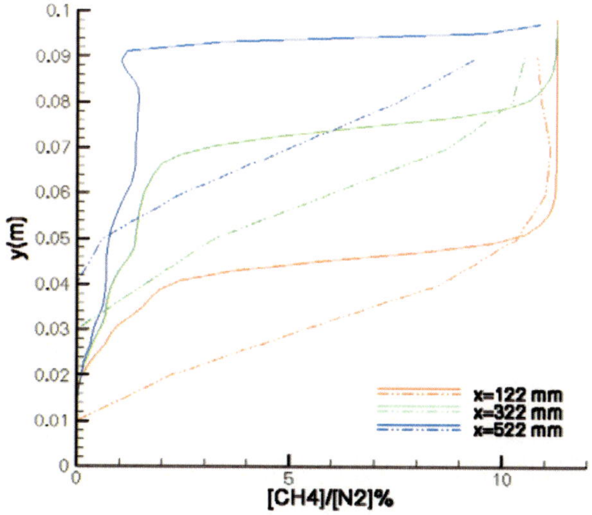

Figure 9: Measured and predicted radial profiles of CH_4 concentration.

8 Conclusion

In this paper, a reaction progress variable combustion model is presented that takes into account the conversion of fuel by the mean and variance of the progress of CH_4 conversion, and the conversion of the intermediate hydrogen by a mean progress variable. Carbon monoxide is considered to be in equilibrium. The chemical reaction source terms were based on a detailed reaction mechanism. With this model the combustion of CH_4 in gas turbine environments can be simulated. By implementing the combustion model into a standard CFD code and its turbulence model, the combustion model is compared with experimental data from Moreau and Boutier. The combustion model is incorporated in the CFD code using combustion scalars. All chemistry data needed for the simulation is calculated before the flame simulation and is stored in averaged form in a thermo chemical database.

The model is validated with experimental data of a premixed CH_4 flame stabilized with a hot flow of combusted CH_4. Comparison was done with laser velocimeter measurements and CH_4 concentration profiles of the turbulent flame. It was seen that after scaling the *t* source term S_t with a factor 10, the FIRST model is capable to predict the velocity field, and hence the density and temperature field correctly. The progress of the composed CH_4 variable is predicted to be much faster than measured in the experiment. This is however inconsistent with the good comparison with the velocity field, that depends strongly on the combustion rate. As the velocity field is measured with greater accuracy than the methane concentration, the methane measurement data have to be discarded.

Appendix

Detailed chemical reaction mechanism for methane/air
(Peters, N., personal communication, 1992)

Nr.	Reaction	A [cm, mol,s]	B [-]	E [kJ/mole]
1	CH_4+H↔CH_3+H_2	1.30 A 10^{04}	3.00	33.60
2	CH_4+OH=CH_3+H2O	1.60 A 10^{07}	1.83	11.60
3	CH_3+O=CH2O+H	8.43 A 10^{13}	0.00	0.00
4	CH_2O+H=HCO+H_2	2.30 A 10^{10}	1.05	0.00
5	CH_2O+OH=HCO+H_2O	3.40 A 10^{09}	1.20	-1.90
6	HCO+H=CO+H_2	9.00 A 10^{13}	0.00	0.00
7	HCO+M=CO+H+M	7.10 A 10^{14}	0.00	70.30
8	HCO+O_2=CO+HO_2	3.00 A 10^{12}	0.00	0.00
9	H+O_2=OH+O	2.00 A 10^{14}	0.00	70.30
10	O+H_2=OH+H	5.06 A 10^{04}	2.67	26.30
11	H_2+OH=H_2O+H	1.00 A 10^{06}	1.60	13.80

Nr.	Reaction	A [cm, mol,s]	B [-]	E [kJ/mole]
12	$OH+OH=H_2O+O$	1.50 A 10^{09}	1.14	0.42
13	$H+O_2+M=HO_2+M$	2.30 A 10^{18}	-0.80	0.00
14	$H+OH+M=H_2O+M$	2.20 A 10^{22}	-2.00	0.00
15	$H+HO_2=OH+OH$	1.50 A 10^{14}	0.00	4.20
16	$H+HO_2=H_2+O_2$	2.50 A 10^{13}	0.00	2.90
17	$OH+HO_2=H_2O+O_2$	6.00 A 10^{13}	0.00	0.00
18	$CO+OH=CO_2+H$	6.00 A 10^{06}	1.50	-3.10

Rate constants: $k_f = A\,T^b \exp(-E/RT)$;
Third body efficiencies:

$$M = 6.5\,CH_4 + 0.75\,CO + H_2 + 0.4\,O_2 + 1.5\,CO_2 + 6.5\,H_2O + H + 0.4\,N_2;$$

References

[1] Kok, J.B.W., Louis, J.J.J., Modeling turbulent combustion in a CO/H_2 diffusion flame using reaction progress variables, *Combust. Sci. and Tech.* **(131)** (1998) 225-249.

[2] Louis, J.J.J., Kok, J.B.W., Klein, S.A., Modelling and measurements of a 16 kW turbulent non-adiabatic syngas diffusion Flame in a cooled cylindrical combustion chamber, *Combust. Flame* **(125)** (2001) 1012-1031,.

[3] Kok, J.B.W., Louis, J.J.J, and Yu, J.H., The IRST Model for Turbulent Premixed Non-Adiabatic Methane Flames, *Combust. Sci. and Tech.* **(149)** (1999) 225-247.

[4] Moreau, P., Boutier, A., "Laser velocimeter measurements in a turbulent flame.", *ONERA*, Toulouse, Report nr ONERA TP 1976-126, France, (1976).

[5] Janicka, J., Ph.D. Thesis, *Techn. Hochschule Aachen*, (1979).

[6] Janicka, J., Kollmann, W., The calculation of mean radical concentrations in turbulent diffusion flames, *Combust. Flame* **(44)** (1982) 319-336

[7] Correa, S.M., *A model for non-premixed turbulent combustion of CO/H2 jets, Archivum Combustionis* **(5:3-4)** (1985) 223-243.

[8] Correa, S.M., Gulati, A., Measurements and modeling of a bluff body stabilized flame, *Combust. Flame* **(89)** (1992) 195-213.

[9] Dixon-Lewis, , G., Goldsworthy, F.A., Greenberg, J.B., Flame structure and flame reaction kinetics IX. Calculation of properties of multi-radical premixed flames, *Proc. R. Soc. London A* **(346)** (1975) 261-278.

[10] Williams, F.A., Combustion Theory, 2nd ed., Addison-Wesley Publ. Comp., Redwood City Ca (1985) p 556.

[11] Eggels, R.L.G.M., Louis, J.J.J., Kok, J.B.W., De Goey, L.P.H., Comparison of conventional and intrinsic low dimensional manifold methods to reduce CO-H2/air

mechanisms for turbulent flame modeling, *Combust. Sci. and Tech.* **123** (1997) 347-362.

[12] Yu, J.H., Combustion of natural gas in gas turbine combustors, PhD thesis University of Twente (NL), (1996).

[13] Skevis, G., Chrissanthopoulos, A., Goussis, D.A., Mastorakos, E., Derksen, M.A.F., Kok, J.B.W., Numercial investigation of methane combustion under mixed air-steam turbine conditions – FLAMESEEK, *Appl. Thermal Eng.* **24**, 1607-1618, (2004).

[14] CFX International, CFX-4.3 solver manual, Harwell, Oxfordshire, AEA Technology, (1997).

In: New Devel. in Combustion Research
Editor: William J. Carey, pp. 227-256

ISBN: 1-59454-826-9

Chapter 7

STRATEGY FOR MULTISCALAR RAMAN DIAGNOSTICS IN HIGH-PRESSURE HYDROGEN FLAMES

Jun Kojima
Ohio Aerospace Institute, 22800 Cedar Point Rd,
Cleveland, OH 44142
Quang-Viet Nguyen
NASA Glenn Research Center, 21000 Brookpark Rd,
Cleveland, OH 44135

Abstract

We present a detailed strategy for performing quantitative spatially-resolved multiscalar measurements of major species concentration and temperature in high pressure H_2-air combustion systems using spontaneous Raman scattering (SRS). This strategy can also be extended to include CO, CO_2 and hydrocarbons (such as CH_4) in both high pressure and atmospheric pressure hydrocarbon-air flames. Central to this strategy is the Raman calibration matrix formalism which is often used to account for spectral interferences resulting from inter-species spectral crosstalk. We describe in detail, the theory, instrumentation, facilities, and data processing techniques used to generate the Raman calibration matrix, and its implementation to obtain quantitative multiscalar measurements. The Raman calibration matrix is realized with the aid of a novel high-pressure fully-premixed calibration burner, an optically accessible high-pressure combustion facility, and a state-of-art 532 nm wavelength laser SRS system. This system and data processing methods, collectively, provide an empirically based, but analytic formulation of the temperature-dependent Raman calibration matrix elements based on the assumption of chemical equilibrium at the measured temperature in the post-flame products of a premixed H_2-air calibration burner operated at a pressure of 10 atm. At pressures up to 30 atm, we find the inclusion of a pressure-dependent term in the calibration matrix unnecessary for the relatively low, but practical, spectral resolution of the optical setup, mainly because the pressure broadening (or narrowing, if any) effect is minor relative to the instrumental resolution (approx. 1.5 nm) . The high signal-to-noise ratio SRS signals yielded accurate (< 2% uncertainty) temperature measurements via the rotational H_2 and vibrational N_2 spectra, and together, enable the generation of a high quality Raman calibration matrix. The Raman calibration matrix determined in this study can be used as a transferable SRS standard for use with other 532 nm wavelength laser-Raman setups to probe H_2-air flames since the signal response was calibrated using traceable wavelength and intensity standards.

1 Introduction

Quantitative measurements of species concentration and temperature in elevated pressure combustion environments are critical to the validation of computer models used in the development of chemical propulsion systems such as gas turbine and internal combustion engines. Laser-excited spontaneous Raman scattering (SRS) spectroscopy is one of the few practical techniques that provide spatially and temporally resolved multi-species information in a simultaneous or single-shot fashion with a high level of accuracy [1-3]. For high pressure combustion studies, SRS has the added benefit of providing stronger signal levels than those performed at atmospheric pressure, because the signal scales linearly with the number density. The choice of using SRS for flame diagnostics also provides a practical benefit: the ability of working with only a single optical access port in the back-scattered collection mode. This is often useful for industrial applications where the number of optical windows needs to be kept to a minimum. Since SRS is a linear optical process, the measurements are relatively simple, less expensive, and more versatile when compared to coherent anti-Stokes Raman spectroscopy (CARS), or stimulated Raman spectroscopy. These facts have motivated a recent renewed interest in fundamental studies on SRS diagnostics in high-pressure combustion [4,5].

Despite the promising capabilities of Raman scattering diagnostics as an inherently linear process, the routine processing of Raman spectral data to yield quantitative values is not yet a maturely developed area. A primary task that permits the accurate processing of Raman scattering data is a comprehensive signal *calibration* process [2,3,6-9]. Without an effective methodology and database for an accurate calibration, Raman measurements in high pressure flames are not suitable for use in validating computational models of combustion. Such calibration procedures for SRS are challenging due to the problematic effects of inter-species spectral interferences or 'crosstalk' that are a function of temperature, species concentration, and possibly pressure.

2 Fundamental Concept

Earlier studies have shown that the SRS calibration procedure is best implemented using a calibration matrix formalism for discrete wavelength detection channels [2,3,6-8]. Even in the case of spectrally resolved detection using a multi-channel array or CCD camera type detector, data interpretation is often implemented by integrating the vibrational band of each species, resulting in a 'superpixel' [9]. This technique efficiently distills spectrally-resolved data into a compact and manageable form, and effectively results in the generation of discrete channel data which lends itself to the calibration matrix formalism. The calibration matrix equation relates the spectrally integrated SRS intensity response S_i for a specific molecular species i, to the number density by the following phenomenological relation:

$$S_i = \Lambda E_{\mathrm{L}} \sum_j k_{i,j}(T) N_j \,, \tag{1}$$

where Λ is the optical collection efficiency of the apparatus that is easily calibrated (being a function of the integrated absolute spectral response of all optical devices involved, the solid angle of detection, and the control volume); E_{L} is the energy of the excitation laser pulse; N_j is the number density of the species j; $k_{i,j}$ is the so-called Raman calibration matrix element (constant) which is a function of temperature. From Eq. (1), if we solve for the number density N_j, we see that it can be calculated from the *inverse* of the calibration matrix as follows:

$$\begin{pmatrix} N_1 \\ N_2 \\ \vdots \\ N_n \end{pmatrix} = \frac{1}{\Lambda E_{\mathrm{L}}} \begin{bmatrix} k_{1,1} & k_{2,1} & \cdots & k_{n,1} \\ k_{1,2} & k_{2,2} & & \vdots \\ \vdots & & \ddots & \vdots \\ k_{1,n} & \cdots & \cdots & k_{n,n} \end{bmatrix}^{-1} \begin{pmatrix} S_1 \\ S_2 \\ \vdots \\ S_n \end{pmatrix}. \tag{2}$$

where the on-diagonal terms of the Raman calibration matrix $k_{i,j}$ are the direct-contribution calibration constants while the off-diagonal terms represent the spectral crosstalk-contribution of species j onto species i. Clearly, from Eq. (2), the calibration matrix $k_{i,j}$ is the key parameter in making accurate measurements of species concentrations using SRS.

Logically, before attempting any multiscalar measurement in a flame of interest, an acceptable value of the matrix $k_{i,j}$ has to be determined by executing a series of calibration experiments conducted using a well-characterized calibration burner accompanied with known species number density and temperature information for combustion products of the calibration burner, and a well-defined Raman instrumentation system. Given that these requirements are satisfied, the Raman matrix element $k_{i,j}$ can be determined by plotting the reference SRS intensity response S_i^{ref} normalized by the reference number density N_j^{ref} for each channel i as a function of temperature. By executing a series of such calibration experiments over various different flame conditions, which give different temperature (and species concentrations) settings, each individual element in the matrix $k_{i,j}$ can be defined as a function of temperature as illustrated in Fig. 1. The temperature dependence of the matrix element has been shown, in general, not to behave linearly, but rather, is typically represented by polynomial relation of more than second order [6,7]. This nonlinearity is due to the fact that, upon a fixed bandwidth factor (i.e., wavelength area for a pixel integration or a detection channel) [1], the matrix $k_{i,j}$ is governed by multiple factors including the Boltzmann distribution, the spectral line-shape function, the instrumental function, and spectral overlap, etc. Note that acquiring calibration flame data in both fuel-lean and fuel-rich equivalence ratios is critical in providing a complete picture of the calibration matrix elements. For example, in fuel-lean flames, we can assume that there is no H_2 present and O_2 should be present; in fuel-rich flames, we can assume that there is no O_2 present and H_2 present. In this way, we can distinguish between the different types of spectral crosstalk. This wide range of equivalence ratio is also important from thc standpoint of studying non-premixed flames where varying degrees of turbulent mixing can occur at any point and time.

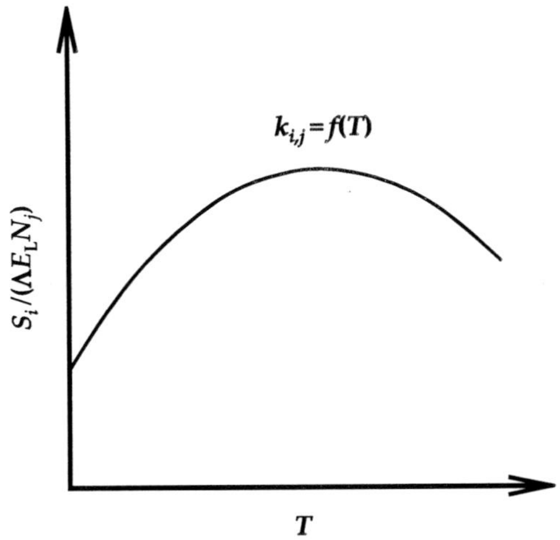

Figure 1: Conceptual illustration of the Raman crosstalk-calibration matrix element. See text for legends used here and explanation.

Despite the numerous studies that use SRS in combustion studies, we are unaware of a previous work that is aimed at providing the Raman calibration matrix applicable to species measurements in *high-pressure* flames and relating it to quantitative measurements. Since collisional broadening and even line-mixing effects can occur, such measurements in high-pressure flames may be required to augment a crosstalk matrix obtained at atmospheric pressure. Therefore, the pressure dependence on Raman spectra has to be carefully investigated over the range of temperature and pressure of interest.

In this chapter, an overall strategy for deriving the Raman calibration matrix is described through a comprehensive high-pressure flame experiment confirmed by detailed spectroscopic simulations. The scope of this chapter is to give an overview and specific detailed procedures for implementing the Raman calibration matrix formalism with an example application. We focus our attention in this work to hydrogen–air flames for the following reasons: chemically, a H_2–air flame provides the simplest combination of combustion products in order to facilitate the spectral analysis, yet retains spectral features that are important to more complicated fuel-oxidizer systems that will be studied at a later time; research data on a H_2–air flames are important in a recent increasing interests in hydrogen-based low-emissions propulsions and power systems as well as rocket combustors or pulse detonation engines concept.

3 Modeling of Raman Scattering

In order to aid our understanding of the nature of the Raman crosstalk at elevated pressure, we developed a theoretical model of the SRS spectra [10] following a semi-classical anharmonic oscillator formalism. Note that while a Raman spectral simulation software for combustion applications [11] is publicly available, having our own program adds flexibility to investigate pressure broadening effects and crosstalk study in a various flame conditions.

3.1 Raman Cross-Section

According to Placzek's polarization theory of the Raman effect, the corresponding differential Stokes-shifted Raman scattering cross-section per molecule at thermal equilibrium is given by [1,12]:

$$\left(\frac{\partial\sigma}{\partial\Omega}\right)_{i\to f} = q(\nu_0 - \nu_k)^4 g \frac{\exp(-E_i / kT)}{Q(T)} \Phi \,. \tag{3}$$

Here, q is the molecule-independent constant; ν_0 [cm^{-1}] is the excitation light frequency [cm^{-1}]; ν_k [cm^{-1}] is the molecular frequency associated with the rotation-vibration, or pure rotation transition from the initial state i to the final state f; g is the total degeneracy; E_i [cm^{-1}] is the energy of the initial state; k is the Boltzmann constant; T [K] is the gas temperature; $Q(T)$ is the total partition function of the molecule at temperature T, which may be written as the sum of the rotational and vibrational partition function [13]; Φ is the space-averaged squares of the transition polarizability tensor components for diatomic molecules (molecule-dependent temperature-independent invariant), which is tabulated in Ref. 12 for each branch of vibrational and pure rotational Raman scattering. Note that Eq. (3) should be multiplied by the factor $(v + 1)b_v^2$ for vibrational Stokes scattering of diatomic molecules. Here, v is the vibrational quantum number and b_v^2 is defined as $h/8\pi^2 c\,\nu_k$ where h [J·s] is the Plank constant, and c [cm·s^{-1}] is the speed of light in vacuum. For calculating a pure rotational H_2 Raman spectrum, the correction factor for the rotation-vibration interactions was further multiplied by Eq. (3).

The Raman spectral line positions are given by the molecular rotational or vibrational frequency ν_k, and are calculated on the basis of the anharmonic oscillator model for the diatomic molecular structure [13]. The energy in a vibrational-rotational state (v, J) is expressed as the sum of the vibrational energy term, $G(v)$ and rotational energy term, $F(v,J)$. These terms can be expressed by the Dunham expansion, which is the conventional power series in (v + ½). The non-rigid rotator model was also applied to take the influence of centrifugal force due to vibration into account. Thus, with the appropriate spectroscopic constants [14], the energy terms for H_2 were effectively calculated. The Q-branch of vibrational Raman line is generally much stronger than the O- and S-branches because the mean polarizability tensor greatly exceeds the anisotropy invariant. Yet the contributions from the O- and S-branches, in general, are not negligible as spectral interferences. We then calculated the relative intensities of the O-, Q-, and S-branches in vibrational Raman scattering for H_2 and N_2 molecules with the present model.

The Raman scattering spectrum of water vapor appears over the range of frequency from 3400 to 4130 cm^{-1}, and results mainly from the OH-stretch vibrational mode. Generally, the temperatures encountered in combustion environments makes the ν_1 band (100)-(000) transition dominant. For the H_2O Raman spectrum, the classical anharmonic-oscillator model is not suitable for calculating the vibrational energy because of the asymmetric top nature of the water molecule that shows a relatively complicated structure. Based on the wavefunctions and the polarizability derivatives, Avila *et al* [15] calculated the vibrational Raman cross-sections of H_2O in the OH-stretching region with great accuracy up to about 1000 K. In the current work, we then used the results of the Raman line frequency and cross-sections which are tabulated in Ref. 15 to simulate the H_2O Raman spectra in our high-pressure flames.

Finally, the Raman line frequencies and intensities of H_2 (rotational and vibrational), N_2, and H_2O were theoretically determined using the above-mentioned model. In order to permit comparisons of the calculated Raman spectra for each species with experimental data, we calculated the relative (but quantitative) intensity by setting the function Φ at an arbitrary value. In the sections below, *vib*-H_2 indicates vibrational H_2 Raman spectra and *rot*-H_2 indicates pure rotational H_2 Raman spectra.

3.2 Pressure Broadening

In spontaneous Raman scattering pressure-broadening linewidths have less impact to the overall spectral shape, unlike with the CARS technique. Yet, it is important to take pressure broadening effects into account for spectral analysis of SRS over the range of pressure in the present work. Hussong *et al* [16] provided an effective, polynomial equation for the H_2 collisional broadening coefficient including the inhomogeneous contribution (nonlinear dependence on the perturber concentration) in the binary system H_2-N_2:

$$\tilde{\gamma}_J^0(T,C) = (\gamma_0^{\text{self}})_J + (\tilde{\gamma}^{self})_J T + (a_1)_J C + (a_2)_J C^2 + (a_3)_J CT + (a_4)_J C^2 T \quad (4)$$

where $\tilde{\gamma}_J^0$ [cm^{-1}/bar] is the total broadening coefficient as a function of temperature, T and concentration of perturber, C. Here C is defined as N_2 fraction to H_2 under the assumption of H_2-N_2 binary system and J is the rotational quantum number. All other parameters, $(\gamma_0^{\text{self}})_J$, $(\tilde{\gamma}^{self})_J$, $(a_i)_J$ are tabulated for J = 0 to 5 in Ref. 16. As Hussong *et al* mentioned in their reports [16,17], the influence of H_2O as an additional collider plays an important role at higher temperature in terms of speed-inhomogeneous effects, thus a ternary system H_2-N_2-H_2O should be considered for better accuracy over the range of flame temperature. However, comprehensive studies on the collisional broadening of ternary mixture including H_2O at high temperature have not yet been reported in the literature, although an alternative strategy adapting the H_2-N_2-He system has been proposed [17]. Since additional collisional broadening coefficients were not available for the ternary system (including water), we calculated the H_2 broadening coefficient for each flame condition using Eq. (4) for the present work.

On the other hand, molecular spectroscopic information (including the collisional broadening coefficients) of N_2 or H_2O are available in an atmospheric database, HITRAN [18]. We then used HITRAN to find spectroscopic information for the collisional broadening coefficients of N_2 and H_2O. In the database, the collisional broadened halfwidth is written as

$$\tilde{\gamma}_J^0(T,C)^{\text{HITRAN}} = \left(\frac{T_0}{T}\right)^n \left[\gamma^{\text{air}}(P_0,T_0)C + \gamma^{\text{self}}(P_0,T_0)(1-C)\right], \quad (5)$$

where $\gamma^{air}(P_0,T_0)$, and $\gamma^{self}(P_0,T_0)$ are the air-broadened, and the self-broadened width at the reference temperature, T_0 = 296 K and the reference pressure, P_0 = 1 atm, and n is the coefficient of temperature dependence.

In an effort to simplify the effects of temperature and J-number dependence of collisional broadening for hydrogen, we use the broadening coefficient of a centered line for all other lines. Note that if this single-value coefficient assumption would reduce the Voigt-a parameter by less than 10% at combustion temperature but this order of variation does not cause a significant error on the spectral shape, particularly for a diagnostic application. In the below section, experimental spectra will be compared with calculated ones based on our assumption regarding the binary system and the single-value coefficient.

3.3 Line-Shape Model Boundary

Spectral line shapes of Raman scattering spectra at atmospheric pressure are typically modeled by a Gaussian profile, and those at higher pressures are usually modeled by a Voigt profile [19] which is the convolution of Gaussian and Lorentzian collisional broadening distributions. In some cases that require higher accuracy, such as in a laser absorption technique, motionally-narrowed Galatry profiles [20] could be also employed. These line-shape models were developed for a single isolated transition. In practice however, multiple overlapping transitions are common and the effect of their overlap may become important at high pressures. Thus, consideration of the line-narrowing due to the interaction between different transitions, the so-called line-mixing effect, may be required to properly model the Raman spectra for quantitative measurements.

Nagali [21] discussed the applicability of the major line-shape models and introduced an effective approach for choosing an appropriate model for neutral species within the temperature and pressure range of interest. Following the approach described by Nagali [21] but include N_2 concentration-dependent collisional broadening coefficients to the analysis, we are able to determine an appropriate collisional broadening model for simulating Raman spectra of H_2, N_2, and H_2O gases over a range of pressure, temperature, and species concentration in high-pressure flames. Here through a molecular collision-scale analysis with molecular vibration frequencies [10,21], we defined three criteria for modeling Raman spectra, i.e., A) Motional narrowing (in which the Galatry profile is considered); B) Impact approximation (in which the Voigt profile is valid); and C) Line mixing (in which the additive approximation is not valid):

$$P = \frac{\Delta\nu_D/2}{\tilde{\gamma}_J^0(C,T)} \text{ : Galatry boundary – A} \tag{6}$$

$$P = \frac{0.079\Delta\nu_{\text{imp}}}{\tilde{\gamma}_J^0(C,T)} \text{ : Voigt boundary – B} \tag{7}$$

$$P = \frac{\nu_{\mathrm{sep}}}{\tilde{\gamma}_J^0(C,T)} \text{ : Line-mixing boundary – C} \tag{8}$$

where $\Delta\nu_D$ [cm^{-1}] is the molecule-dependent Doppler width; $\Delta\nu_{imp}$ is the spectral region in which the impact approximation is accurate and can be written as [10,21]

$$\Delta\nu_{\mathrm{imp}}(T,C) = \frac{0.1}{2\pi c t_{\mathrm{coll}}}. \tag{9}$$

t_{coll} [s] is the molecular collision duration, t_{coll} [s] and can be expressed as [10,21]

$$t_{\mathrm{coll}}(T,C) = \frac{1}{\nu_{\mathrm{mol}}}\sqrt{\frac{c\cdot\Delta\nu_c^0}{\nu_{\mathrm{mol}}N_j}}, \tag{10}$$

where $\Delta\nu_c^0$ [cm^{-1}] is the collisional broadening fullwidth which is twice the total broadening coefficient $\tilde{\gamma}_J^0$ at room temperature and atmospheric pressure; c [m/s] is the speed of light; and ν_{mol} [m/s] is the mean speed of the gas molecule. ν_{sep} is the averaged separation distance in frequency between all interacting (consecutive) rovibrational lines for practical convenience. With these boundaries, the region with $P\tilde{\gamma}_J^0 \le \Delta\nu_D/2$ indicates the necessity of considering the motional narrowing effect, i.e. use of the Galatry profile; that with $\Delta\nu_D/2 < P\tilde{\gamma}_J^0 \le 0.079\Delta\nu_{\mathrm{imp}}$ indicates the use of the Voigt profile for entire line (additive approximation is valid); that with $0.079\Delta\nu_{\mathrm{imp}} < P\tilde{\gamma}_J^0 < \nu_{\mathrm{sep}}$ supports the use of the Voigt profile, with the caveat, that there is a possible breakdown in the far-wing (additive approximation is likely valid); and that with $P\tilde{\gamma}_J^0 \ge \nu_{\mathrm{sep}}$ indicates the necessity of considering the line mixing effect.

3.4 Target Flames

Before choosing appropriate line-shape models, we calculated thermodynamic properties for five different flames under the assumption of complete combustion. The chemical equilibrium compositions of several different hydrogen-air flames, which hereafter we call "target flames", were calculated using the CEA chemical equilibrium code [22] for assigned equivalence ratios, pressure (30 atm), and an initial gas temperature (300 K). Table 1 shows the calculated mole fractions of H_2, N_2, and H_2O and the adiabatic temperatures. The temperatures, pressure, and N_2 mole fraction perturber concentrations were used as input parameters to calculate the collisional broadening coefficients, and the line-shape diagrams as well as for the Raman relative cross-section.

Table 1 Adiabatic temperatures and chemical-equilibrium species compositions calculated for hydrogen-air combustion system at assigned pressures and equivalence ratios.

Case #	1	2	3	4	5
P [atm]	30	30	30	30	30
ϕ	1.5	2	3	4	5
T [K]	2262	2065	1770	1560	1402
H2 [Mole frac.]	0.147	0.257	0.409	0.509	0.58
H2O [Mole frac.]	0.295	0.257	0.205	0.17	0.145
N2 [Mole frac.]	0.55	0.479	0.381	0.317	0.271

3.5 Choosing Line-Shape

To choose the best line-shape function to model Raman spectra in the target flames over a range of pressure, temperature, and perturber concentration we introduce a three-dimensional diagram that is consists of the boundaries defined in the above section. Figure 2a shows the diagram designed for the pure rotational H_2 spectrum (center wavelength ν_0 = 1035 cm^{-1}). This diagram directs one to choose an appropriate line-shape model for spontaneous Raman spectra simulation in any flame of interest with a prior knowledge of pressure, temperature, and N_2 concentration. The locations of the five target flames (labeled case # 1 to # 5 in Table 1) are marked in the diagram to determine which line-shape model is the most appropriate for each flame condition. All five cases here are between boundary A and B as shown in Fig. 2a, which indicates the entire line shape can be modeled by the Voigt profile with the additive approximation. Note that the same is true for other major *rot*-H_2 lines of higher J number while the Galatry boundary (boundary A) slightly moves toward higher pressure depending on the transition center frequency ν_0.

The diagram for the *vib*-H_2 spectrum (ν_0 = 4160 cm^{-1}) is shown in Figure 2b with points representing the target flames. All the target flames in Fig. 2b are also between boundary A and B, indicating the entire line shape can be also modeled by the Voigt profile with the additive approximation. This trend is also true for the other species over the flames of our interest [10]. The line shape for the N_2 Q-branch can be appropriately modeled by the Voigt function with the assumption of the additive approximation in the near-wing region, but their line shape could break down in the far-wing regions. The impact and additive approximation are, for the most part, appropriate for the entire region of the H_2O spectrum in the target flames.

In general, the Voigt profile was found to be a reasonable model for simulating the Raman spectrum over the range of pressure up to 30 atm, and for temperatures from approximately 1500 K to 2500 K. While there may be a slight effect of breakdown at the far wings of the N_2 Q-branch, the effect is small and is ignored in favor of Voigt function's simplicity, and the widespread availability of fast algorithms for its rapid computation. Hence, overall profiles of Raman spectra were calculated by summation of Voigt-shaped lines based on the additive approximation assumption. Note that in calculations of the Voigt-function *a* and *x* parameter we used the overall convolution Gaussian width including the instrumental function.

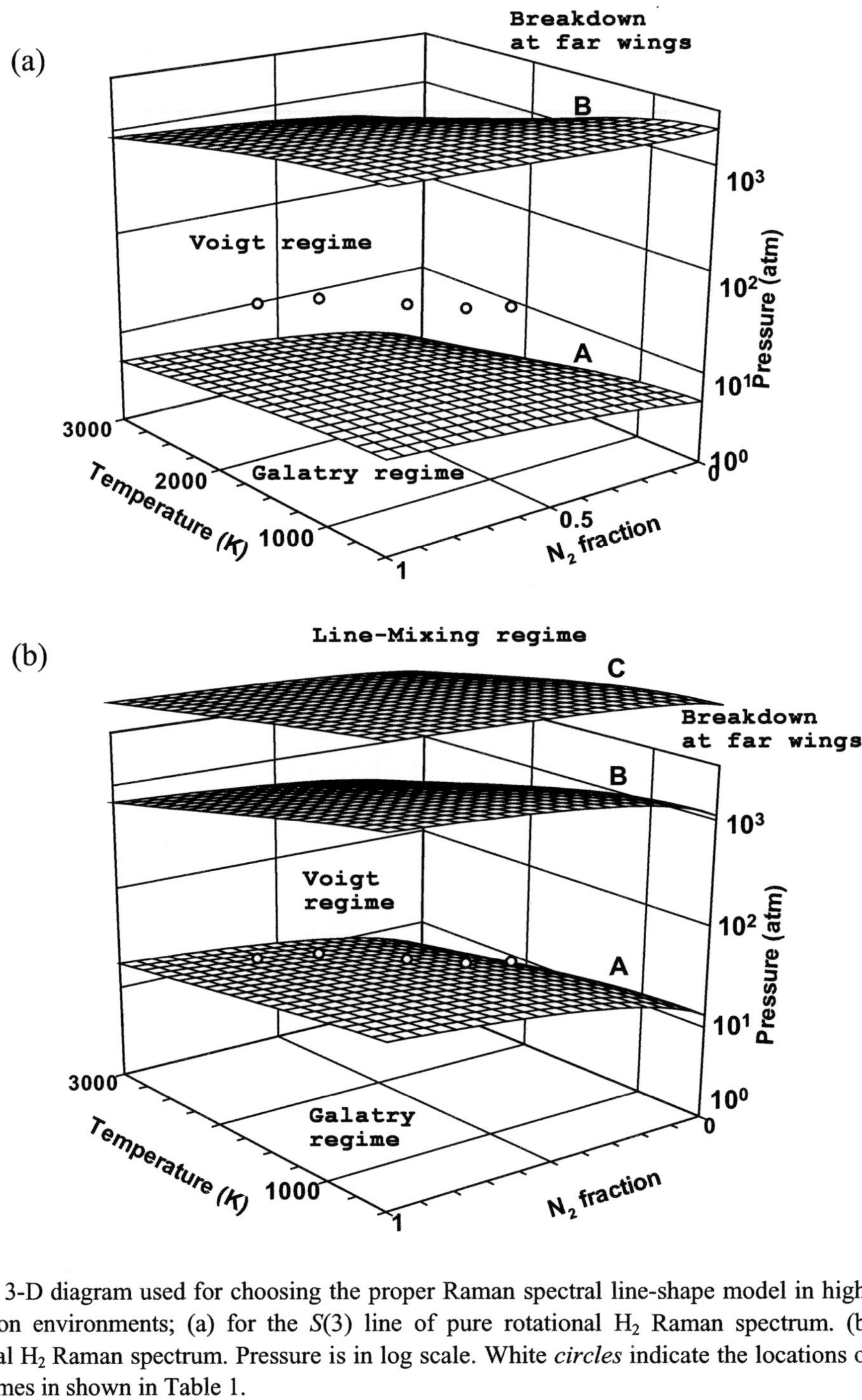

Figure 2: 3-D diagram used for choosing the proper Raman spectral line-shape model in high-pressure combustion environments; (a) for the *S*(3) line of pure rotational H_2 Raman spectrum. (b) for the vibrational H_2 Raman spectrum. Pressure is in log scale. White *circles* indicate the locations of the five target flames in shown in Table 1.

The effect of pressure-broadening on the vibrational N_2 Raman spectrum is calculated using the approach described above, and is shown in Fig. 3. Note that the pressure-broadening simply smoothes the narrower rotationally-resolved lines, but does not alter the overall area or shape of the vibrational band. As evident in Fig. 3 and our previous measured data [23], pressure broadening (or narrowing) is not a dominant factor for Raman scattering measurements in hydrogen flames (up to 30 atm) that use vibrational-band-integrated techniques for spectral detection.

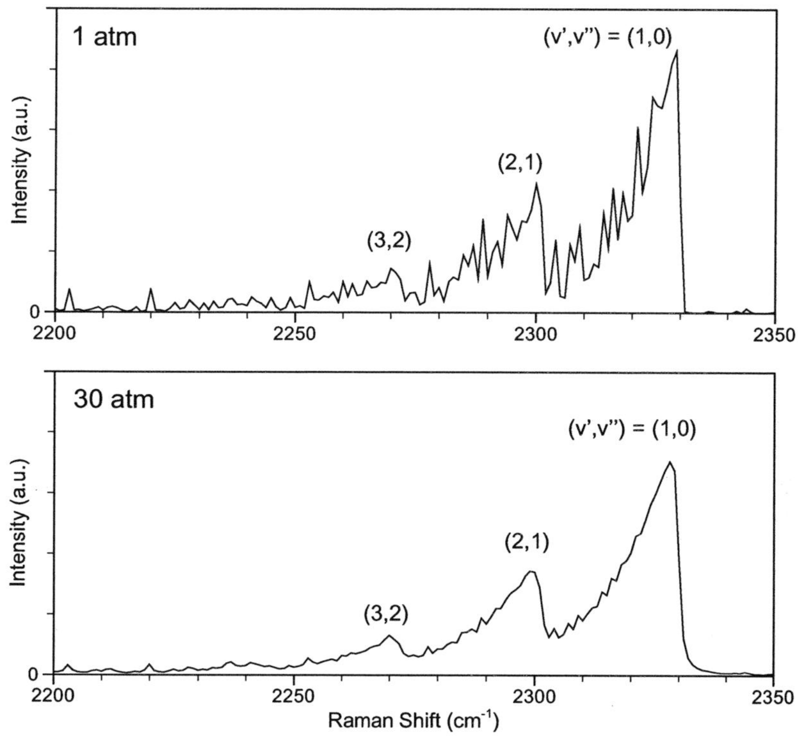

Figure 3: Simulated high-resolution N_2 Raman spectra (vibrational *O*-, *S*-, and *Q*-branches) in atmospheric (1 atm) and high-pressure (30 atm) hydrogen combustion. The instrumental resolution is 0.7 cm^{-1}. Strong pressure dependence on detail spectral shape is evident here while the overall spectral regions show little difference. Similar results due to pressure line shape broadening are also observed experimentally.

3.6 Visualizing the Raman Crosstalk

Since simple deconvolution techniques are not suited for analyzing complex spectral interferences such as those that result from multi-species interferences, a direct simulation of spectrum compared with measured data is an effective method to reveal and understand the nature of the crosstalk. From Fig. 4, we can see that the crosstalk results in a spectral overlap or signal that is difficult to analyze due to ambiguity. In Fig. 4a (target flame #1), the *S*(10) Stokes line of *rot*-H_2 around 605 nm is completely overwhelmed by the N_2 *Q*-branch. There is an interference between the N_2 *O*-branch (the blue-side tail) and the *rot*-H_2 *S*(8), or *S*(9) lines; and there is also an interference between the N_2 *S*-branch (the red-side tail) and the *rot*-H_2

$S(11)$ line. As also seen in Fig. 4b (target flame #5), the *vib*-H_2 $O(0,3)$ line around 657 nm significantly interferes with H_2O in highly H_2-rich flames. We confirmed these spectral interferences by comparing simulated spectra and data measured in 10-atm fuel-rich H_2–air flames (the experimental apparatus is described in the following sections) as clearly shown in Fig. 5. The simulated spectra for the combustion species show excellent agreement with the experimental data in terms of the transition frequency, intensity distributions and line shape. These spectral interferences (crosstalk) are important because they affect determinations of the species number density and/or temperature in fuel-rich regions of a flame.

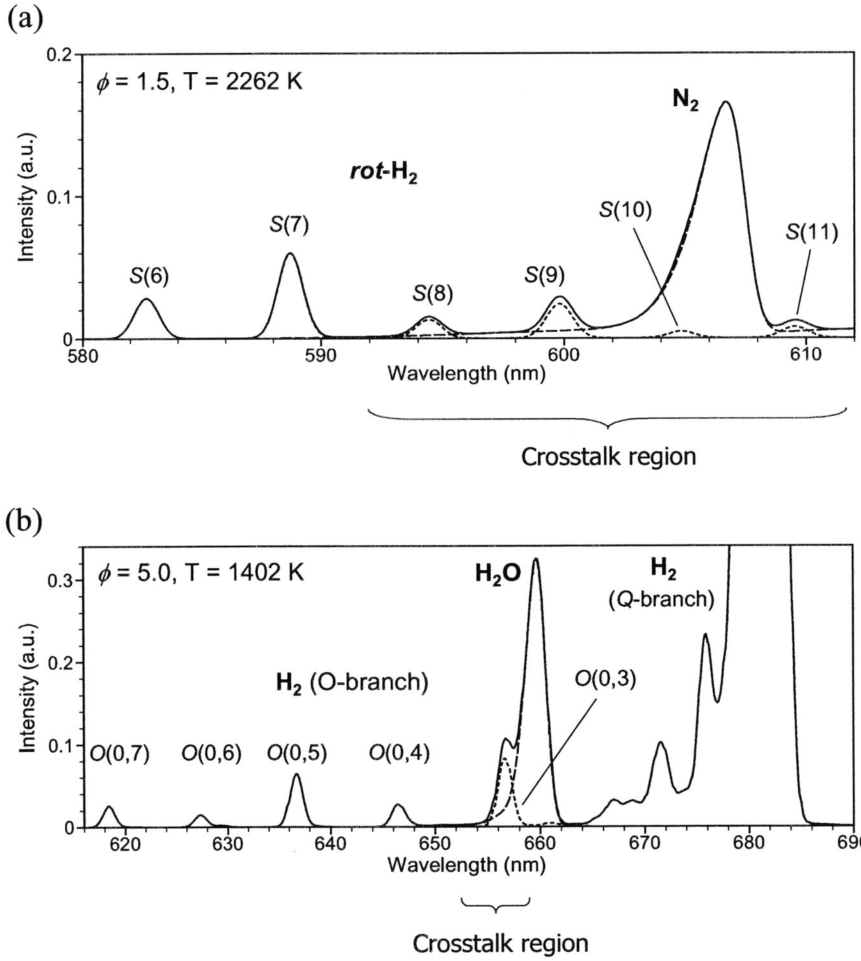

Figure 4: Simulated low-resolution Raman spectra in 30-atm H_2-air flames, showing the spectra from different species interfere each other. (a) Crosstalk between pure rotational H_2 lines and vibrational N_2 line. (b) Crosstalk between vibrational H_2 and H_2O. Here, the *solid line* represents the summation of the two spectra. Equivalence ratios and temperature of flames are shown. Note that due to the spectral overlap resulting from the crosstalk, a quantitative determination of the contribution of each species is very difficult without use of either the simulation or a calibration flame database.

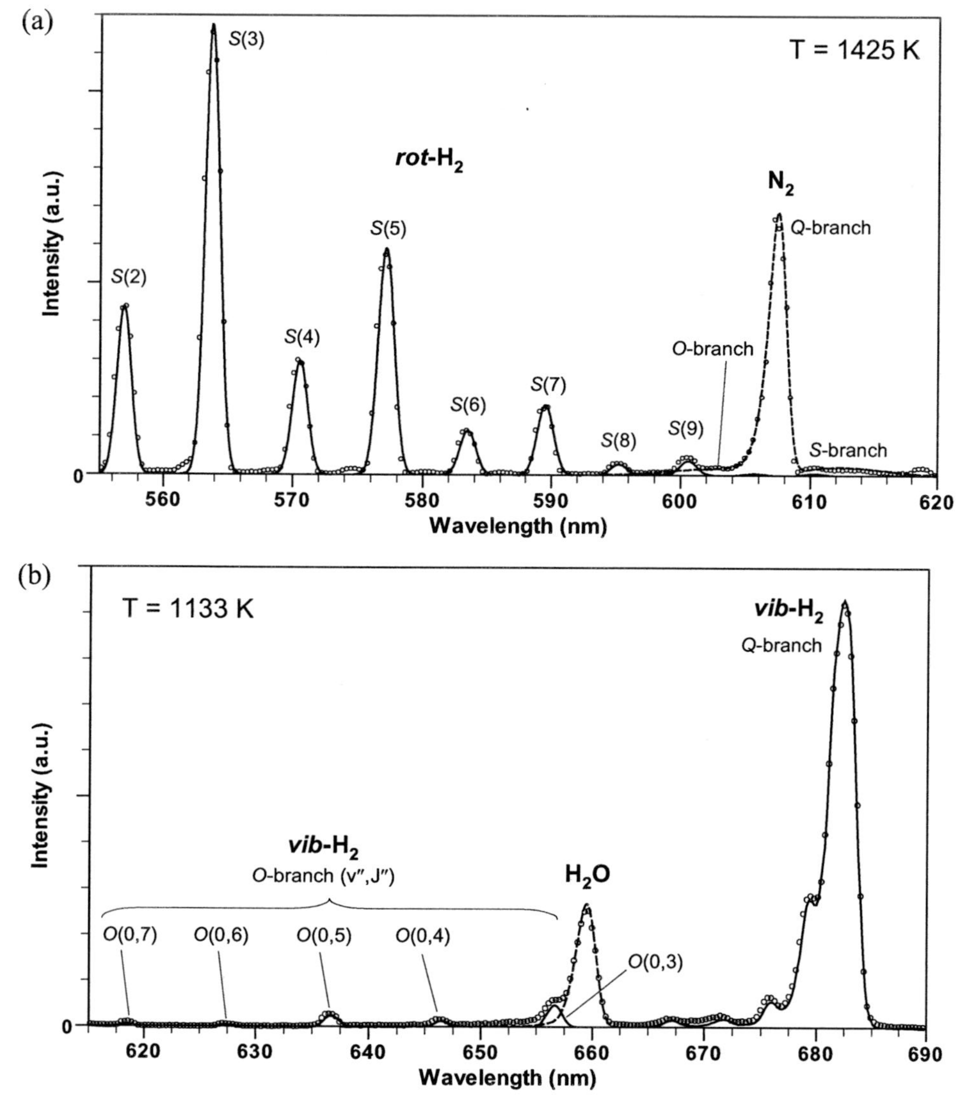

Figure 5: Comparison between measured and simulated Raman spectra in fuel-rich 10-atm H_2-air flames. (a) N_2 and rotational H_2; (b) H_2O and vibrational H_2. The simulations were made at the measured temperatures (shown in the upper corners of the figures) with an instrumental resolution of 35 cm^{-1}, which corresponds to the spectral resolution of the spectrometer used in the measurement. The experimental data is represented by the *points* (200-shot averaged, 0.1 mm spectrometer slit). Two calculated lines (*solid* and *dashed lines*) are shown separately but not superimposed.

The utility of accurate spectral simulations for predicting and analyzing spectral crosstalk in H_2-air flames is clearly evident. However, for the best accuracy, the empirically-based Raman calibration matrix formalism still has to be used when more complicated species such as H_2O and CO_2 are considered due to the difficulty of accurately simulating polyatomics at high temperatures.

4 Combustion Apparatus

4.1 High-Pressure Calibration Burner Concept

In order to obtain the Raman calibration matrix for high-pressure flame diagnostics, a high-pressure burner that can serve as a precision calibration standard would be of great use. Such a burner should provide a stable and repeatable source of combustion products over a wide operational range. In the past, water-cooled sintered-metal porous plate flat-flame burners (so-called McKenna burners) have been used for generating high-pressure laminar hydrocarbon flames [5,9,24,25]. This type of burner was reported to work well in providing stabilized high-pressure flames at pressures up to 60 atm for hydrocarbon flames, and is useful for optical calibrations at reduced flame temperatures [9]. This type of burner, however, suffers from flame temperatures that decrease with increasing pressures due to a reduction in burner surface-to-flame distance (thus, higher heat loss factor) caused by the higher pressures [5]. Furthermore, the low flow velocities through the sintered plate produce a flame that is susceptible to the increased effects of buoyancy at high pressures. As for hydrogen flames, sintered bronze-matrix burner have been proposed to provide a 15-atm H_2-air laminar premixed flame for the calibration of CARS diagnostics, albeit with the risk of thermal meltdown even in fuel-rich ($> \phi = 3.5$) conditions [17]. Despite these drawbacks, the McKenna burner has been a popular burner for high pressure studies. However, if higher flame temperatures with higher pressure (higher heat capacity) are required, then a different type of burner must be used.

Another common system is the so-called Hencken burner operated in a non-premixed mode with multiple fuel tubes is generally used for optical calibration in H_2 flames [9]. This type of burner works well at atmospheric pressure conditions and is able to provide near adiabatic flames over a wide range of equivalence ratios although flow-field uniformity and the uncertainty in flow rates are still an issue [9]. The Hencken burner does not likely survive a high pressure hydrogen-air flame due to thermal meltdown, and is hardly operated in a premixed mode with hydrogen due to flash-back.

The burner capable of satisfying the above requirements without flash-back is indeed challenging. The design of a burner that we employed as an alternative, is similar to the Hencken burner concept but is able to run in a premixed mode. It is a versatile high-pressure burner that can operate over a wide range of temperature, pressures, equivalence ratios, and with different fuels and oxidizers.

4.2 Calibration Burner

The burner is based on a premixed concept that utilizes back-side impingement cooling and fuel-air mixing in a "micro" chamber at high velocities to overcome these challenges. Figure 6 shows a schematic of the high-pressure premixed burner design. Basically, the burner consists of an array of closely-spaced premixed fuel/oxidizer jets that issue from the burner face and quickly combine downstream of the face to form a uniform flow pattern. The fuel/oxidizer premixing occurs just upstream of the burner face in a thin cavity just before the flows exit the burner face. An array of 7 x 7 fuel tubes and 8 x 8 oxidizer tubes approximately 1 mm in diameter in a staggered arrangement provides the flows to the thin premixing cavity.

The oxidizer flow first impinges on the backside of the burner face and then turn 90 degrees to effect the mixing with the fuel jets. The mixing is also enhanced by the large shear forces induced in the sudden directional change of the oxidizer flows. The premixed fuel/oxidizer jets then exit through an array of 7 x 7 premix holes that are coaxially aligned with the fuel-tube supply array. These premixed jets have an approximately 1:5 expansion ratio to slow the jet velocities down from the high velocities through the nozzle throat required to prevent flash-back. The premixed jet array is approximately 18 x 18 mm in size. These 49 jets then continue to mix and combine to become a single uniform flame downstream of the burner face. By mixing the fuel and oxidizer just upstream of the burner face in a region of high gas velocity, flashback into the thin premixing chamber is avoided if the bulk velocity is kept above the laminar flame speed of the fuel-air mixture (which can be as high as 10 m/s for high pressure H_2-air flames).

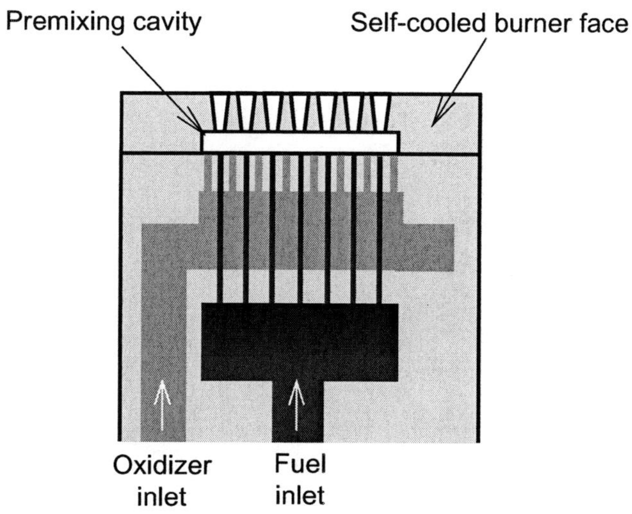

Figure 6: Schematic of multiple-tube burner nozzle to provide a premixed flame at high pressures. The fuel and oxidizer mix rapidly in a small cavity prior to exiting the burner face where they burn in a premixed mode.

By impingement cooling the burner face with the oxidizer flow, the burner can be operated without any water cooling, and this helps to keep the flame temperatures high. The burner produces a region of combustion products directly downstream of a premixed flame with a uniform flow pattern over an approximate 5 x 5 mm zone. The copper burner face also is cooled to a certain degree by convection to the gases in the mixing holes as well as the direct conduction to the main burner body. At equivalence ratios below 0.6 the burner can be operated indefinitely; at higher equivalence ratios, the burner can be operated for typically 2 minutes before the burner face temperatures (sensed by thermocouples) get too hot (920 K). However, this is not a problem as only about 30 seconds of run time is needed to acquire the Raman scattering data. Note that no preheat air-fuel mixture was fed to the burner even for a lean flame in this experiment.

4.3 High-Pressure Rig Facility

Figure 7 shows a schematic of the high-pressure burner rig and gas flow system. Figure 8 shows a photograph of the facility with the laser pulse-stretcher in use. In Fig. 7, ambient temperature air from a facility compressor (not shown) provided 30.4 bar air for cooling. The cooling air was introduced at the bottom of the rig for liner cooling (0.11 kg/s max.), and also at the upper side as "quenching" of hot exhaust gas (0.09 kg/s max.). These airflows were controlled by remotely operated regulators using calibrated venturi flow meters. Approximately 10% or less of the total cooling flow rate of the facility air was used as a purge-air for the optical windows during experiments to prevent water vapor condensation. The flow rate of the purge-air was set via a sonic venturi. The pressure of the burner rig was maintained via the back-pressure valve mounted at the top of the chamber, which is remotely operated by an auto-feedback process controller. The rig pressure was varied from 1.2 atm to 30 atm in the present work although the burner rig can operate up to 60 atm. The pressure was stabilized within ± 1 % accuracy for each condition. For optical access, the burner rig has four 44 mm thick ultraviolet grade fused silica windows with a clear aperture of 85 mm. A burst disk (65 atm) was placed between the pressure chamber and exhaust pipe in case of a hazardous pressure buildup.

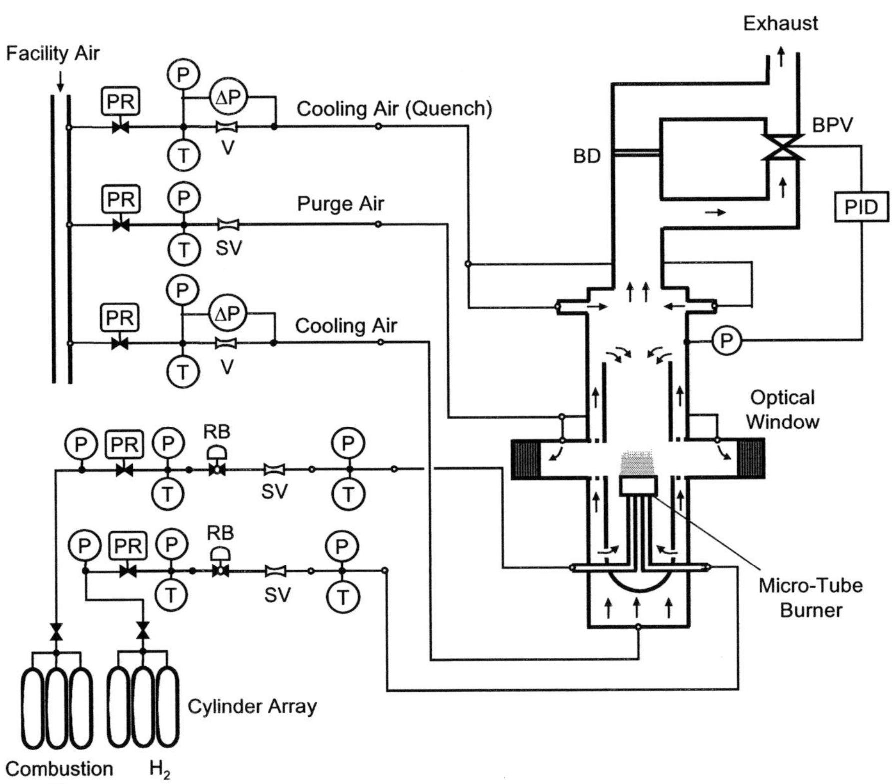

Figure 7: Schematic of high-pressure gaseous burner facility flow systems. Legends — P: pressure transducer; T: thermocouple; PR: remotely operated regulator; RB: remotely operated ball valve; V: venturi; SV: sonic venturi; BPV: back-pressure valve; PID: process controller; BD: burst disk.

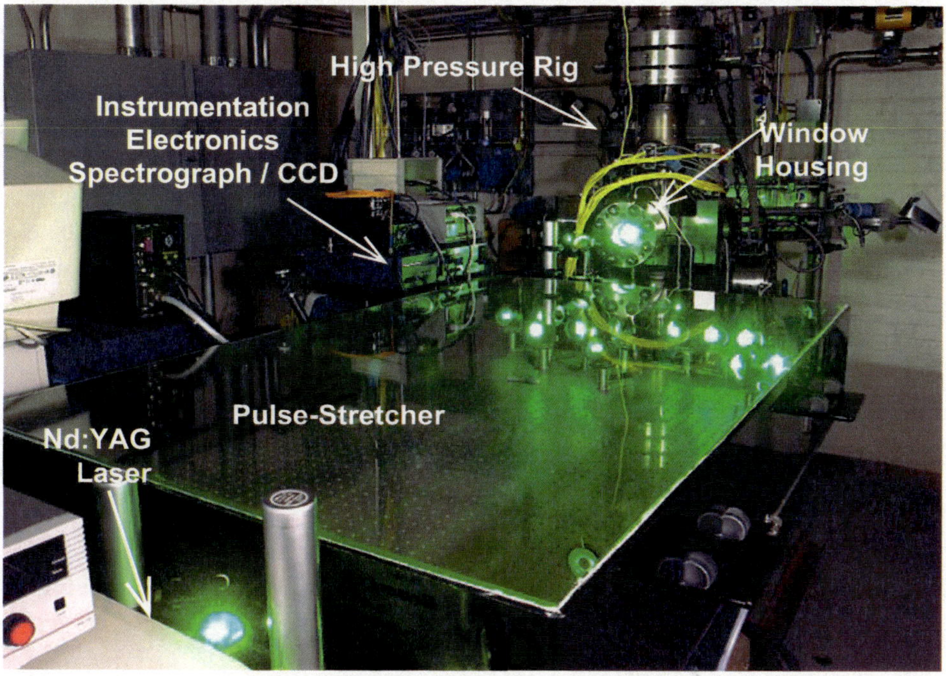

Figure 8: A photograph of the high-pressure burner rig and the instrumentation. The system resides in a test cell separated from the control room by a blast-proof wall for safety. The entire apparatus is remotely controlled using a combination of closed-circuit television cameras and computer-controlled automation. The laser pulse-stretcher (inside enclosure) can be clearly seen in the foreground. The stretched-pulse is directed into the window housing using a mirror periscope.

The premixed burner was mounted inside a high-temperature liner casing inside the rig. Note that the pressure inside the casing was the same as for the rig. A small amount of cooling air was fed into the casing to avoid building up of combustion products around the burner. The oxidizer air and the fuel H_2 were provided by 12-pack cylinder arrays at 150 atm pressure. The flow rates of the air and fuel were precisely controlled within ± 1% accuracy using a bank of three critical-flow venturis (not shown in Fig. 7 for simplicity), fitted with pressure transducers and computer operated precision pressure regulators and valves; the bank of flow meters allows a wide dynamic range of flow rates to be metered with high accuracy. The sonic flow venturis also serve to limit the maximum flow of the H_2 in case of a downstream drop in rig pressure (as in the case of a burst disc rupture event). The use of critical flow venturi meters ensures that the mass flow rate stays constant regardless of pressure fluctuations in the rig.

A retractable, high-energy surface-discharge aircraft style igniter system is used to ignite the burner. The igniter is inserted to a radial location above the burner using a geared stepper motor, and a 10-spark sequence is then initiated prior to opening the H_2 flow valves to introduce the fuel flow.

Our first demonstration using this high-pressure burner [26] indicated that a condition close to chemical equilibrium was reached over the range of equivalence ratios at the measurement location, although at temperatures below the adiabatic temperature by more than 100 K, possibly due to radiation and conduction back to the burner face. Direct observations of flame luminescence showed that the burner generated multiple small hydrogen premixed flames just above the burner face. The burned gases combined downstream to deliver a homogeneous zone of combustion products. The luminous zone appeared different in size and shape depending on the equivalence ratio. The flow rate of the gases was not fixed in the current experiment, so the mean flow velocity at the burner port was different for each equivalence ratio. Also our preliminary analysis of the burner element using a comprehensive computational fluid dynamics code with finite-rate H_2-air chemistry indicates that the primary flame zone may be located as low as 1 mm above the burner face [27]. Since the measurements are performed at a point (12 mm above the burner face) located well down-stream of the predicted and observed flame zone, we use our high-pressure premixed calibration burner as a source of hot combustion products that is closely approximated by chemical equilibrium at a known temperature. The temperature of the chemical equilibrium condition is determined unambiguously with spectroscopic temperature measurements (described in the following sections) to avoid inaccuracies that would result from assuming an adiabatic equilibrium temperature condition.

5 Spontaneous Raman Scattering Diagnostics

5.1 Laser Excitation System

Figure 9 shows a schematic of the Raman scattering measurement apparatus. For Raman excitation an injection seeded, Q-switched Nd:YAG laser provided 500-mJ, 532-nm pulses (8.4 ns full-width-half-maximum, or FWHM) at a 10 Hz repetition rate. Although not needed for Raman scattering in general, the injection seeding feature produces a better pulse-to-pulse energy stability with less timing-jitter. The injection-seeded laser line width as stated by the manufacturer is 0.003 cm^{-1}. The pulses were then temporally "stretched" to about 75 ns (FWHM) using a pulse-stretcher [7] based on multiple ring-cavities. The pulse-stretcher has been described in detail elsewhere [28] and consists of sixteen mirrors and three beam splitter-combiners arranged in a sequential 3-cavity setup to produce an output pulse with approximately 10% of the peak power of the input pulse, while retaining 83% energy throughput efficiency. Figure 10 shows the performance of the laser pulse-stretching system. The reduced peak power prevented laser-induced spark/plasma formation at the probe volume due to the breakdown of the gas dielectric properties under the intense electric field that would otherwise be generated by the tightly focused Q-switched laser pulse. In addition, the pulse-stretcher reduced the possibility of damage to the windows. The dielectric breakdown of air under intense laser power is problematic because air has negative pressure dependence for the breakdown threshold versus laser power. Even with a pulse-stretcher, spark or plasma formation was still a problem if dust particles or particulates found their way into the laser probe volume. A 750-mm focal lens focused the output of the pulse-stretcher to the probe volume waist diameter of approximately 1 mm. To effectively double the probe volume energy, the laser beam was back-propagated through the probe volume at a small crossing angle using a 400-mm collimating lens and a right-angle prism. The probe volume was formed at 25.4 mm above the burner nozzle surface on the center axis located downstream of the actual flame zone.

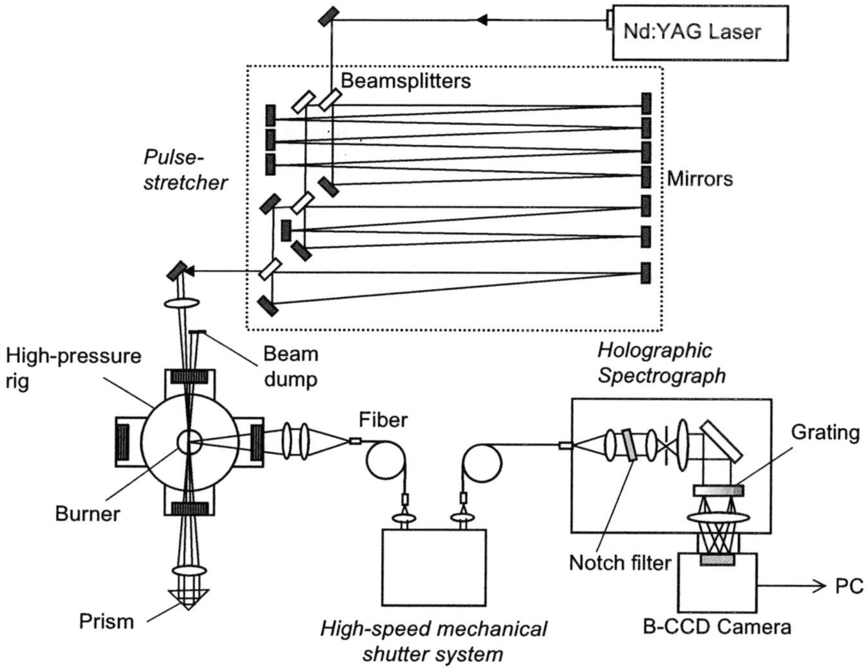

Figure 9: Schematic of high-performance linear Raman scattering apparatus. There are four main components to the system: Nd;YAG laser, pulse-stretcher, high-speed shutter, and spectrograph/backside-illuminated CCD camera.

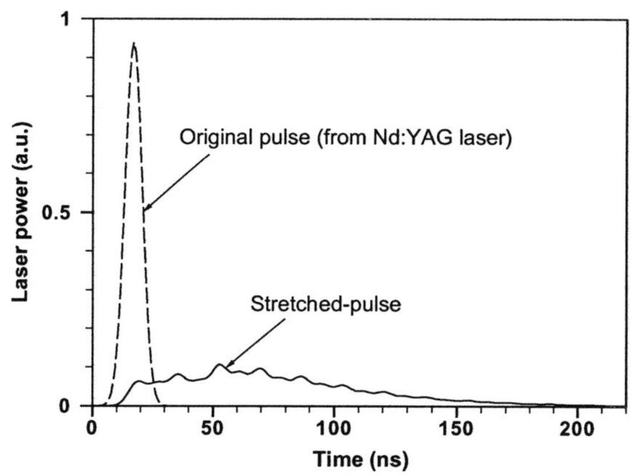

Figure 10: Measured (256-shot average) temporal profile of excitation laser pulse demonstrating the performance of the pulse-stretcher. Using an optimized 3-cavity pulse-stretcher, an original laser pulse is stretched out in time and reduced to 10% of original peak power with an 83% throughput efficiency.

5.2 Signal Detection

The vertically polarized Raman scattering signal was collected at a 90-degree angle relative to the beam-propagation direction using a standard 35-mm format camera lens (85 mm f/1.4). Note the actual solid angle of the current collection optics is f/5.33 (400 mm focal length with 75 mm in diameter) which is dictated by the maximum window aperture. The collected light was then focused onto a silica optical fiber (400 μm core diameter) that is sheathed in a flexible metal conduit. The diameter of the collection fiber core and the lens magnification defines a probed region of approximately 1.9 mm in length along the laser beam waist. An electro-mechanical high-speed shutter [29] synchronized to the 10-Hz laser pulses optically gated the light emerging from the fiber. Although capable of operating at gate widths as short as 9 μs, the shutter system in this study provided a 24 μs gate time (FWHM) with 0.4 μs jitter through a 12 x 0.8 mm clear aperture with an overall 55% optical throughput including fiber transmission losses. The high-speed shutter system reduced the effects of background light interferences by a factor of about 1000 compared to a conventional electro-mechanical leaf shutter. The optically gated light from the shutter was fiber-optically directed to the spectrograph using a custom made fiber bundle consisting of 7 x 100 μm core silica optical fibers arranged in a circular pattern to match the 400 μm diameter core of the input fiber into the shutter. The output of the fiber bundle was arranged in a 7 x 1 linear array and aligned to match the 100 μm wide input slit of the spectrograph. The Raman signal was dispersed with an axially transmissive visible-wavelength spectrograph (f/1.8) fitted with a holographic notch filter (optical density > 6.0, spectral edgewidth < 4.0 nm) to attenuate the Rayleigh scattering component of the signal. A volume holographic transmission grating dispersed the signal into a spectrum which was detected by a non-intensified, liquid-nitrogen cooled (-120 °C), back-illuminated charge-coupled-device, or CCD camera (1340 x 100 array, 20 x 20 μm pixels).

The back-illuminated CCD camera and electro-mechanical shutter combination were selected over an intensified CCD camera because it provides better than 90% detector quantum efficiency (QE) with a 200,000 photoelectron dynamic range that was digitized at 16-bits of resolution. The simultaneous requirements of high detector QE and high dynamic range were critical for accurately capturing the full range of weak and strong Raman signals and its background interferences as they appear in high-pressure flames. The electronic exposure of the CCD was set at 5 ms (but the actual time exposure limited by the shutter was 24 μs), and the data was binned on-chip over 50 vertical pixels and averaged over hundreds of shots on chip to increase signal-to-noise ratio. The spectrograph was calibrated in wavelength using a primary standard (Ne and Hg/Ar discharge lamps). The spectral resolution of the standard holographic grating was determined from the calibration lamp lines (which are much narrower than the spectrograph resolution) over the spectral range of interest. The spectral resolution of the standard grating was approximately 35 cm^{-1}/pixel in a visible wavelength range (540 nm to 700 nm) using a 100 μm wide slit. We calibrated the spectral intensity response of the CCD detector and the entire collection optics train into an absolute value using a calibrated tungsten halogen lamp. This is essential for the accurate determinations of vibrational and rotational population distributions used for temperature measurements.

6 Application and Analysis

Laser Raman scattering measurements in combustion environments often suffer from background interferences such as flame luminosity and/or laser induced fluorescence of radicals. Removing these interferences is the key to a quantitative analysis. Even though Raman scattering was measured using a pulsed laser with a high-speed shutter system to minimize spontaneous flame emissions and luminosity, a certain level of background was observed in the Raman scattering spectra. To remove this, the flame luminosity was measured without laser excitation and was subtracted from the original signal. This technique is effective for steady flows which we have in this study. However, in non-steady flows, single-shot measurements can utilize background subtraction if the background can be represented as a smooth curve that can be numerically fitted based on intensity values in between Raman features of interest. Figure 11 shows the averaged flame-luminescence-subtracted Raman spectra of H_2–air combustion gases in lean to rich conditions at 10 atm. Relative intensities of *rot*-H_2, and *vib*-H_2 increased significantly when the flame became richer while the vibrational Raman spectra of N_2, and H_2O were observed in all conditions. Obviously, the O_2 Raman spectrum was seen only in lean flames. Note that the spectra acquired at higher pressures (20, and 30 atm), but not shown here, had the same tendency but they had higher signal-to-noise ratio because of higher density of molecules, thus higher Raman scattering intensities.

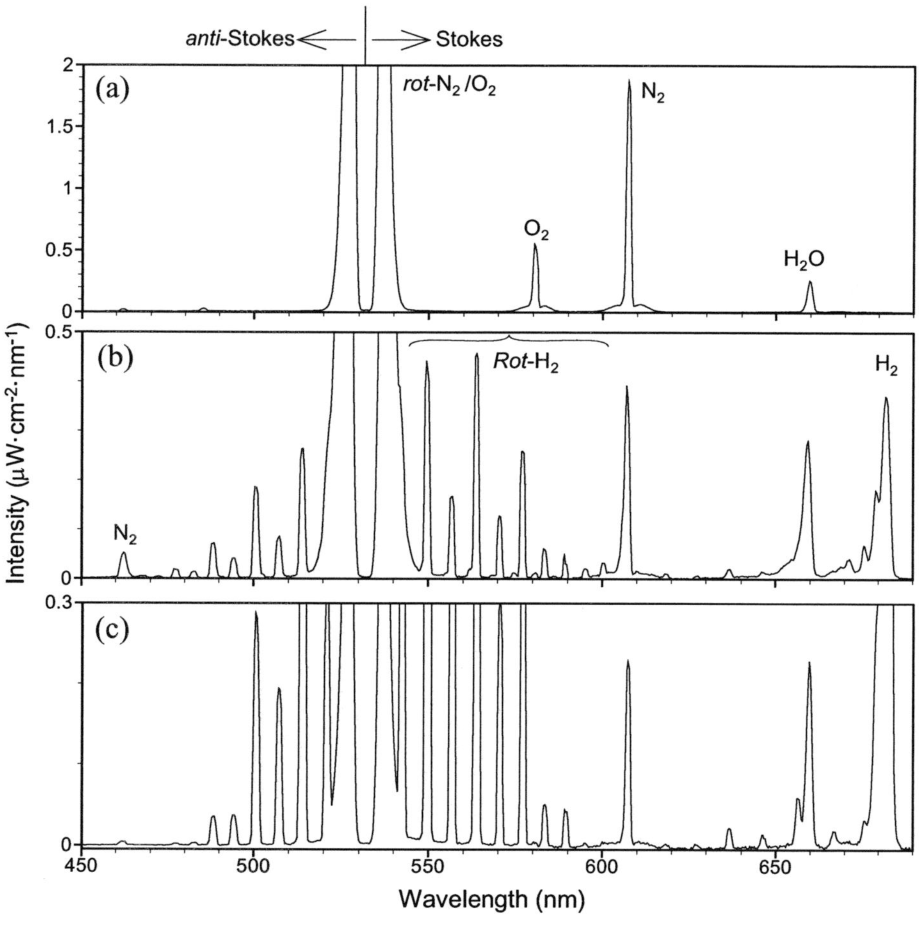

Figure 11: Measured spectra of spontaneous Raman scattering from high-temperature gaseous molecules in 10-atm premixed hydrogen air flames at different equivalence ratios (200-shot averaged): (a) $\phi = 0.15$, (b) $\phi = 1.4$, (c) $\phi = 5.1$.

6.1 Temperature Determination

There are a several ways to determine the flame temperature from SRS spectra. The Raman thermometry using the ratio of the Stokes to anti-Stokes intensity of the vibrational N_2 *Q*-branch (S-AS method) [30] or spectral line-shape fitting to the N_2 *Q*-branch [31] are well known. The H_2 Raman thermometry [32,33] using the line intensity distribution of the *rot*-H_2 is a good alternative in H_2-rich flames since the *rot*-H_2 has a relatively strong cross-section, and each rotational spectrum is discriminated very well due to relatively large rotational level spacing. Intensity-integration calculations can also be effectively implemented to increase the signal-to-noise ratio. For the present work, temperature at each equivalence ratio was calculated by either the rotational H_2 thermometry or the N_2 S-AS method.

The H_2 Raman thermometry uses the measured relative intensities of pure rotational Raman transitions, which is governed by the relation [33]:

$$\ln\left[f(J)S(J)\right] = -hcF(J)/kT + \text{const}, \tag{11}$$

where $S(J)$ is the rotational Raman intensity (integrated), $F(J)$ [cm^{-1}] is the rotational energy. The term $f(J)$ contains J-dependent terms for the Raman intensities, including correction of the rotation-vibration interactions which are significant in hydrogen [32,33],

$$f(J) = \frac{2J+3}{g_s(J+1)(J+2)} \cdot \frac{1}{1+A(J^2+3J+3)} \left(\frac{\nu_0}{\nu_k(J)}\right)^4 \quad : S\text{-branch}, \tag{12}$$

$$f(J) = \frac{2J-1}{g_sJ(J-1)} \cdot \frac{1}{1+A(J^2-J+1)} \left(\frac{\nu_0}{\nu_k(J)}\right)^4 \quad : O\text{-branch} \tag{13}$$

where g_s is the nuclear spin degeneracy factor (g_s = 1 for even J; g_s = 3 for odd J); A is the molecule-dependent spectroscopic constant for H_2, and equals 4.02×10^{-3} [33]; $\nu_k(J)$ here is the experimentally observed Raman line frequency.

The N_2 S-AS method is based on the formula that can be expressed as:

$$\frac{I_{AS}}{I_S} = \left(\frac{\nu_0+\nu_k}{\nu_0-\nu_k}\right)^4 \exp\left(-\frac{hc\nu_k}{kT}\right), \tag{14}$$

where I_S and I_{AS} are the Stokes and anti-Stokes intensities (integrated) of the vibrational N_2 Raman spectra; and ν_k here is the molecular vibrational frequency particularly for Strokes *Q*-branch of N_2 (i.e., Raman shift).

Figure 12 shows the temperatures measured via the above-mentioned two different spectroscopic methods for 10-atm H_2–air flames over a range of equivalence ratios from 0.15 to 5.1, which were produced by our calibration burner. The measurement uncertainty of H_2 rotational temperature is less than 2%. As observed in Fig. 12, temperatures based on the different techniques agree well on the fuel-rich side, which indicates the higher accuracy of the measurements. Since the uncertainty of the *rot*-H_2 temperature increases toward the stoichiometric condition due to the lower H_2 signal intensity, the discrepancy between the vibrational (N_2) and the rotational (H_2) temperatures at ϕ around 1.5 to 2.5 do not necessarily imply a non-thermal equilibrium between two molecules. The non-thermal equilibrium condition, generally, is not likely in high-pressure combustion due to the small collisional time scales which provide rapid thermalization [1,34]. Based on the signal-to-noise ratio, the N_2-based temperature is more accurate for flames with $\phi < 1.5$, whereas H_2-based temperature is more accurate for flames with $\phi > 1.5$. The selected temperatures indicated in Fig. 12 are then used for a further analysis on the Raman calibration matrix in the below section.

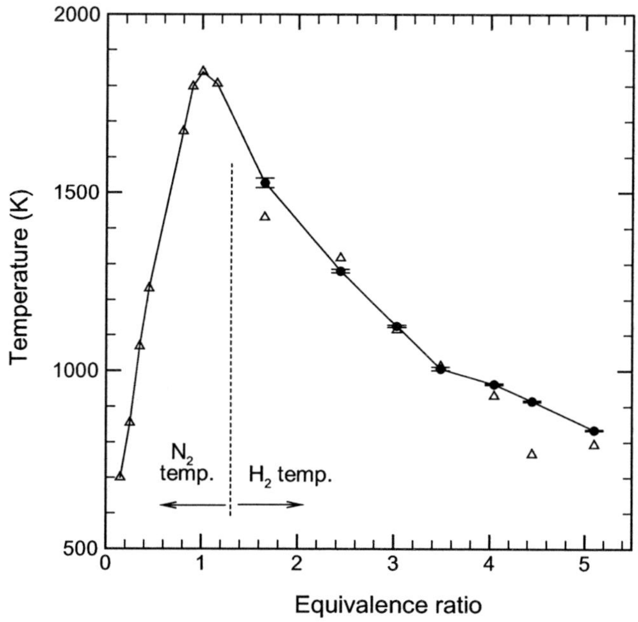

Figure 12: Spectroscopically determined temperature derived from Raman spectral data (such shown in Fig. 5) measured in the 10-atm premixed H_2–air flames. *Triangle* : temperature via the vibrational N_2 Raman spectra (ratio of anti-Stokes band to Stokes band). *Circle*: temperature via the population distribution of the pure rotational H_2 Raman spectra (for fuel-rich region only), with measurement uncertainties (*error bars*). Appropriate thermodynamic temperatures are selected (*Line*) out of two values depending on the region where each has the best accuracy.

6.2 Calibration Matrix: Quantitative Multiscalar Measurement

The Raman calibration matrix provides a means to relate the signal to the number density while simultaneously compensating for the spectral crosstalk. If one were to simply process the Raman data by merely using the raw but calibrated intensity for determining number densities of combustion products accounting for the crosstalk, considerable errors between the measured and true values would result. Thus, it is necessary to apply the crosstalk correction method described in this Chapter to obtain a more precise value. The Raman signal can be calibrated, as described in Section 2, by the calibration matrix formalism which includes the direct-contribution (on-diagonal elements) and crosstalk-contribution terms (off-diagonal elements). The calibration process determines the *elements* of the Raman calibration matrix as a function of temperature. Since these matrix elements have temperature dependence, the first step is to empirically find the appropriate functional forms for the matrix elements by analyzing the correlation between the amount of crosstalk and temperature for each species of interest.

Our analysis reveals that the Raman calibration matrix for H_2-air combustion can be formulated as follows:

$$\boldsymbol{k}_{i,j} = \begin{pmatrix} f(T)_{O2} & 0 & 0 & 0 \\ 0 & f(T)_{N2} & f(T)_{H2 \to N2} & 0 \\ 0 & 0 & f(T)_{H2} & 0 \\ 0 & 0 & f(T)_{H2 \to H2O} & f(T)_{H2O} \end{pmatrix}, \tag{15}$$

where $f(T)$ represents a temperature-dependent function. The functions for the on-diagonal elements, each of which represents one species, is determined by fitting an analytical function that best describes the density-normalized Raman responses in the following expression:

$$f(T^{\text{ref}})_i = \frac{1}{\Lambda E_{\text{L}}}\left(\frac{S_i^{\text{ref}}}{N_i^{\text{ref}}}\right), \tag{16}$$

whose units are [$\mu W \cdot cm^{-2} \cdot nm^{-1}$/molecules$\cdot cm^{-3}$] (irradiance/number density) in the present work. Here T^{ref} is the reference temperature measured for the calibration burner via Raman thermometry. For example, the N_2 on-diagonal element, i.e. $f(T)_{N2}$, can be calculated from Eq. (16) using the SRS data measured in the fuel-*lean* H_2–air calibration flames ($\phi < 1.0$) that should receive no crosstalk effect from the other species at those equivalence ratios.

The matrix $k_{i,j}$ also includes the off-diagonal elements to correct the crosstalk of the rotational hydrogen spectrum onto nitrogen (indicated by the subscript $H_2 \to N_2$); or the crosstalk of the vibrational hydrogen spectrum onto water vapor (indicated by the subscript $H_2 \to H_2O$). By expanding Eq. (15) and (16), the off-diagonal elements are given by the expression:

$$f(T^{\text{ref}})_{j \to i} = \frac{1}{\Lambda E_{\text{L}}}\left(\frac{S_i^{\text{ref}} - f(T^{\text{ref}})_i N_i^{\text{ref}}}{N_j^{\text{ref}}}\right) \tag{17}$$

by assuming that the residual after subtracting the self or direct-contribution calculated from Eq. (16), is the crosstalk contribution from a nearby spectral line. For example, the N_2 off-diagonal element, i.e. $f(T)_{H2 \to N2}$, can be calculated from Eq. (17) using the SRS data measured in the fuel-*rich* H_2–air calibration flames ($\phi > 1.0$), where the values should be a summation of the N_2 self-contribution *and* a certain amount of crosstalk from the rotational H_2 spectrum from the fuel-rich side.

Applying Eqs. (16) and (17) to the spectral data taken in the experiments described in Section 4 & 5, we derived the temperature-dependent functions of the Raman calibration matrix elements for the major species of the 10-atm H_2–air flame. The results are shown in Fig. 13. Here we used the SRS spectra (partially shown in Fig. 11) measured in our calibration burner over a range of equivalence ratios from 0.15 to 5.1 for the reference intensity S_i^{ref}; and the measured temperature shown in Fig. 12 as the reference temperature T^{ref}. The reference number density N_i^{ref} was given by the chemical equilibrium numbers calculated by the CEA code with the assigned temperature (T^{ref}) and the pressure of 10 atm. Figure 13 confirms that the nonlinear behavior of the Raman calibration matrix elements over the temperature range.

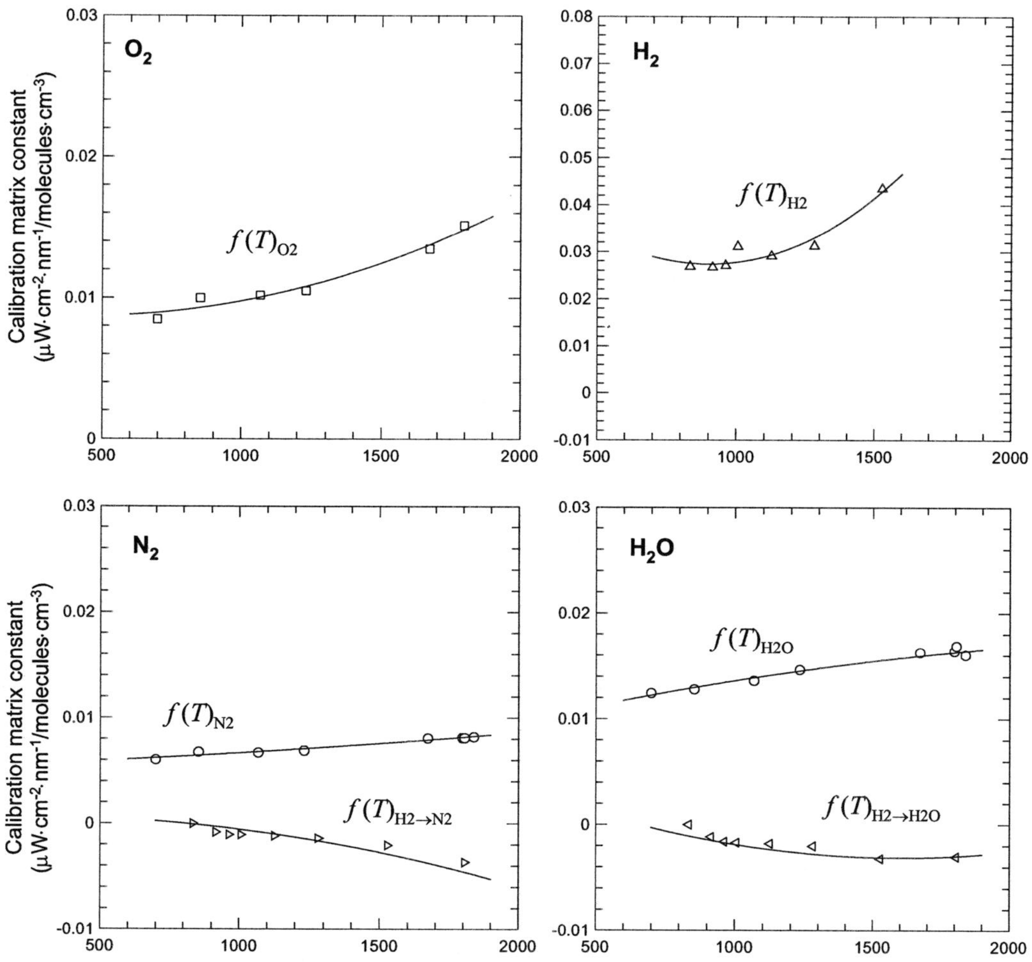

Figure 13: Empirically determined Raman calibration matrix elements. These curves as a function of temperature directly provide the elements shown in Eq. (15) in the text. The functions were determined by the polynomial fitting (¡Â 3rd order) to the processed data.

Finally, the *inverse* matrix $k_{i,j}^{-1}$ was calculated from the original matrix $k_{i,j}$ consisting of the polynomial functions determined in Fig. 13 and substituted into Eq. (2) with inputs of the reference temperature vector $\overrightarrow{\boldsymbol{T}^{\mathrm{ref}}}$ and the reference intensity vector $\overrightarrow{\boldsymbol{S}_i^{\mathrm{ref}}}$. Figure 14 shows the resulting crosstalk-corrected quantitative number densities of N_2, H_2, O_2, and H_2O. The reference temperature is also plotted in Fig. 14 for a comparison. The multiscalar data determined here clearly demonstrates the quality of the Raman calibration procedure in the present work. The excellent self-consistent agreement between the data points and the lines which represent the chemical equilibrium predicted number density indicates that the strategy described in this chapter has great promise for future studies in high pressure H_2-air flames.

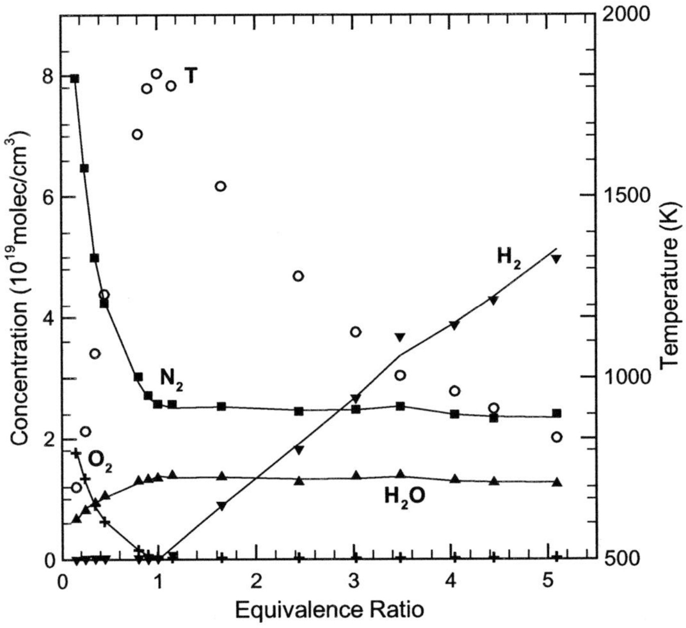

Figure 14: Calibrated and crosstalk-corrected number densities of species O_2, N_2, H_2, and H_2O for the experimental data obtained in 10-atm hydrogen flames. *Points* represent the resulted data. *Lines* represent the values obtained from a chemical equilibrium assumption.

7 Conclusion

In this chapter, we presented the theoretical and experimental background, and practical strategy for making quantitative multiscalar measurements in high pressure H_2-air combustion systems. The foundations of this technique can also be extended to multiscalar measurements in hydrocarbon flames that involve CO, CO_2 and hydrocarbons (such as CH_4) in both high pressure and atmospheric pressure flames. At the core of this strategy, the Raman calibration matrix formalism which includes both direct-contribution calibration terms and crosstalk-contribution terms was described with the aid of a Raman spectroscopic model and simulation. The Raman calibration matrix was realized with the aid of a novel premixed high-pressure calibration burner, an optically accessible high-pressure combustion facility, and a state-of-art spontaneous Raman scattering system. This collective apparatus was used to provide empirically based, but analytic formulations of the temperature dependent Raman calibration matrix elements for each species based on the assumption of chemical equilibrium at the measured temperature in the post-flame products of the calibration burner. The resulting Raman calibration matrix $k_{i,j}$ (and thus the inverse matrix $k_{i,j}^{-1}$) for the species H_2, O_2, N_2, and H_2O was then successfully determined from series of data measured in 10-atm H_2-air flames over a wide range of equivalence ratios from 0.15 to 5.0. From the calibration matrix, the number densities of the major species were calculated from the Raman signal vector and showed obvious and excellent agreement with the values calculated from chemical equilibrium at the measured temperatures.

High SRS signal-to-noise ratios were enhanced by the high pressure conditions and helped to yield highly accurate (< 2%) temperature measurements via the rotational H_2 and vibrational N_2 spectra. These accurate temperatures contributed greatly to the quality of the calibration matrix values. At pressures up to 30 atm, we found that the calibration matrix $k_{i,j}$ does not require the inclusion of a pressure-dependent term at the low (but practical) spectral resolution of the optical setup, mainly because the pressure broadening (or narrowing, if any) effect is minor compared with the instrumental function and the band factor (approx. 1.5 nm).

The matrix $k_{i,j}$ determined in this study provides a transferable SRS calibration standard for use with other experimental setups to probe H_2-air flames since the signal response was absolutely calibrated in wavelength (nm) and intensity (radiance) using traceable calibration lamp standards. In order to use this calibration matrix with other systems (at pressures up to 30 atm), all that is required is that the instrumental spectral resolution of the systems be matched, that the spectral response of the detector array be normalized using a traceable calibration standard, and that a '1-point' system response calibration be performed to account for the differences in optical collection efficiencies and variations in the probed volume size and geometry. Such a 1-point calibration can be conveniently performed by taking the spectrum of N_2 and O_2 in ambient air at a known pressure and temperature.

Our ongoing efforts include the extension of the Raman calibration matrix to more complex combustion systems that include hydrocarbons, as well as applications in practical combustion systems such as swirl-mixed non-premixed flames. Our preliminary test results in 10-atm H_2/CO–air premixed flames confirms that the matrix elements for the four major species considered in this chapter is consistent and quite reasonable for the generation of additional elements calibration matrix elements for CO and CO_2.

Acknowledgements

This work was supported principally by the Ultra-Efficient Engine Technologies (UEET) Project at NASA John H. Glenn Research Center at Lewis Field. The authors acknowledge Mr. Gregg Calhoun, Mr. William Thompson, Mr. Raymond Lotenero, and Mr. Gary Lorenz for their assistance in the construction and operation of the facilities.

Permissions

Pressure Fuel-Rich Hydrogen-Air Flames, 439-466, Copyright (2005), with permission from Elsevier.

Figure 2 Reprinted from J Quant Spectrosc Radiat Transf 94, Kojima, J., Nguyen, Q.-V., Quantitative Analysis on Spectral Interference of Spontaneous Raman Scattering in High-Pressure Fuel-Rich Hydrogen-Air Flames, 439-466, Copyright (2005), with permission from Elsevier.

Figure 5 Reprinted from J Quant Spectrosc Radiat Transf 94, Kojima, J., Nguyen, Q.-V., Quantitative Analysis on Spectral Interference of Spontaneous Raman Scattering in High-Pressure Fuel-Rich Hydrogen-Air Flames, 439-466, Copyright (2005), with permission from Elsevier.

Figure 7 Reprinted from Meas Sci Technol 15, Kojima, J., Nguyen, Q.-V., Measurement and simulation of spontaneous Raman scattering in high-pressure fuel-rich H_2–air flames, 565-580, Copyright (2004), with permission from Institute of Physics.

Figure 9 Reprinted from Meas Sci Technol 15, Kojima, J., Nguyen, Q.-V., Measurement and simulation of spontaneous Raman scattering in high-pressure fuel-rich H_2–air flames, 565-580, Copyright (2004), with permission from Institute of Physics.

Figure 10 Reprinted from Meas Sci Technol 15, Kojima, J., Nguyen, Q.-V., Measurement and simulation of spontaneous Raman scattering in high-pressure fuel-rich H_2–air flames, 565-580, Copyright (2004), with permission from Institute of Physics.

References

[1] Eckbreth, A. C., Laser Diagnostics for Combustion Temperature and Species, 2nd ed. Amsterdam, Gordon and Breach Publishers, 1996, 209-273.

[2] Hassel, E. P., Linow, S., Laser Diagnostics for Studies of Turbulent Combustion, *Meas Sci Technol* **11** (2000) R37-R57.

[3] Nooren, P. A., Versluis, M., van der Meer, T. H., Barlow, R. S., Frank, J. H., Raman-Rayleigh-LIF measurements of temperature and species concentrations in the Delft piloted turbulent jet diffusion flame, *Appl Phys B* **71** (2000) 95-111.

[4] Gu, Y, Zhou, Y, Tang, H, Rothe, EW, Reck, GP, Pressure Dependence of Vibrational Raman Scattering of Narrow-Band, 248-nm, Laser Light by H_2, N_2, O_2, CO_2, CH_4, C_2H_6, and C_3H_8 as High as 97 bar, *Appl Phys B* **71** (2000) 865-871 .

[5] Cheng, T. S., Yuan, T., Lu, C.-C., Chao, Y.-C., The Application of Spontaneous Vibration Raman Scattering for Temperature Measurements in High Pressure Laminar Flames, *Combust Sci Technol* **174** (2002) 111-128.

[6] Masri, A. R., Dibble, R. W., Barlow, R. S., The Structure of Turbulent Nonpremixed Flames Revealed by Raman-Rayleigh-LIF Measurements, *Prog Energy Combust Sci* **22** (1996) 307-362.

[7] Nguyen, Q.-V., Dibble, R. W., Carter, C. D., Fiechtner G. J., Barlow R. S., Raman-LIF Measurements of Temperature, Major Species, OH, and NO in a Methane-Air Bunsen Flame, *Combust Flame* **105** (1996) 499-510.

[8] Mansour, M. S., Chen, Y.-C., Line-Raman, Rayleigh, and Laser-Induced Predissociation Fluorescence Technique for Combustion with a Tunable KrF Excimer Laser, *Appl Opt* **35** (1996) 4252-4260.

[9] Barlow, R. S., Carter, C. D., Pitz, R. W., Multiscalar Diagnostics in Turbulent Flames. In: Applied Combustion Diagnostics, Ed. Kohse-Hoinghaus K., Jeffries J. B., Taylor & Francis, New York, 2002, 384-407.

[10] Kojima, J., Nguyen, Q.-V., Quantitative Analysis on Spectral Interference of Spontaneous Raman Scattering in High-Pressure Fuel-Rich Hydrogen-Air Flames, *J Quant Spectrosc Radiat Transf* **94** (2005) 439-466

[11] Dreizler, A., RAMSES: Software for Calculation of Raman Scattering Spectra, Technical University of Darmstadt, http://www.tu-darmstadt.de/fb/mb/ekt/main.html.

[12] Long, D. A., Raman Spectroscopy, McGraw-Hill, London, 1977, CRS1-16.

[13] Hertzberg, G., Molecular Spectra and Molecular Structure - I. Spectra of Diatomic Molecules: 2nd Ed., Krieger Publishing Company, Florida, 1950, 66-141.

[14] Huber, K. P., Hertzberg, G., Molecular Spectra and Molecular Structure - IV. Constants of Diatomic Molecules, Van Nostrand Reinhold Company, New York, 1979, 250.

[15] Avila, G., Fernández, J. M., Maté, B., Tejeda, G., Montero, S., Ro-vibrational Raman Cross Sections of Water Vapor in the OH Stretching Region. *J Mol Spectrosc* **196** (1999) 77-92.

[16] Hussong, J., Stricker, W., Bruet, X., Joubert, P., Bonammy, J., Robert, D., Michaut, X., Gabard, T., Berger, H., Hydrogen CARS Thermometry in H_2-N_2 Mixtures at High Pressure and Medium Temperatures: Influence of Linewidths Models, *Appl Phys B* **70** (2000) 447-454.

[17] Hussong, J., Lückerath, R., Stricker, W., Bruet, X., Joubert, P., Bonammy, J., Robert, D., Hydrogen CARS Thermometry in High-Pressure H_2-Air Flame. Test of H_2 Temperature Accuracy and Influence of Line Width by Comparison with N_2 CARS as Reference, *Appl Phys B* **73** (2001) 165-172.

[18] Rothman, L. S., Rinsland, C. P., Goldman, A., Massie. S. T., Edwards, D. P., Flaud, J.-M., Perrin, A., Camy-Peyret, C., Dana, V., Mandin, J.-Y., Schroeder, J., McCann, A., Gamache, R. R., Wattson, R. B., Yoshino, K., Chance, K. V., Jucks, K. W., Brown, L. R., Nemtchinov, V., Varanasi, P., The HITRAN Molecular Spectroscopic Database and HAWKS (HITRAN Atmospheric Workstation), 1996 ed. *J Quant Spectrosc Radiat Transfer* **60** (1998) 665-710.

[19] Schreier, F., The Voigt and Complex Error Function: A Comparison of Computational Methods, *J Quant Spectrosc Radiat Transfer* **48** (1992) 743-762.

[20] Varghese, P. L., Hanson, R. K., Collisional Narrowing Effects on Spectral Line Shapes Measured at High Resolution, *Appl Opt* **23** (1984) 2376-2385.

[21] Nagali, V., Ph.D. Dissertation, Mech. Eng. Dept., Stanford University, *TSD report* **117** (1998) 8-35.

[22] Gordon, S., McBride, B. J., *Computer Program for Calculation of Complex Chemical Equilibrium Compositions and Applications* – I. Analysis, NASA RP-1311, (1994). http://www.lerc.nasa.gov/WWW/CEAWeb/xHowCEA.html.

[23] Kojima, J., Nguyen, Q.-V., Measurement and simulation of spontaneous Raman scattering in high-pressure fuel-rich H_2–air flames, *Meas Sci Technol* **15** (2004) 565-580.

[24] Schulz, C., Sick, V., Meier, U. E., Heinze, J., Stricker, W., Quantification of NO A-X(0,2) Laser-Induced Fluorescence: Investigation of Calibration and Collisional Influences in High-Pressure Flames, *Appl Opt* **38** (1999) 1434-1443.

[25] Bessler, W. G., Schulz, C., Lee, T., Jeffries, J. B., Hanson, R. K., Strategies for Laser-Induced Fluorescence Detection of Nitric Oxide in High-Pressure Flames. I. A-X(0,0) Excitation, *Appl Opt* **41** (2002) 3547-3557.

[26] Kojima, J., Nguyen, Q.-V., *Development of a High-Pressure Gaseous Burner for Calibrating Optical Diagnostic Techniques*, NASA/TM 2003-212738 (2003). Available electronically at http://gltrs.grc.nasa.gov.

[27] Private-communication with Iannetti, A. of RTB, NASA Glenn Research Center, combustor simulations were performed using NASA's National Combustor Code, (2003).

[28] Kojima, J., Nguyen, Q.-V., Laser Pulse-Stretching Using Multiple Optical Ring-Cavities, *Appl Opt* **41** (2002) 6360-6370.

[29] Nguyen, Q. V., *High Speed Electromechanical Shutter for Imaging Spectrographs*, NASA Disclosure of Invention and New Technology, (2001).

[30] Rabenstein, F., Leipertz, A., Two-Dimensional Temperature Determination in the Exhaust Region of a Laminar Flat-Flame Burner with Linear Raman Scattering, *Appl Opt* **36** (1997) 6989-6996.

[31] Schoenung, S. M., Mitchell, R. E., Comparison of Raman and Thermocouple Temperature Measurements in Flames, *Combust Flame* **35** (1979) 207-211.

[32] Drake, M. C., Rosenblatt, G. M., Rotational Raman Scattering from Premixed and Diffusion Flames, *Combust Flame* **33** (1978) 179-196.

[33] Breiland, W. G., Coltrin, M. E., Ho, P., Comparisons between a Gas-Phase Model of Silane Chemical Vapor Deposition and Laser-Diagnostic Measurements, *J Appl Phys* **59** (1986) 3267-3273.

[34] Dring, I. S., Devonshire, R., Meads, J., Boysan, H. F., Greenhalgh, D. A., Observation of Non-Equilibrium Effects in Free-Convective Flows Using CARS: Recirculating Flow Relaxation Spectrometry, *Chem Phys Lett* **132** (1986) 283-290.

In: New Devel. in Combustion Research
Editor: William J. Carey, pp. 257-312
ISBN: 1-59454-826-9

Chapter 8

EXPERIMENTAL, THEORETICAL AND NUMERICAL INVESTIGATION OF SURFACE FOREST FIRE

Paul-Antoine Santoni*, Albert Simeoni and Frédéric Morandini
SPE – UMR CNRS 6134, Université de Corse, Campus Grossetti,
B.P. 52, 20250 Corte, France Fax : (33) 495 450 162

Abstract

The study of wildland fire is a growing field bringing important improvements to our understanding of heat and mass transfer and combustion. Indeed, forest fire is a complex phenomenon in which the levels of description cover a huge range, from the details of the kinetics of gaseous combustion and thermal degradation of fuels, up to the physical-chemistry characterization of the flame and the vegetation cover as a fuel. This paper addresses experiments, modelling and simulation of surface forest fires. Experimental techniques and results are first displayed to investigate fire behaviour. Then modelling tools and simulation results devoted to understand the mechanisms involved in such fires are presented. Different methods and measuring structures build in order to improve the knowledge on the fundamental physical mechanisms that control the propagation of wildland fire are detailed. The basic properties of flames encountered in forest fire are presented along with the major tools developed for measuring their characteristics at both laboratory and field scales. Indeed, although the modelling of forest fire spread has made considerable progress over the last ten years, however there remains a lack of reliable experimental results concerning fire front properties for the validation of the different existing models. Then, a theoretical model for the spreading of a surface fire is proposed incorporating the effects of wind and slope. The aim of our ongoing research is to propose a model for the spreading of a forest fire to be integrated in an operational management tool. To proceed, we carry out the reduction of a detailed multiphase set of equations that describe the finest mechanisms involved in forest fire. This reduction allows us to develop a simplified model by considering mainly the buoyancy effect induced by combustion in the flaming zone. The contributions of both convection and radiation impinging on the top of the fuel-bed ahead of the fire-front are included. Finally, model predictions are compared to laboratory-scale experiments. To this end, dedicated fires were carried out for horizontal spreading in still air across beds of pine needles, in order to measure flame and fire fronts properties, thermocouples and heat flux sensors. Experiments

* E-mail address: santoni@univ-corse.fr

conducted under wind and slope conditions are also considered for the validation of the model.

Nomenclature

Latin Symbols	
a	absortivity
C_p	specific heat at constant pressure
C_{xy}	cross-correlation function between signal x and y
d	diameter of hot junction of a thermocouple
d_l	thickness of the fuel layer
DL	digital level
e	total energy
$\vec{e}$	radiant intensity direction
$\vec{g}$	acceleration due to gravity
h	enthalpy
h^0	heat of formation
h_a	reduced convection coefficient
H	convection coefficient
K	reduced thermal diffusivity
K_T	McCaffrey's correlation coefficient for temperature
k_w	reduced advection coefficient
k_w^*	constant in the k_w expression
L	heat of vaporization
L_c	continuous flame lenth
L_m	maximum flame lenth
L_f	mean flame lenth
L^Ω	radiant intensity
$\dot{m}$	masse rate
m_{eq}	surface thermal mass of the medium equivalent to the litter
$\dot{M}$	mass flux
N	sample of size N
p	pressure
q	heat flux
Q	heat release rate
Q_{eq}	reduced combustion enthalpy
r	distance between two probes
R	reduced radiant flux
S	surface element
s	surface mass
s^2	variance
t	time

T	temperature
u	internal energy
$\vec{V}$	velocity
$\vec{V}_\infty$	maximal wind velocity
w	vertical component of the velocity
x, y, z	coordinates of a point in space
Y	mass fraction of a chemical species
Greek Symbols	
α	volume fraction
β	parameter defined in equation (15)
χ	drag forces coefficient
δ	parameter defined in equation (13)
ε	emissivity
ϕ_{sl}	slope angle
γ	gain coefficient
γ_c	combustion time constant
Γ	rate of production of a chemical species at the solid / gas interface
η	linearization parameter
φ	angle from the normal of the surface
κ	Mc Caffrey's correlation coefficient
λ	thermal conductivity
$\overline{\overline{\pi}}$	stress tensor in the gas
$\vec{\Pi}$	stresses at the solid / gas interface
θ	time shift
ρ	density
ρ_{xy}	correlation coefficient between signal x and y
σ_B	Stefan-Boltzmann constant
σ^S	surface to volume ratio
τ	time constant
$\dot{\omega}$	species mass rate of production
Ω	solid angle
ξ	linearization function
ψ	rate at which the radiant energy from the flame-front is absorbed
ΔH	heat of reaction
Δt	time interval
$\vec{\nabla}_s$	surface divergence vector
Subscripts and superscripts	
$1,2$	thermocouple 1, 2
a	ambient
ash	ash
$char$	char

conv	convection
fl	flame
g	gas
gk	interface exchange between gas and solid
H_2O	moisture
i	sample *i*
ig	ignition
j	chemical species *j*
k	solid phase
pyr	pyrolizate gases
rad,abs	absorbed by radiation
rad,lost	lost by radiation
s	surface component of a vector
surf	surface regression
th	thermocouple
w	wind
0	initial condition
Diacritical	
[ ]	source term
$\overline{\ }$	time average
$\overline{\overline{\ }}$	average property over the fuel depth
$\Vert\ \Vert$	euclidean value

Introduction

Devastating fires regularly damage all continents over the world. The ones that have occurred over the last few years (USA, China, Portugal, France...) involved some human losses and huge environmental and goods damages. In front of these catastrophic events, forest fire research is a young field that started in North America and Australia fifty years ago [1-9]. Researchers investigated both modelling and experimental areas. Those experiments are necessary in order to understand and model fire spread.

Numerous fire-spread experiments were conducted across beds of fuel at laboratory scale [9-19]. This kind of experiments allows conducting each run under controlled conditions and their instrumentation is quite easy to implement. These works were mainly devoted to study the effect of fuel moisture content, slope and wind velocity on the fire spread and flame properties. They were very useful to understand the main phenomena involved in fire spread and to obtain the first models. They remain of interest particularly to dissociate phenomena in order to understand them and to derive sub-models (like wood pyrolysis or soot production in flames). Our team carried out horizontal fire spreads in still air. We particularly developed a method to measure the temperature in the fire plume thanks to double-thermocouple probes [20] and we applied it to the characterization of a fire spread across pine needle fuel beds [21]. We also measured fire front properties using an infrared camera, thermocouples and heat flux sensors [22].

Nevertheless, due to scale effects, such experiments are not sufficient to understand and describe all the phenomena involved in wildland fires, like wind and vegetation effects. Some observations of large-scale forest fires were made [23,24], showing at best the fire shape history and associated meteorological and fuel conditions. Nevertheless, all of these data are not useful for modelling purpose and few experimental studies were conducted up to now because instrumentation in the field is very difficult. Cheney and Gould [25,26] studied fire growth in open grassland and related the rate of spread to the head of fire width. Theses pioneer works confirm that field experiments are the best alternative for the understanding of real fire behaviour. Some research teams [27-29] conducted field experiments recently in order to measure rates of spread, environmental conditions and the influence of parameters like moisture content or structure of vegetation. On the other hand, a few studies [30,31] began to include measurements of thermodynamic quantities in the aim of providing information on the forest fire threat to habitations. However, thermodynamics quantities allowing characterising fire spreads in field experiments from a physical viewpoint were scarcely investigated by researchers [28,30]. Furthermore, a common protocol was not yet defined by the scientific community to describe and characterise such fires apart from common behaviour descriptors such as: flame properties, flame residence time, Byram's fireline intensity [2] and fire front shape.

These considerations have provided the main motivation for our team to generate valuable data at field scale. To proceed, a fire-spread experiment was conducted in Corsica (France), across Mediterranean shrub. We elaborated a protocol and a set of structures in order to improve the knowledge on the physical mechanisms driving the spreading of wildland fires. The collected data were concerned with behaviour descriptors and measurements of thermodynamic quantities. They include the shape and the rate of spread of the fire front, the measurement of the amount of energy impinging ahead of it, the characterization of the vertical distribution of the temperature within the fire plume as well as the measurement of the wind velocity and direction. We reconstructed the fire behaviour and we provided its main properties thanks to the whole data set. Wind and vegetation effects on the fire spread were particularly described. We also investigated the effects of wind and turbulence on the properties of the fire front. Indeed, the effects of turbulence are poorly described in fire spreading [32] whereas it is an essential character of wind at the field scale that significantly affects the flame front.

In parallel of experimental activities, the scientific community has become increasingly involved in the modelling of forest fires and a number of approaches have emerged. Three types of models can be identified [33,34]. The first one includes statistical models that do not take physical information into consideration [7,35-38]. The second one incorporates semi-empirical models, which are based on the principle of energy conservation but which do not distinguish between the different mechanisms of heat transfer [39, 40]. Some simulators like BEHAVE [41] or FARSITE [42] were derived from them. They need huge experimental databases to be efficient but the absence of a real physical description makes them inapplicable with other fuel and environmental conditions than the ones used in the experiments. They are particularly inappropriate in areas with high variability of fuel and environmental conditions like in the Mediterranean area. Finally, the last type of models includes physical models, which are the most numbered. They describe all of or part of the mechanisms of heat transfer and production in order to predict fire spread [43-59]. Among these models, the multiphase approach, which takes the finest mechanisms involved in fire spread into consideration, is the most detailed modelling that was developed so far [54-59]. There are few simulators based on physical models and they are not validated by experiments. For instance, AIOLOS-F simulator, which is a simplified physical model [52] and FIRETEC model [53], which is a coupled atmospheric transport/wildfire spread model based on sound physics, are just compared with the final contours of a few wildland fires. The detailed multiphase models are inappropriate to develop simulators because they mainly describe the fire as an infinite straight line at small or medium scale (length scale < 100 m) and they neglect the atmospheric influence. Hence, the most relevant data available in the literature for field experiments such as the fire front shape cannot be used to validate the simulation of such models whereas it is a significant variable influencing fire spread [25,26].

The interest in modelling the spread of forest fire is twofold. Firstly, forest managers need models to manage forests. They require models that are necessarily accurate to allow dimensioning forest works (like fire break lines or pruning heights for instance). Detailed physical models seem to be particularly interesting to meet this kind of requirements. However, they need to be improved to circumvent the weaknesses previously mentioned. The second interest of modelling fire spread concerns fire fighting. Creating an operational management tool, capable of describing the spread of a forest fire in order to help fire fighters make the appropriate decisions is a challenge. That necessitates the development of a model which possesses a short calculation times. Approaches based on fundamental conservation equations are computationally intensive and cannot be used. Furthermore, as the main features to predict are the fire rate of spread, the fire-front geometry and the temperature field in the fuel layer, a satisfying solution can be found in simplified models based on both modelling and experimental data. Review of the literature confirms that models combining empiricism and theory are probably better than using either independently and should therefore be preferred [60,61]. Some modelling approaches were developed in this way [62,63] but these formulations assume the fire front as an infinite straight line and only consider heat transfers in the fuel bed ahead of the fire font.

In previous works [64,65], we have developed a two-dimensional fire spread model based on a diffusion-reaction equation with the aim to integrate it in an operational management tool for fire fighters. It allowed us to determine, from a single equation, the main characteristics of a surface fire at the laboratory-scale under no wind and no slope conditions [66]. Calculation times were lower than real time thanks to the development of timesaving algorithms [67,68]. In a second step, we improved the model to better take into account radiant transfer [69-71]. Although this model was able to predict both strong slope and combined slope and low wind effects, it failed to describe fire behaviour at increasing wind velocity. Then, a theoretical study was conducted [72] based on the reduction of a multiphase model [56] that proved the necessity to add an advection term in our model. A simplified flow was further derived from the same approach [73,74] to calculate the gas velocity inside the fuel bed. The model including advective and radiant transfers was validated thanks to laboratory experiments, providing valuable rates of spread, temperature distributions and heat fluxes [22]. The model predictions for the fire front shapes over time were validated only under slope conditions because of a lack of experimental data under wind conditions [16,17]. The model remains to be validated at the field scale.

In the present chapter, we will present the contribution of our research team to experimental and modelling activities in the frame of forest fire research. The first section presents our experimental activities at both the laboratory and the field scales to determine the basic properties of flames and fire fronts. The second section deals with the modelling of surface forest fires. We present the reduction of the multiphase approach leading to a two-dimensional model including radiative and convective transfers. The last section presents the validation of the model predictions thanks to experimental data at laboratory scale.

Major Properties of Flames and Fire Front in Forest Fires

This aim of this section is twofold. First it concerns the presentation of experimental techniques developed to study the fire spread and the related phenomena at both laboratory and field scales. Secondly, results from experiments conducted at those scales are provided and discussed. With regard to the fire spreads experiments conducted at the laboratory scale [20-22], we measured the temperature and velocity within the flame, the heat fluxes ahead of the fire front as well as the flaming intermittency (based on infrared imaging). All those tests, as well as those performed by other research teams [75-76] involve low fuel load (lower than 1.0 kg/m^2 in general) and small area for the spreading (fuel bed area lower than 10 m^2). Since there is a need of valuable data on fire spread at a scale larger than the laboratory one, we have also performed fire spread experiments in the field. Experimental data from a fire across a Mediterranean shrub are then presented. They concern thermodynamics data as well as common behaviour descriptors previously mentioned [25-27] that will be useful for the validation of physical models of wildland fire spread.

Fire Behaviour and Fire Front Characteristics at the Laboratory Scale

The flames encountered in free burning forest fuels are typical diffusion flames in which one can distinguish three behavioural regions. The lower part of the fire plume is the continuous flame region close to the source of the fire. Over that region, there is the intermittent one where balls of flame come away from the source, and the upper part is the conventional thermal plume. In this section, we investigate the lower and intermittent regions for a fire spreading across pine needles. A reliable and robust experimental procedure was developed in [20] using a double thermocouple probe for reconstructing the fluctuations in gas temperature. This method is applied to measure the temperature distribution within the fire plume. Then upward gas velocity measurements are provided from the method developed by Cox [77], which considers the cross-correlation of random thermal fluctuations. The reconstruction for the gas temperature is as an important operation to perform before to determine the gas velocity using the cross-correlation technique.

Measurement Method for Fluctuating Temperature in a Fire Plume

The method used to measure the temperature consists of compensating the thermocouple measurements. A thermocouple acts as a low-pass filter on the gas temperature and its measure should be corrected in order to restore the dynamic of the signal. A technique using a double thermocouple-probe was developed to reconstruct the gas temperature.

Model of the Thermocouple's Bead Temperature

A model was first elaborated for the temperature of a thermocouple bead immersed in a flame. This model was then applied to reconstruct the gas temperature by using two exposed junction thermocouples with different beads diameters. The instantaneous heat balance for the bead of a thermocouple immersed in a continuous flame region can be formulated using the following assumptions:

- the bead is supposed to be spherical and homogeneous,
- catalytic heat transfers are not considered,
- conduction losses along the wires are assumed negligible,
- convective heat flux is modeled by using Newton's law

$$q_{conv} = -H\left(T_{th} - T_g\right), \tag{1}$$

by assuming that the bead behaves like a gray body, the heat flux lost by radiation is given by

$$q_{rad,lost} = -\varepsilon_{th}\,\sigma_B\,T_{th}^4, \tag{2}$$

if we further assume that the region of the flame surrounding the bead behaves like an isothermal homogeneous semi-transparent medium, the heat flux absorbed in the bead by radiation can be evaluated by

$$q_{rad,abs} = \varepsilon_{th}\,\sigma_B\,\varepsilon_g\,T_g^4. \tag{3}$$

The instantaneous heat balance is thus

$$\frac{dT_{th}}{dt} = -\frac{6H}{\rho_{th}C_{p_{th}}d_{th}}\left(T_{th} - T_g\right) - \frac{6\varepsilon_{th}(1-\varepsilon_g)\sigma_B T_g^3}{\rho_{th}C_{p_{th}}d_{th}}T_g - \frac{6\varepsilon_{th}\sigma_B}{\rho_{th}C_{p_{th}}d_{th}}\left(T_{th}^4 - T_g^4\right) \tag{4}$$

where d_{th} is the bead's diameter. Let assume that a parameter η can be defined so that

$$\left|\frac{T_g - \eta T_{th}}{T_g}\right| << 1 \tag{5}$$

we have

$$T_{th}^4 - T_g^4 \approx \xi(\eta)\,T_g^3\,(T_{th} - T_g)$$

with $\xi(\eta) = 1 + \frac{1}{\eta} + \frac{1}{\eta^2} + \frac{1}{\eta^3}$. In the following, the parameter η will be omitted for simplicity and ξ will be written instead of $\xi(\eta)$. The heat balance for the thermocouple becomes

$$\tau(\xi)\frac{dT_{th}}{dt} + T_{th} = \gamma(\xi)T_g \tag{6}$$

with

$$\tau(\xi) = \frac{\rho_{th} C_{p_{th}} d_{th}}{6(h + \xi \varepsilon_{th} \sigma_B T_g^3)}, \tag{7}$$

and

$$\gamma(\xi) = \frac{H + \varepsilon_{th}(\xi - 1 + \varepsilon_g)\sigma_B T_g^3}{h + \xi \varepsilon_{th} \sigma_B T_g^3}. \tag{8}$$

The three unknowns H, ε_g and η vary because of large fluctuations of temperature, velocity and gas composition in flame. And consequently τ and γ also fluctuate. Instead of identifying h, ε_g and η, a method was developed to determine the two fluctuating parameters τ and γ. Indeed, the use of mean parameters results in inaccurate temperature measurement in turbulent combustion as demonstrated before for τ [78].

Parameter's Identification

Let consider two thermocouples, different only in diameters, which are immersed in the continuous region of a flame. If T_{th1} and T_{th2} denote the temperatures of the thin and thick thermocouples, respectively, their governing equations are

$$\begin{cases} \tau_1 \frac{dT_{th1}}{dt} + T_{th1} = \gamma_1 T_{g1} \\ \tau_2 \frac{dT_{th2}}{dt} + T_{th2} = \gamma_2 T_{g2} \end{cases} \tag{9}$$

The identification of the parameters necessitates that the thermocouples are close enough to ensure firstly that the velocity and temperature fields of the surrounding flows are identical and secondly that there is no mutual perturbation of their boundary layers. These conditions are necessary for the sensors to measure the same temperature T_g. The identification is managed in two stages: the first one is devoted to the time constants determination, and the second one deals with the gain coefficients identification.

Time Constant Identification

To identify the time constants τ_1 and τ_2 the variance

$$s^2 = \frac{1}{N}\sum_{i=1}^{N}\left(\left(\gamma_2 T_{g_2}^{\,i} - \gamma_1 T_{g_1}^{\,i}\right) - \frac{1}{N}\sum_{i=1}^{N}\left(\gamma_2 T_{g_2}^{\,i} - \gamma_1 T_{g_1}^{\,i}\right)\right)^2 \qquad (10)$$

was minimized for a number of N samples recorded by the thermocouples during a time interval Δt as follow

$$\begin{cases} \dfrac{\partial s^2}{\partial \overline{\tau_1}} = 0 \\ \dfrac{\partial s^2}{\partial \overline{\tau_2}} = 0 \end{cases} \qquad (11)$$

Since during Δt, τ_1 and τ_2 are assumed constant, we note them from now on $\overline{\tau_1}$ and $\overline{\tau_2}$.

Gain Coefficient Identification

Two more relations are needed to identify the gain coefficients. The averaged gain coefficients $\overline{\gamma_1}$ and $\overline{\gamma_2}$ are considered below since they depend on $\overline{\tau_1}$ and $\overline{\tau_2}$. Considering equations (7) and (8), a first relation between the gain coefficients was derived

$$\overline{\gamma_1} - \delta\overline{\gamma_2} = 1 - \delta \qquad (12)$$

with

$$\delta \approx \frac{\overline{\tau_1}}{\overline{\tau_2}}\frac{d_{th2}}{d_{th1}}. \qquad (13)$$

The second relation is obtained from equations (9) since $T_{g1} \approx T_{g2}$

$$\overline{\gamma_1} - \beta\overline{\gamma_2} = 0 \qquad (14)$$

with

$$\beta = \frac{\dfrac{1}{N}\sum_{i=1}^{N}\left(\overline{\tau_1}\dfrac{dT_{th1}^{i}}{dt} + T_{th1}^{i}\right)}{\dfrac{1}{N}\sum_{i=1}^{N}\left(\overline{\tau_2}\dfrac{dT_{th2}^{i}}{dt} + T_{th2}^{i}\right)} \qquad (15)$$

$\overline{\gamma_1}$ and $\overline{\gamma_2}$ are obtained by solving equations (12) and (14). Finally, the gas temperature is reconstructed from equation (9).

Measurement Method for the Upward Gas Velocity in a Fire Plume

The method used to obtain the upward gas velocity within the flame is based on the measurement of the transit time of a thermal fluctuation between two points of the flow [77]. The statistical properties of the signals are employed to recognise a fluid fluctuation pattern and to time its movement. It is assumed that this pattern changes little for a small separation r between two probes. The transit time θ_0 is that value of time shift for which the delayed version $x(t-\theta)$ of signal $x(t)$ from the first measuring point is almost equal to the signal $y(t)$ from the second point, *i.e.* when the cross-correlation function

$$C_{xy}(r,\theta) = \lim_{\chi\to\infty}\frac{1}{\chi}\int_0^{\chi} x(t-\theta)\,y(t)\,\mathrm{d}t \tag{16}$$

has a maximum value. This function is often expressed in normalised form as the correlation coefficient $\rho_{xy}(r,\theta)$ where

$$\rho_{xy}(r,\theta) = \frac{C_{xy}(r,\theta)}{\left[\overline{x(t-\theta)^2}\right]^{\frac{1}{2}}\left[\overline{y(t)^2}\right]^{\frac{1}{2}}} \tag{17}$$

with

$$\overline{x(t-\theta)^2} = \lim_{\chi\to\infty}\frac{1}{\chi}\int_0^{\chi} x(t-\theta)^2 dt \tag{18}$$

For $r = 0$, $y(t) = x(t)$ and the cross-correlation function becomes the auto-correlation function of $x(t)$

$$C_{xx}(0,\theta) = \lim_{\chi\to\infty}\frac{1}{\chi}\int_0^{\chi} x(t-\theta)\,x(t)\,\mathrm{d}t\,. \tag{19}$$

The auto-correlation function describes the relationship between a signal and its time-shifted versions, and serves as an indicator of the validity of Taylor's assumption of frozen turbulence [77], which states that fluctuations are frozen into the mean flow. It means that the cross-correlation function is identical to the auto-correlation function with a time-shift θ_0: $C_{xy}(r,0) = C_{xx}(0,\theta_0)$. However, in reality this is not the case due to viscous dissipation of the eddies. The cross-correlation coefficient is poorer than the auto-correlation one. A measure of the departure from this assumption will be considered.

Experimental Procedure

The experimental set-up is displayed in figure 1. The bench of combustion was composed of an air-entrained concrete plate (1.2 m long and 0.5 m wide). It was protected by sand and supported by a larger table (1.5 m long and 1 m wide) to allow lateral inflows of air. Pure oven-dried pine needles of *Pinus pinaster* were spread as evenly as possible on a portion (0.7 m long and 0.4 m width) of the total area of the combustion table. This gave a homogeneous structure, representative of a typical layer found in Corsican forests with a load of approximately 0.5 kg/m^2 on a dry basis. The needles area to volume ratio and density are 4550 m^{-1} and 680 kg/m^3, respectively. The depth of fuel was 2.3 cm, giving a bulk density of 21.74 kg/m^3 and a packing ratio of 0.032. The needles were dried in an oven at 60°C for 24 hours before each run. The properties of the ambient air were measured for each test; the temperature ranged between 25 to 30°C and the relative humidity varied from 40 to 50%.

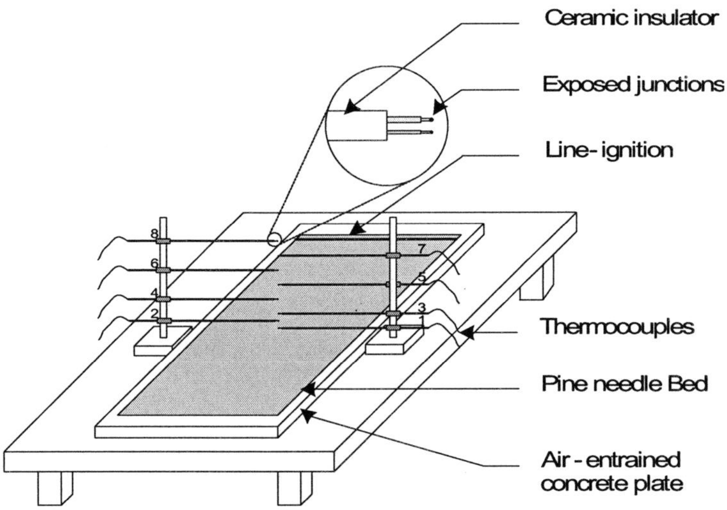

Figure 1. Laboratory experimental set-up.

Three experiments were performed to observe the spread of the fire under no slope and without any wind for line-ignition. A line was ignited using alcohol at one small edge of the fuel-bed, as shown in figure 1. Two arrays of double thermocouple probes were installed above the fuel at approximately 0.35 m before its end, *i.e.* far enough from the ignition line for a fire to achieve a stable spread-rate, before it passed through the thermocouples. Each thermocouple-probe is composed of a two-hole insulator made of recrystallized pure alumina ceramic containing over 99.8% Al_2O_3. The insulators were 28.5 cm long (o.d. 3.1 mm; i.d. 1 mm). The thermocouples were threaded through these holes and their junctions were set at a distance of about 0.3 mm from each other. Those insulators protected the thermocouples' stainless steel sheaths. Thermocouple-probes were placed in the following way. The first probe was placed at the top of the litter. The second one was at 2 cm from the first one, as well as the third one from the second one. The others were placed 4 cm apart (see Table 1). The thermocouples used in these experiments were mineral-insulated integrally metal-sheathed (MIMS) pre-welded type K (chromel-alumel) pairs of wire with an exposed junction.

Table 1. Positions of the double thermocouple-probes.

Probe	1	2	3	4	5	6	7	8
Height (cm)	4	6	8	12	16	20	24	28

The MIMS form of thermocouple consisted of matched thermocouple wires surrounded by insulating material compacted by rolling until the sheath is reduced in diameter. In this work, we used cheap stainless steel sheathed (304 SS) and M_gO insulated standard thermocouples (Omega Engineering). The exposed junction consists of an extended junction going beyond the protective metallic sheath to give accurate fast response. The double thermocouple probe consisted of two 30 cm long sensors with sheaths of 250 and 500 μm in diameter for both thin and thick thermocouples. At the exposed hot junctions the wires were 50 and 100 μm in diameter, respectively. The exposed junctions were nearly perfect spheres, whose diameters varied from 100 to 225 μm and from 250 to 425 μm, respectively. The wire diameters were chosen since they constituted a compromise between their robustness and their rapidity (low time constant). Furthermore, the sizes of the wires adopted here limit the sagging of the junction due to thermocouple expanding. The sampling frequency was 700 Hz; the precision in temperature measurement was 0.5°C.

Results and Discussion

Temperature

Figure 2 shows the temperature curves versus time of two thin thermocouples. The first one (thermocouple 1) takes place at 4 cm from the ground and the second one (number 6) at 20 cm (see Table 1). The lower thermocouple, which takes place on the top of the litter, is firstly heated in a more regular and less fluctuating way than the upper one by the spreading fire front. We can notice that a few seconds before the ignition of the litter, the temperature of the thermocouple number 6 increases more significantly. That points out the growth of the fire plume radius with height, which heats the upper thermocouple before the lower one.

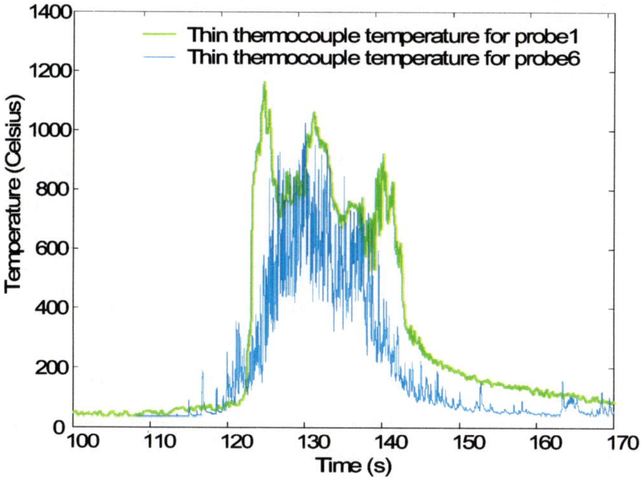

Figure 2. Temperatures for thin thermocouples located at 4 cm and 20 cm height.

Then, when the litter ignites the two thermocouples are immersed in the flame. During this stage, the temperature profiles show two properties of the gas flow. The temperature fluctuates less on the top of the litter than above it. This reveals the laminar feature of the flow in the ejection zone of the pyrolyzate gases, before it becomes turbulent with height.

Figure 3 displays the temperature curves of a double thermocouple probe before and after the numerical compensation. The reconstruction for the gas temperature provided here was performed for thermocouple number 5 located in the fully developed flame zone. Solely a portion of the reconstructed temperature is displayed, which corresponds to the time interval when the temperature fluctuates around the maximum temperature. In this time interval, the reconstructed temperature curves have a correlation coefficient of more than 99%. The maximum temperature compensated on the whole experiments was $\approx 1250°C$. This result is not absurd, considering the adiabatic combustion temperature ($\approx 1330°C$) in air for formaldehyde COH_2, an equivalent substance to the volatiles released by the pyrolysis of wood [79].

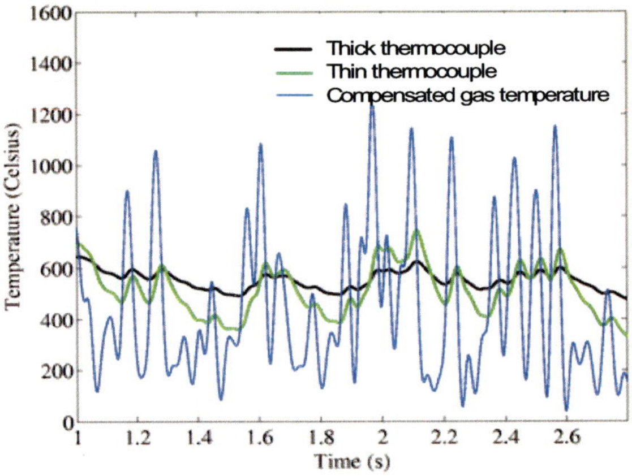

Figure 3. Compensation for the gas temperature at probe 5.

Afterwards, the compensation was applied for the whole set of double thermocouple probes. The compensated temperatures were averaged on the time interval associated with the zone of maximum temperature, which corresponds to the flame residence time. This was performed to examine the effect of the compensation on average temperature values. Figure 4 shows the mean profiles versus height of these two temperatures. We can observe that the numerical compensation slightly modifies the averaged values of the thin thermocouples, while keeping the shape of the temperature profiles. The interest of the compensation technique relies more on the rendering of the dynamics of the signals (as shown in figure 3) than in the correction of its mean value. This advantage will become important in the next section for the calculation of the upward mean gas velocity. Like [80, 81] we can observe two behaviours. The first one concerns the temperature increase just above the litter. In this region there is an oxygen deficit due to the ejection of the pyrolyzate gases, so combustion is incomplete. Then the pyrolyzate gases burn (when they reach the oxygen) and the mean temperature reaches its maximum (here about 1050°C). The second behavioural zone corresponds to the typical cooling of the fire plume which is mainly due to radiative losses in the lower part and entrainment and mixing with ambient air in the upper part.

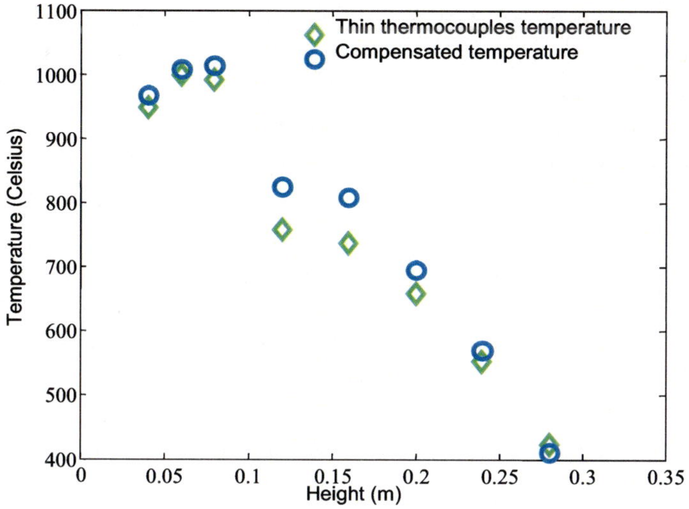

Figure 4. Measured and compensated mean temperature profiles along the fire-plume axis.

Velocity

The reconstruction of the temperature signals performed in the previous section lets restoring their dynamics which was lost by the thermocouple, (acting as a low-pass filter). Figure 5 presents such compensated signals of temperature for two probes 4 *cm* apart on a short time interval.

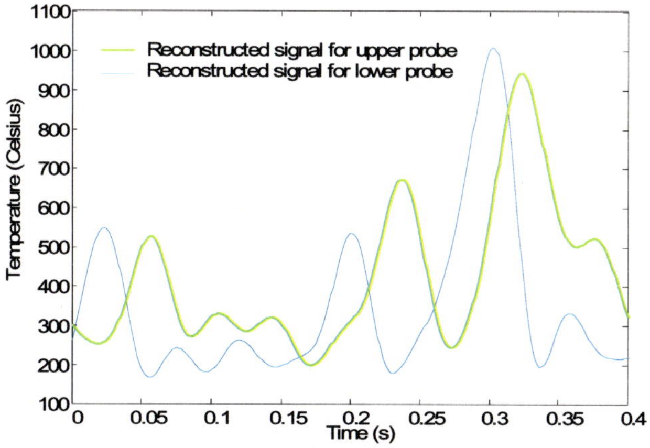

Figure 5. Compensated temperature for two probes 4 cm apart.

Each fluctuation of the compensated temperature of the lower probe is observed with a time shift on the compensated temperature of the upper one. The reconstruction allows evaluating easier the transit time of a thermal fluctuation between two parts of the flow to determine the upward gas velocity. Conversely to [82], it is not necessary to apply a high-pass filter to the signal in order to carry out the main value to keep only the "high" frequencies to calculate the gas velocity. To implement this method, it is necessary to verify the assumption of frozen turbulence, however. This assumption was investigated and appears reasonable. For example, for the two successive probes 4 and 5 (located at 0.12 cm and 0.16 cm from the ground, respectively), the cross-correlation coefficient $\rho_{45}(0.04, 0) \approx 0.36$ is near to the auto-correlation coefficient $\rho_{44}(0, \theta_0) \approx 0.38$ for probe number 4. $\theta_0 = 0.0364$ s is the time shift corresponding to the peak of cross-correlation coefficient. The delayed correlation is poorer because of viscous dissipation of eddies.

Upward gas velocities versus height, determined by the cross-correlation technique, are plotted in figure 6. The velocities were calculated for each pair of successive probes in the same time interval as those previously mentioned for the calculation of the mean maximum temperature. We can note that the profiles follow the same tendency for the three experimental runs. In the lower part of the fire plume (height < 0.1 m), the measured velocities are rather different. This is due to the laminar pattern of the flow as previously mentioned. The cross-correlation technique lies in the measure of the transit time of thermal fluctuation. Obviously, the velocity calculation is less efficient in the near fire plume than in the above regions since the temperature signals present less fluctuations. Conversely, in the upper part of the flow the measured velocities are very close at each probe interval for each experimental run. The maximum relative error on the determination of the gas flow velocity was under 10%.

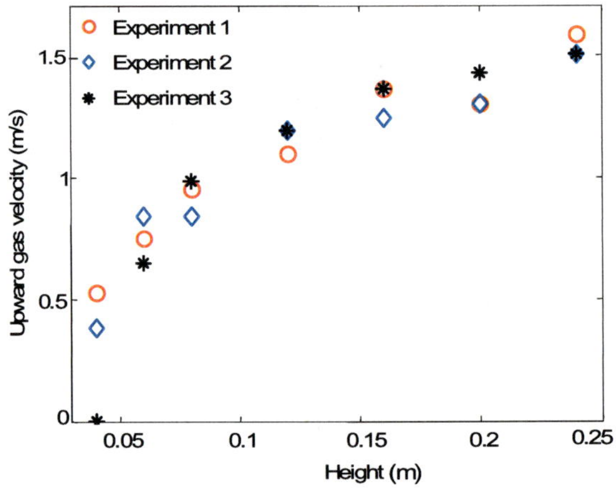

Figure 6. Mean gas velocity profiles along the fire-plume axis.

Infrared Thermography in Fire Applications

The measurements of fire front properties are of great interest for the understanding of the fire spread behaviour. In this aim, non-intrusive techniques have to be developed since they are the only ones that could be implemented in the field for measurements during forest fires. In a fire, the radiation is the significant heat transfer process and very high temperatures are encountered. Thus, the infrared thermography appears to be the more appropriate and accurate treatment to determine some characteristics of the fire. The infrared images from fire spread across bed of pine needles were recorded with a focal plane array (FPA) camera operating in the MIR band. This band is a part of the electromagnetic spectrum located from 3 to 5 μm. It was chosen for fire applications since for temperature of about 1000°C, approximately one third of the radiated power is emitted in the MIR band. Furthermore, at that temperature the wavelength of maximum emission is 4 μm, which is in the centre of the

MIR band. The infrared images are composed of 256×256 pixels, which digital levels are quantified with a 14-bit dynamic range. The digital level of a pixel is proportional to the radiative intensity emitted by the fire. The adjustable integration time of the camera, which can be set from 10 to 1000 μs, allows accommodating a wide range of flux levels. The integration time is the duration that the FPA is allowed to collect infrared photons. Thus, short integration times are used to measure thermal scenes with high temperatures, like fire presence phenomena, without saturation problems. The acquisition rate of the camera allows capturing up to 25 images per second. There are two main contributions to the infrared emission of a fire: that of the emitted gases (mainly CO_2 and H_2O in the flame) and that of the hot solid particles (embers). The use of specified filters, ahead of the detector of the infrared camera, allows focusing either on the combustion processes occurring in the gaseous phase or on those occurring in the solid phase. Two applications of the infrared thermography will be provided hereafter. The first one concerns the determination of the characteristic length scale of a flame front. These results are then used to estimate the centreline temperature distribution within the flame front. The second one deals with the temperature measurements within the hot solid particles.

Flame Characteristic Length Scales

During a spreading fire like depicted in figure 1, we disposed the infrared camera to face the flame front. It was placed at 1 meter from the end of the bed and focussed on the measuring section where the thermocouples were located. Figure 7 illustrates the position of the camera with an example of the field of view (the supporting rods of thermocouples, which will be used for the validation of the correlations predictions are visible ahead of the flame front). At that distance the square pixel size is about 0.14 mm.

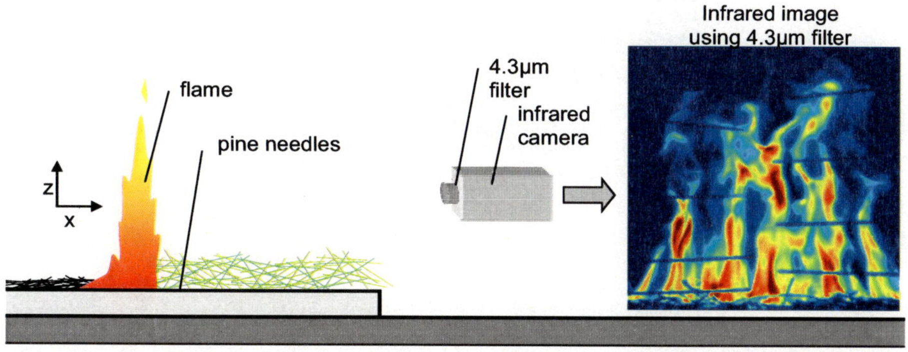

Figure 7. Location of the infrared camera for the study of the flame front.

During combustion processes in the gaseous phase the more abundant products of the reaction is carbon dioxide. The major part of these combustion products emits radiation more intensively in the 4.3 μm band due to the asymmetric stretch vibration of CO_2 species. Measurements of the geometrical properties of the flames were done by using a filter for CO_2 species, which have a specified wavelength near 4.3 μm. The integration time of the infrared camera was set to 33 μs, to avoid image saturation problems when the flame is recorded. The infrared camera was firstly used to determine the characteristic length scales of the flame front. These flame lengths will be used in the following with empirical correlations to determine the temperature distribution along the flames.

The continuous flame length, L_c and the maximum flame length, L_m determine the limits between three regions: the continuous flame region is found below L_c, the intermittent is located between L_c and L_m and the plume region is found above L_m. The flame characteristic length scales, namely the continuous and maximum flame lengths, were obtained using an image processing technique. This treatment of infrared images is derived from the one used for video recordings by [83]. The image processing technique is based on a binarization of the infrared images. It consists in assigning a threshold value of the digital level of the pixel DL_b below which the flame is considered to be absent. For this work the value of DL_b of 5000 is considered to describe best the flame contour. This digital level, compared to thermocouple measurements, corresponds to temperature of about 500°C. Below DL_b the flame is considered to be absent and a value of 0 is assigned to the pixels; above this threshold flame is considered to be present and a value of 1 is assigned to the pixels. This technique was applied to 150 images of the fully developed buoyant diffusion flame when the flame front enters in the measuring section. By averaging all binarized images, the intermittency of flame length [84], which represents the flame presence probability contours, is obtained (figure 8).

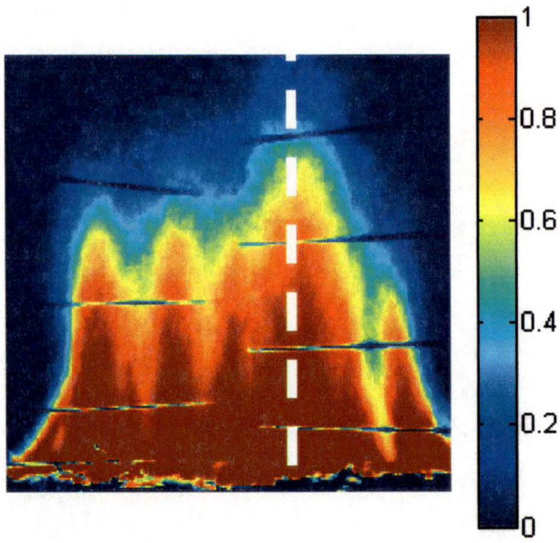

Figure 8. Flame presence probability contours.

The continuous flame length L_c is defined as the distance from the bed of fuel at which the flame intermittency is 0.95 and the maximum flame length L_m is defined as the distance from the bed of fuel at which the intermittency is 0.05. The continuous flame region is found below L_c, the intermittent region is found between L_c and L_m and the plume region is found above L_m. Another important characteristics length scale defined by [85] is the average flame height, L_f which is defined as the height for which intermittency is 0.50. The presence probability along the vertical direction for the flame zone located at the centre of the fuel bed is provided in figure 9. The values of L_c and L_m, for the flame zone located at the centre of the fuel bed are 0.11 and 0.31 m, respectively. The mean flame height L_f is found to be about 0.21 m. It should be noticed that, for fire spread across bed of fuels, the intermittent flame region occupies the major part of the flame. In this case, it represents here about 65% of the maximum height of the flame.

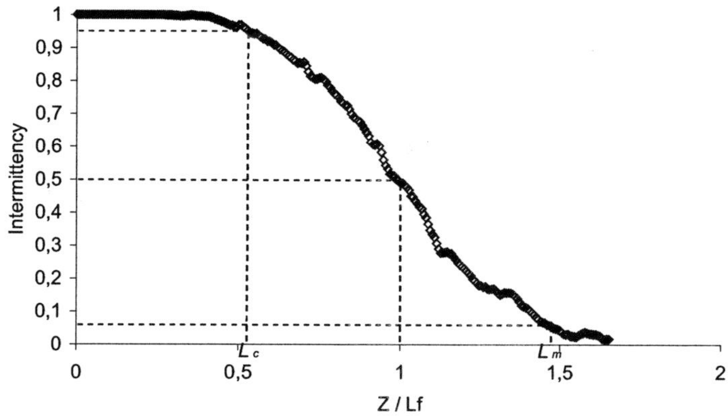

Figure 9. Intermittency of flame at the centreline of the flame front.

Centreline Temperature Distributions

The emissivity of a reacting gaseous medium such as flame is not easy to determine. In particular, it depends on gas composition and soot generation processes. Thus the infrared images obtained are not calibrated and the temperature of the flame can not be derived directly from them. However, the centreline temperature in the flame along the vertical direction is of great interest. Many authors [81,84,86] have identified this distribution as being the main characteristic of the fire. In particular, McCaffrey [84] established the following empirical correlation for diffusion pool flame, to describe the centreline temperature distribution in the three flame regions (continuous flame, intermittent flame and plume):

$$\frac{2g\left(T-T_a\right)}{T_a}=K_T\left(\frac{z}{Q^{2/5}}\right)^{2\kappa-1} \qquad (20)$$

Equation (20) will be used here to determine the temperature distribution on the basis of the continuous and intermittent flame height obtained previously. This correlation was established for square burners but experimental works conducted by [86,87] have demonstrated that the McCaffrey structure is preserved for a wide range of geometrical configurations and experimental conditions. Furthermore, Hasemi and Nishita [88] determined that an important parameter affecting the flame structure is the width/depth ratio. They stated that for a ratio greater than 4 the centreline temperature distribution is that of a line fire (which is the present case) and remains invariant with further increase of the aspect ratio. Although the coefficient K_T provided here is not the same as that proposed by McCaffrey, this relation is still applicable for a line fire since the flame structure is preserved. The parameters K_T can be determined in each region from characteristic length scales, namely L_c and L_m. The value of this empirical coefficient is provided in Table 2. The heat release rate, Q is given by

$$Q=\dot{m}\,\Delta H \qquad (21)$$

where $\dot{m}$ represents the mass rate of the fuel gas and ΔH is the heat of reaction. By assuming the mass rate of the fuel gas to be equal to the mass loss rate measured during experiments, namely 0.44 10^{-3} kg/s and with ΔH=21000 kJ/kg, we obtain Q = 9 kW.

Table 2. Empirical coefficients for McCaffrey correlations.

Flame zone	η	K_T
Continuous	$\frac{1}{2}$	$\left(\frac{C_C}{C_B}\right)^2$
Intermittent	0	$\frac{2g(T_c - T_a)}{T_a} \times \frac{L_c}{Q^{2/5}}$
Plume	$-\frac{1}{3}$	$\frac{2g(T_c - T_a)}{T_a} \times \frac{L_c}{Q^{2/5}} \times \left(\frac{L_m}{Q^{2/5}}\right)^{2/3}$

where C_C is coefficient for center line correlations (C_C = 6.9) and C_B is the buoyancy constant (C_B = 0.9).

The temperature distribution is presented in figure 10. The agreement between predicted centreline temperature distribution and experimental data is good in the continuous and intermittent flame regions. The characteristics slope coefficient κ, defined by McCaffrey is conserved for a line fire spread across pine needles bed. The maximum probe height (0.28 m) was not sufficient to allow measurements in the plume region, which is located above the maximum flame length (0.31 m).

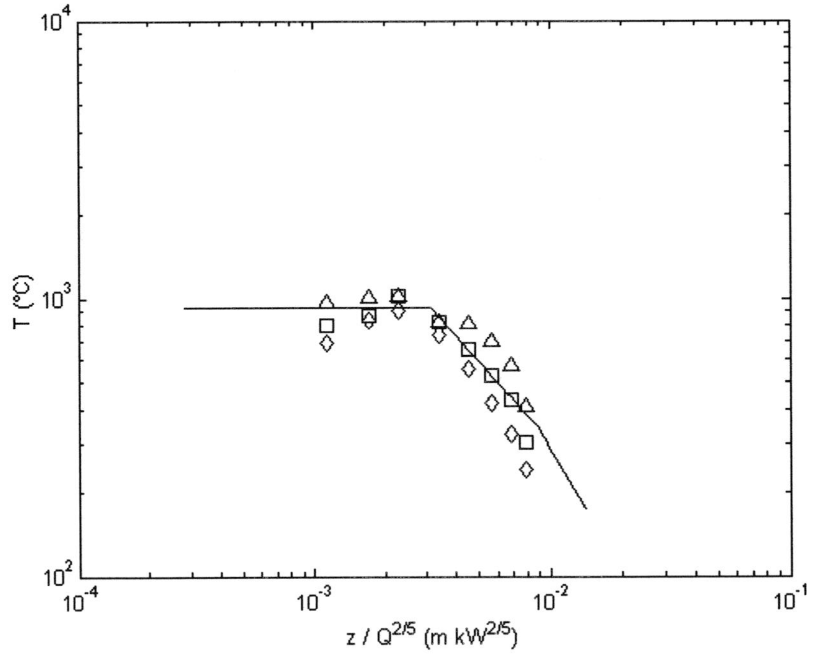

Figure 10. Centreline temperature distribution for fire spread across a bed of pine needles.

Solid Phase Temperature

In the present section a quantitative application of the infrared thermography is provided. The infrared camera is used to make non-intrusive measurements of the temperature of the hot solid particles. These measurements were done using a flame filter to eliminate the CO_2 and H_2O radiation mainly present within the flame. This filter allows making high temperature measurements of the fuel bed particles (specified wavelength near 3.9 μm). With such flame filter and an integration time of 50 μs, the camera is capable to make no contact temperature measurements in the range of 200°C to 1100°C. Indeed, the calibration relies on a linear fit between radiative intensity and digital level on each pixel, measured with the camera viewing a blackbody source at different temperatures. The calibration curve for a 50 μs integration time is provided in figure 11.

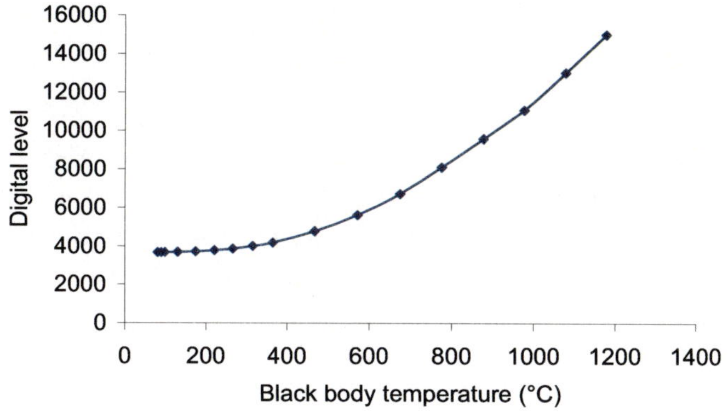

Figure 11. Calibration curve of the infrared camera (50 μs integration time and 3.9 μm filter).

The pine needles were assumed as black-body (emissivity value equal to 1). The camera was located at a height of 2 meters and set perpendicularly to the main fire spread direction. Figure 12 illustrates the position of the camera with an example of the field of view. The location of the fire front is provided on the left side of the figure.

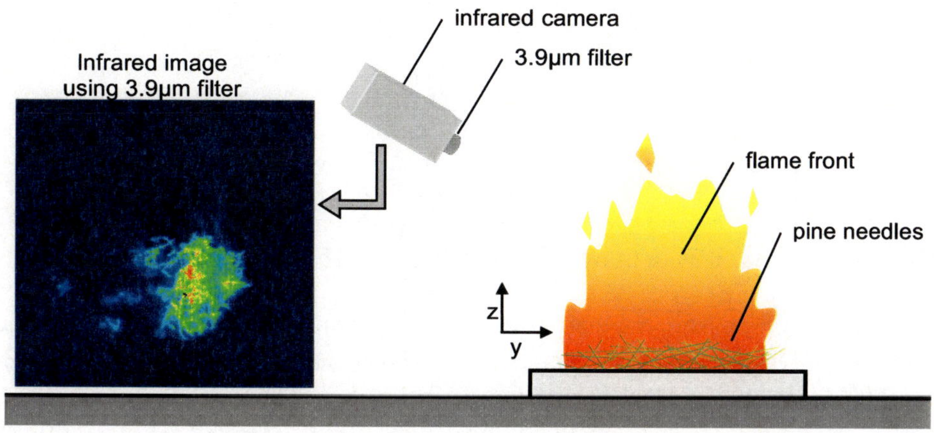

Figure 12. Location of the infrared camera for measurements of the needles temperature.

Figure 13 compares the temperature measured by the camera with that provided by a 100 μm thermocouple located inside the fuel bed. As previously mentioned, due to the integration time of 50 μs taken for the camera, it was not possible to record accurate temperature under 200°C (for instance, an integration time of 200 μs, would have allow measurements in the range of 0°C to 250°C). Hence, in the following we will only discuss the results for temperature greater than 200°C. The infrared images and the temperature measured with the thermocouple allow a better understanding of the fire spread behaviour across the fuel bed. These measurements show the presence of two fire fronts: a flaming zone delimited by the devolatilization front (first temperature peak) and a char combustion zone (second temperature peak). It should be noticed that the presence of the two fronts can only be measured with very fine wires thermocouple and this phenomenon is filtered with bigger one [13,75]. Between these two high temperature zones, there is a colder one. In this last zone, char combustion is limited since there is an oxygen deficit. The thermocouple measures the gas temperature whereas the camera measures the temperature of the fuel bed. Thus we can say that during the devolatilization, the thermal equilibrium between gaseous and solid phases is not achieved as depicted by figure 13, in which the temperature increase differs between the two measurements. Both temperature match during char combustion and thermal equilibrium seems to occur at that time. During the cooling phase, the comparison is meaningless because the infrared camera measures the temperature of the combustion bench (there is little ash) heated by the fire spread while the thermocouple measures the air temperature above it.

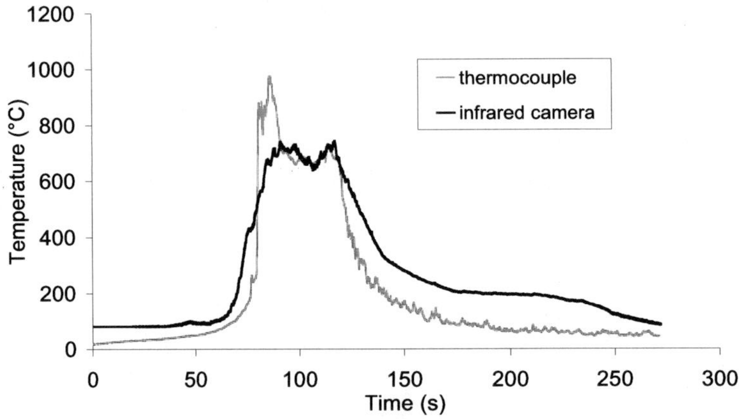

Figure 13. Comparison between temperatures measured by thermocouples and infrared camera.

Fire Behaviour and Fire Front Characteristics at the Field Scale

In this section we propose a method and a set of measuring structures built in order to improve the knowledge on the fundamental physical mechanisms that control the propagation of fire at field scale. Wind and vegetation effects on fire behaviour, which are mainly responsible of the hazards encountered in forest fires, are addressed. The effects of wind on the thermal properties of the flame front are also investigated.

Site Description

The test site was selected in order to get simultaneously some of those conditions for which high intensity fires occurred in summer in south Corsica, *i.e.* western windy condition (greater than 12 m/s), high ambient temperature (greater than 30°), low relative humidity (lower than 30%), low fuel moisture content and drought conditions. The aim was to reproduce through a small-scale field experiment, a fire spreading representative of severe fires occurring in summer, although being aware that it could not duplicate actual wildfire. The layout of the plot was designed by taking into account the experimental requirements, the terrain configuration and the safety conditions. Situated on an uneven terrain, it was in the shape of a rectangle, 30 meters wide perpendicular to the slope with a northwestern outlook on that side and 80 meters long parallel to the slope (Figure 14). The 30 meters wide upper limit of that rectangle was delimited by a fuel break. Along the biggest sides, the first 50 meters were under upslope condition while the 30 remaining meters were on a flat terrain. For each plot, a safety belt of 20 to 30 meters wide was build with a bulldozer.

Figure 14. General view of the plot.

Before the burning, we took an inventory of the vegetation located on the parcel. We built a statistical description by species and by class of vegetation and we realized a cartography. We measured the location of each species and their heights. Vegetation was classified in a flora. The shrub layer represented the most part of the vegetation cover. Dominant species were olive tree (*Olea Europaea*), strawberry tree (*Arbutus Unedo*) and white heath (*Erica Arborea*). There were also notable amounts of cistus (*Cistus Monspeliens*), golden-chain (*Cytisus Triflorus*) and oak (*Quercus Ilex*). The mean height of vegetation ranged between 2 to 3 meters. The herbaceous layer was sparsely distributed on the parcel mainly due to the presence of a dense shrub layer. Vegetation was also classified in strata: upper shrub layer (> 2 m), lower shrub layer (< 2 m) and herbaceous layer (see Table 3).

Table 3. Vegetation heights and moisture contents.

Species	Minimal height (m)	Maximal height (m)	Moisture content (%)
Olea Europaea	2.13	3.26	84
Quercus Ilex	1.45	2.89	67
Arbutus Unedo	1.11	2.29	108
Cistus monspeliensis	0.93	2.01	78
Cytisus triflorus	0.58	1.50	100

The community studied showed a total cover of practically 100%. The upper shrub layer represented 15% of the cover of the parcel, the lower one 85% and the herbaceous layer 5% (located under the two shrub layers). We also measured moisture content for each species (see Table 3). To proceed, we collected samples of leafs in the vegetation neighbouring the parcel. These samples were oven-dried at 60°C for 24 h and the fuel moisture was expressed as a percentage of dried weight. We can note high values for moisture contents. This was mainly due to the rainy conditions in spring before the experiment.

Weather variables, which are of extreme importance, were surveyed in order to plan both the day and the time of the experimental burns. A weather Campbell® automatic station was installed in the study area two months before the experiments in order to gather data on the local meteorological conditions. It loaded the following parameters every minute: average air temperature, average relative humidity, average wind speed at 2 meters high above ground level and average wind direction. The recorded data revealed two regimes for the wind speed and direction over the plot. The first one, which is the most frequent, is an easterly regime due to a see breeze effect. It is present during daytime in the eastern façade of Corsica. The second one is a westerly regime due to large-scale atmospheric circulation. It takes place mainly during the night and after 16:00, when the sea breeze effects cease. The wind velocity is higher for the westerly regime than for the easterly one, increasing the fire danger. Figure 15 displays the averaged daily dynamics of air temperature and relative humidity recorded on the plot for 16 days before the burn. All these data concurred to select the best period of the day to burn between 16:00 and 18:00, since we expected to have simultaneously high west wind with high air temperature and low relative humidity values.

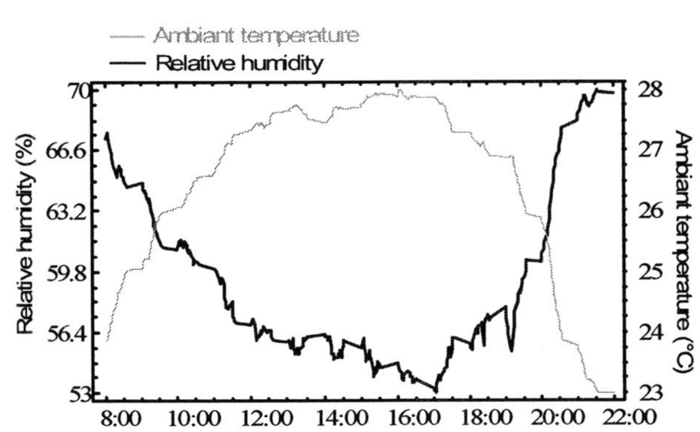

Figure 15. Averaged daily air temperature and relative humidity for 16 days before July 2.

Experimental Method and Apparatus

General Overview of the Experimental Protocol

The experimental protocol was elaborated in order to investigate two sets of fire behaviour variables. The first one concerns the properties of the fire front itself and their changes during the spreading according to the environmental conditions. The second one is related to the effects of the head fire section ahead of it. With regard to the investigation of the properties of the fire front, several devices were located inside the vegetation and along the biggest sides of the plot. The first series of devices, named in the following "characterization lines", were elaborated to determine the fire front perimeter over time. The second ones, called "thermocouple trees" (❶ in figure 16), are measuring structures of six meters high built in order to study the temperature in the fire plume. Three two-dimensional sonic anemometers (❷ in figure 16) were placed around the parcel in order to record the surface wind entering the plot. The first two were located on the right and left sides of the parcel at two meters height while the third one was positioned in the fuel break ahead of the upper limit of the plot. Two video cameras were located on the sides to survey the fire and investigate the flame shape (height, length, depth and tilt angle). Concerning the study of the effects of the head fire section, nine posts (❸ in figure 16) containing each one four heat flux sensors were placed in the fuel break ahead of the upper limit of the plot in order to record the radiant and total heat flux. An infrared camera and a video camera were used to film the fire font reaching the thermocouple trees in a frontal view and a remote video camera served to localize the fire front advance.

Figure 16. General overview of the measuring structures.(❶ Thermocouple trees, ❷ Sonic anemometers, ❸ Flux meter trees).

Characterization Lines

At the field scale, the determination of the fire front perimeter over time is tricky because of the visibility problems induced by smoke and vegetation structure. Thus a device, called "characterization line", was developed in order to detect the fire front advance. It consists of six sensors provided by a 12 V direct power supply. Each sensor is composed of a fuse, which breaks at 300°C and a resistor. The fuses are located within the vegetation 15 cm above the ground and the resistors are buried in the ground, 20 cm deep. As the fire front reaches a fuse, this last one breaks while the corresponding resistor remains intact. A voltage measure, taken at the terminals of a charge resistor R allows to follow the arrival of the fire front at the sensor location.

Thermocouple Trees

Up to know, studies on the vertical distribution of the temperature within spreading [20,21] and static [89] fire plumes were mainly conducted at laboratory scale. Although these previous works brought some insight on flaming combustion of vegetative fuels, conversely they involved low fuel load for spreading fires and small quantities of fuel for static fires. Therefore, their results could not be extrapolated to actual wildland fuel structures. Similar experiments are almost inexistent at the field scale apart from [90], which investigated fire as a heat source to the atmosphere. In the present work, three measuring structures (❶ in figure 16) of six meters height were built in order to evaluate the temperature profile along the vertical direction within the fire plume while the propagation. The aim was to investigate, whether the classical structure of the fire plume revealed by laboratory experiments [21], remains valid at field scale. These "thermocouple trees" were placed in the upper part of the plot, inside the vegetation, and set at a distance of 6 meters from each other. Each mast was equipped with 10 thermocouples placed 0.6 m apart (see Table 4). The thermocouples were mineral-insulated integrally metal-sheathed (MIMS) pre-welded type K pairs of wire with an ungrounded junction. There were 15 cm long with sheaths of 250 μm in diameter and wires of 50 μm in diameter inside. These sensors were chosen since they constituted a compromise between their robustness and their rapidity. Then, each thermocouple was threaded through an insulator (13 cm long) in order to protect the stainless steel sheaths. Both insulators and thermocouples were placed in stainless steel tube devoted to protect the whole from twigs and mechanic contacts. The thermocouples protrude 40 mm from this set-up and were vertically aligned. There were connected to a Hewlett Packard (HP E9811B) data logger. The sampling frequency was 100 Hz.

Table 4. Positions of the thermocouples.

Probe	1	2	3	4	5	6	7	8	9	10
Height (m)	0.6	1.2	1.8	2.4	3	3.6	4.2	4.8	5.4	6

Flux Meter Trees

Nine posts supporting each one two devices containing two radiant and convective heat flux sensors (Captec®) were placed in the fuel break (❸ in figure 16). On each post, the devices were located at two measurement heights (2 m and 4 m) to represent respectively the height of a standing person and the height of the lower edge of a house roof. The posts were positioned in three lines located at a distance of 5, 10 and 15 m from the top of the parcel, in order to investigate the heat flux variation along distance.

Results and Discussion

Due to exceptional rainfalls in spring, live fuel moisture content was greater than usually. We encountered thus some difficulties to ignite the parcel and we had to conduct the fire the July 2, during the wildfire season. It was performed between 16:33 and 16:59, within the period of the day during which high western wind (up to 8 m/s), high air temperature (32°C) and minimum relative humidity (38%) values were recorded.

Time History of the Fire Front

A point ignition was performed on the right side of the parcel along the windward edge in order to exploit the overall measuring section. The fire developed into two distinct head shapes over time. At the beginning, it burned with little variation in direction and few changes in speed. It took a narrow shape at the head. A first change in wind direction caused part of the right flank to become a heading fire with a parabolic shape. The fire reached the limit of the plot and stopped due to unavailable fuel. The time history of this propagation, based upon the remote imagery and "characterisation lines" data analysis, is provided in figure 17. Then after a lull in wind which last about four minutes, the wind speed increased and changed again in direction. This caused part of the left flank to become a heading fire with a broad parabolic shape, like a solid wall of flames, creating very high flames and potential danger of spotting. The fire rate of spread was variable, but generally ranged from 0.1 to 0.4 m/s.

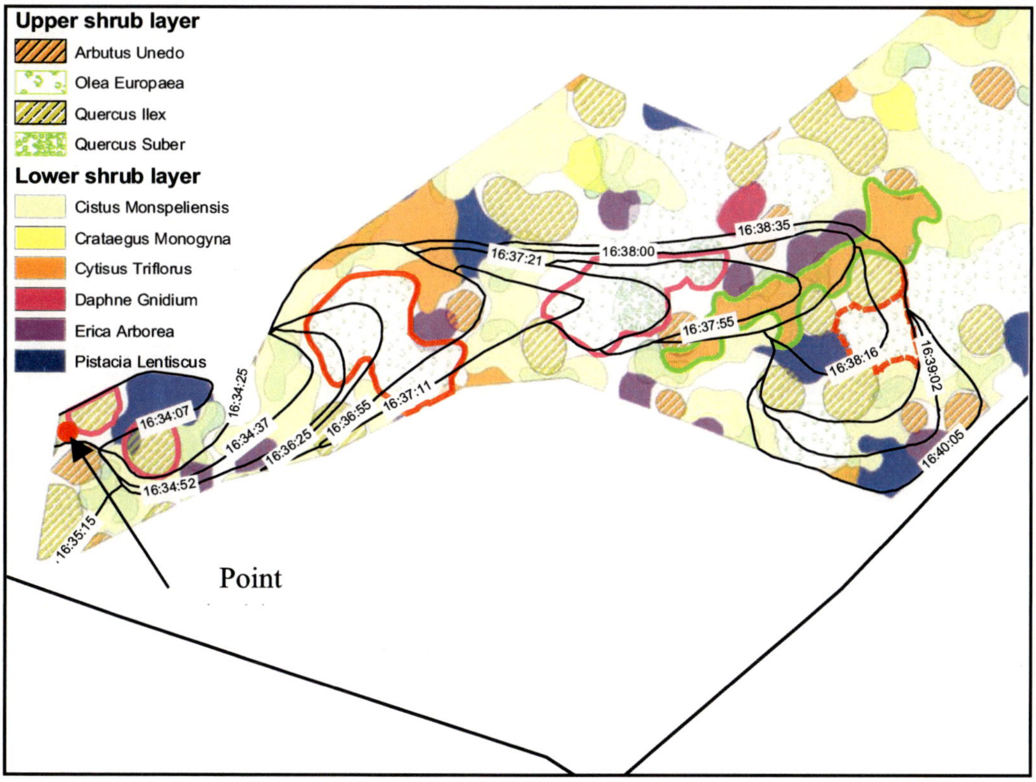

Figure 17. Vegetation cover and diagram of the fire perimeters over time for the first head shape.

Wind Effects on the Fire Behaviour

It is well known that rate of fire spread increases with wind velocity [75, 76]. Combustion is controlled by the oxygen and volatiles supplied to the reaction zone. The arrival of oxygen depends on the wind direction and intensity or/and on the natural convection created by the rising hot plume. The higher the wind speed, the higher the combustion efficiency and the bigger is the heat released by the flame. Furthermore, if wind is blowing in the direction of the fire front advance, flames are tilted forward. There are brought closer to the unburned fuel, increasing the radiation and convection impinging on the fuel and thus the rate of spread. In order to discuss the wind effects on fire behaviour, we consider the measurements of the anemometer located on the right side of the parcel. It was not submitted to the induced wind and smokes generated by the fire, which spread downward and provided information on the wind blowing through the fire front and influencing it. Wind speed and direction from that sensor are presented respectively in figures 18a and 18b (where 0° is for north and 90° for est), over the duration of the spreading displayed in figure 17. We can see that there was a variation on local wind conditions both on speed and direction. Wind velocities were always in the low to moderate range and varied between 2 to 6 m/s at this location. Wind direction shifted from southwest to northwest and reciprocally over time. In order to evaluate the wind influence on fire spread, we have superimposed on figures 18a and 18b, the two periods of time for which the fire stopped (identified from video analysis). The second time interval corresponds obviously to low wind speeds (lower than 3m/s in average) but the first one occurred during moderate wind activity (mean wind speed of 4.5 m/s) for which fire spreads the rest of the time. The reason for this stopping in fire spread is to be found in the vegetation effects and will be detailed in the next section. With regard to the periods of time associated with fire propagation, we can note that highest rate of spread is strongly correlated with highest wind speed (see figures 17 and 18a at time interval [16:36:25, 16:38:00]). At those moments, the surface wind was able to overcome the fire's buoyancy resulting in very inclined flames. This increased the heat transferred towards the unburned fuel, resulting in accelerating the fire rate of spread. Figures 17 and 18b show that the main directions of wind and fire are strongly correlated. During the first period of spreading, the fire is driven by the west/south-western wind. While the second period of fire activity the shifts in wind direction from west/south-west to west/north-west did the fire turned on the right. Vegetation effects further increased this change in fire spread.

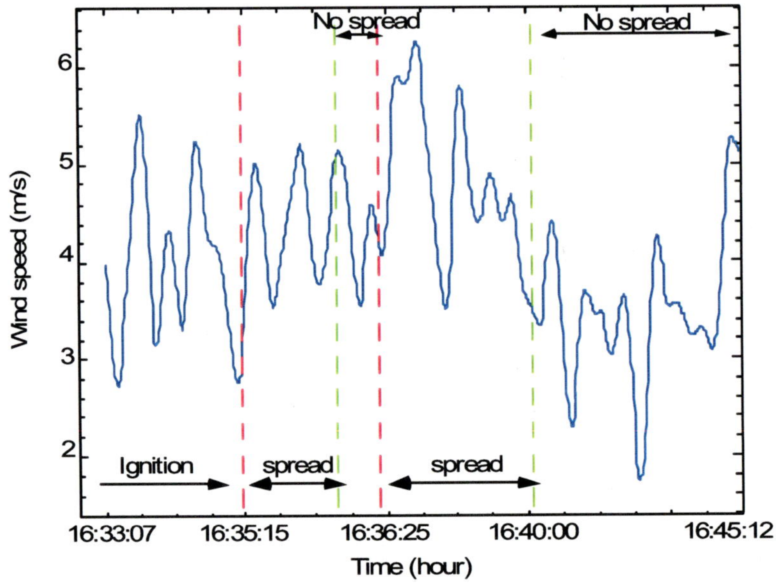

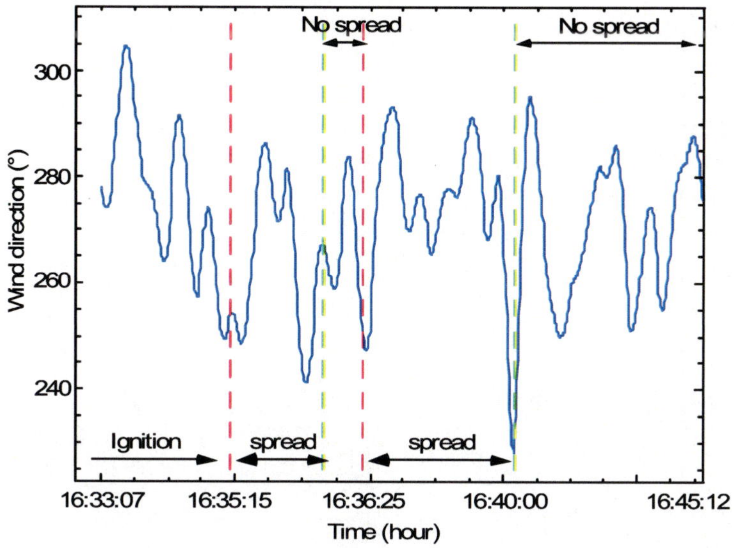

Figure 18. Wind speed (a) and direction (b) during the fire spread.

Vegetation Effects on the Fire Behaviour

We noticed four important effects of vegetation on fire spread. The first two were most probably increased because of the high moisture content of vegetation (rainy spring). The first effet deals with the rate of spread and the second one with the changing in direction. Concerning the first effect, we noted that the first attenuation in fire spread corresponded to the entering of the fire in a grove of Olea Europaea (surrounded in red in figure 17). As noticed previously, we observed a break in fire spread, even if there was a moderate wind activity on the parcel for which fire spreads the rest of the time. We attribute this behaviour to the high moisture content of this species (see table 3) and to the low preheating effect of small flames from the lower shrub layer on its foliage. Indeed, Olea Europaea groves are 1-1,5 m higher than the species of the lower shrub layer (cf. table 3). This layer effect is not usually observed in dryer conditions. Conversely, under other conditions, the fire passed through groves of the upper layer without slowing down. This effect was observed for time intervals [16:37:11, 16:37:55] when the fire passed quickly through a Quercus Suber and Olea Europaea grove (surrounded in purple in figure 17), following the wind direction. These intervals correspond to the higher wind speeds observed during the experiment (see figure 18a). The second effect was observed during the time interval [16:38:00, 16:40:05]. The wind was blowing approximately in the 280° direction (see figure 18b) and the fire followed a direction of 315-320°. We attribute also this behaviour to the presence of an Olea Europaea grove (surrounded in dashed red lines in figure 17) on the left side of the fire front that acted as an inhibition zone. The resulting changing in direction ranged between 35-40° from the wind direction that is substantial. An acceleration of the fire involved also a changing in direction. For instance, we observed that a Cytisus Triflorus grove burned quickly during time interval [16:37:21, 16:38:00] (see figure 17, grove surrounded in green), involving a fire acceleration and a slight changing in direction. The third effect that should be mentioned concerns the species of the upper layer: after an induction phase before ignition we observed a torching effect of groves that induced spotting. The last effect is related to the grass layer. Grass had no influence when fire spread but played a great role in some attenuation intervals as it sustained the fire. It was the single burning layer and the fire grew again from it, thanks to wind squalls.

Flame Front Geometry

The processing of the infrared images allows measuring the characteristic length scales of the flame reaching the boundary of the plot. This technique was applied to 700 images of the fully developed buoyant diffusion flame. The intermittencies of both flame fronts, which represent the flame presence probability contours, are provided in figure 19.

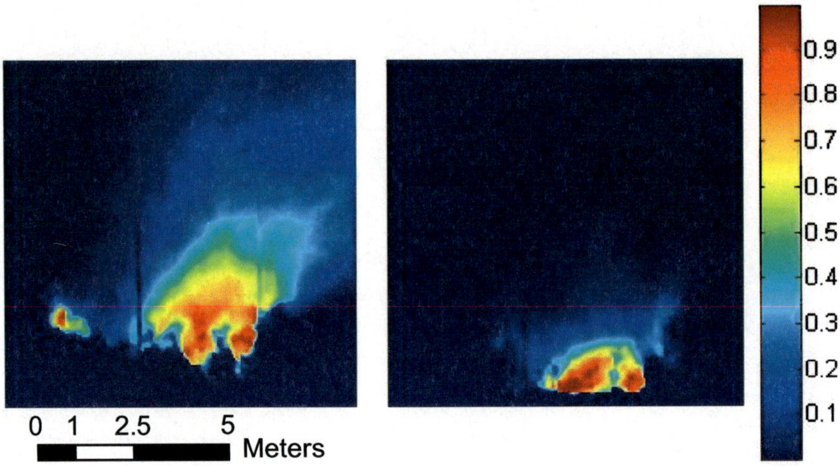

Figure 19. Intermittency of the flame front for fire spread (a) at the upper and (b) lower part of the vegetation.

It should be noticed that the structure of the flame at field scale differs from the one encountered at laboratory scale. In particular for both fire spreads, the intermittent flame region was larger than the one measured in the previous works at the laboratory. This could be due to the effects of the wind and the properties of the vegetation but more experimental fires are required to confirm this result. The ratio of the continuous flame height to the maximum flame height is lower than 0.25. The intermittent flame represents here about 75% of the maximum height of the flame while it represented only 65% at the laboratory scale, as shown previously. The value of the characteristics length scales, for both fire spreads in the different vegetation layers, are provided in Table 5.

Table 5. Characteristic length scales of the flame front measured in upper and lower strata.

	Height of the flame base	**Continuous length, L_c**	**Mean length, L_f**	**Maximum length, L_m**
Upper stratum	0.7 – 1.8 m	2.3 m	4.0 m	7.2 m
Lower stratum	0.0 m	0.7 m	1.3 m	3.0 m

Flame Front Temperature and Structure of the Fire Plume

The "thermocouple trees", built to study the temperature with height during the fire plume formation process, necessitate obviously that the fire enters the measuring section and that the flames remain vertical long enough to observe the relevant phenomena. However, one of the disadvantages inherent to field experiment is the impossibility to have a control over the meteorological conditions. Either a small shift in wind direction or a strong wind speed overcoming the fire plume could respectively make the fire does not get through the measuring structures or stretched the flame and alter the measurements.

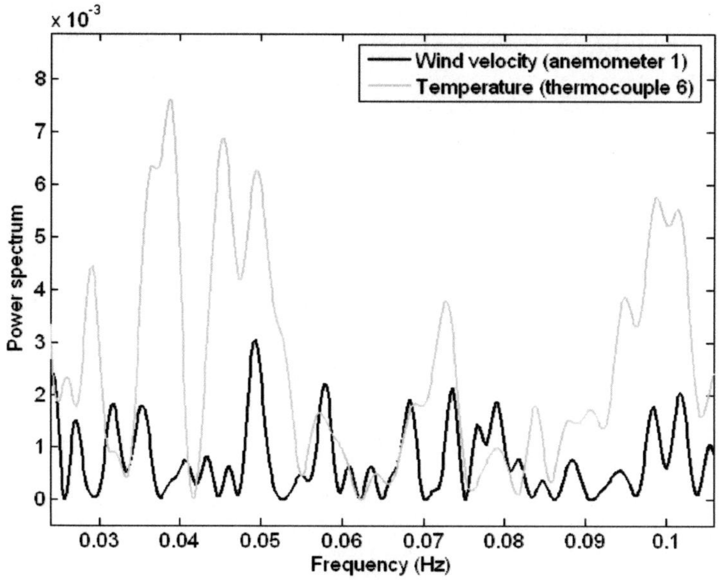

Figure 20. Frequency content, between 0.03 and 0.1 Hz, of wind and temperature of thermocouple 6 at mast 2.

For instance, the first "thermocouple tree" placed on the right side was not reached by the fire. Concerning the second one, located in the middle of the plot, there were variations on local wind conditions when the fire reached it. The thin thermocouples used, with 50 μm wire diameter, allow recording the fluctuations of the temperature when the flame front was in the measuring section. The frequency content of wind velocity and temperature curves (for thermocouple 6 located above the vegetation), between 3.10^{-2} and 11.10^{-2} Hz, is provided in figure 20. These results show that the frequency content of the temperature and wind velocity exhibit similarities within this frequency range. The temperature and wind velocity seems to be correlated and the characteristic duration of the gusts of wind, which affect the temperature, range from 10 to 30 s. It should be noticed that the effect of wind on temperature fluctuations is less significant for the thermocouples located inside the vegetation. The wind hardly penetrates inside the vegetation.

With regard to the third "thermocouple tree" located on the left side of the plot, lull in wind occurred when the fire reached it. The wind speed decreased to less than 3 m/s and the fire plume overcame the surface wind. The flames remained vertical for a long time at that location (up to 20 seconds). Figure 21 shows the distribution of the vertical temperature over time, during part of this stage (time interval [16:48:30, 16:48:42]), for height of the ten available levels (0.6; 1.2; 1.8; 2.4; 3.0; 3.6; 4.8; and 6 meters), simplifying the analysis. The first three levels (thermocouples number 1 to 3) take place inside the vegetation zone while the others (thermocouples 4, 5, 6, 8 and 10) are located above it. Those results show the three classical behavioural regions encountered in free burning fires. The lower part of the fire plume is the continuous flame region containing the vegetation zone where the temperature is maximum reaching 1060°C for thermocouple 1, 2 and 3. Over that region, there is the intermittent one, where balls of flame come away from the source. This zone includes thermocouple 4 and 5 where the temperature grows up to 700°C. During this short time interval it is not visible for thermocouple 5 but other time intervals make it clearer. The upper part is the beginning of the conventional thermal plume. Thermocouple 6 to 10 belong to this region, the cooling of which is mainly due to entrainment and mixing of ambient air.

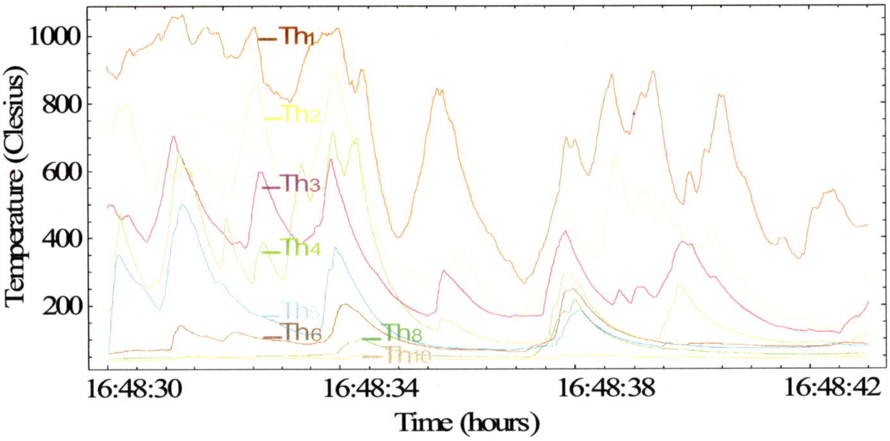

Figure 21. Gas temperature over time, along the fire plume axis at the third "thermocouple tree".

It should be pointed out that important discrepancies appear between those results and data from laboratory scale experiments [21]. In the latter, we could see that the temperature greatly fluctuates above the fuel. This last result was attributed to the turbulent feature of the flow above the ejection zone of the pyrolyzate gases. This pattern was not encountered here. The reason may to be found in the scale of the eddy structures originating from the flames at field level that are bigger than those observed at laboratory scale. Furthermore, the temperature increase just above the vegetation mentioned by different authors [21, 82] for small pine needle fuel beds was not observed. This last characteristic was attributed to an oxygen deficit due to the ejection of the pyrolyzate gases, which makes the combustion incomplete. It was not the case in our experiment where the mixing between fuel and oxidant led to combustion even in the vegetative structure.

Fire Impact

The measurement of the thermal radiation emitted ahead of the flame front can be analyzed thanks to heat fluxes recorded at 5, 10 and 15 meters from the upper limit of the plot. In this study, radiation appears to be the dominant heat transfer mechanism involved in the long-range preheating process of the vegetation. As the flame front spread and got closer to the sensors, the radiant heat flux increased. The comparison between radiant and total heat fluxes measurements show that convection incident beyond 5 m ahead of the flame front is negligible and will not be discussed here. The sensors located at two and four meters high gave similar measurements. The radiant heat flux measurements at 5 m from the flame front, for the two fire spreads ranged from 3 to 8 kW/m^2 and from 2 to 7.5 kW/m^2, respectively. The flux measured during the first spread was greater due to taller flames. The peak value was greater than 6.4 kW/m^2 during more than 8 s, which is the limit of time to feel a pain on an exposed skin [86]. At 10 meters from the plot, the maximum heat flux measured was 3 kW/m^2. This value corresponds to a pain felt on exposed skin after 120 s [86]. This time interval was not observed during that part of the fire spread. At 15 meters, the peak of the heat flux was twice the summer sunshine at that place.

Mathematical Modelling of Surface Forest Fire

This part details the reduction of a multiphase formulation which can be used to develop or to improve simple physical models of forest fire spread. The first section recalls the multiphase concept and the set of governing equations. Then, a simple physical model, to be improved is described in the second section. The multiphase reduction is presented in the third section wherein the resulting improved equation of energy and simplified flow of the simple model are detailed. The fourth section concerns the presentation of that reduced model. The numerical method used to solve it is described in the last section.

Theoretical Balance Equations

We present here the detailed multiphase approach mainly developed by [56]. The fire spread medium is represented as multiphase, reactive and radiative. This medium includes the fluid phase and N solid phases. Each solid phase consists of a set of particles that possess the same geometry and thermochemical properties. As it is impossible to study forest fire behaviour at the particle scale, an elementary multiphase volume is defined to carry out averaged properties of both gaseous and solid phases. This last volume should be smaller than the scale of the phenomenon but greater than the size of the particle. The entire set of multiphase averaged equations is obtained in two steps. Firstly, point equations for the fluid and fuel phases, as well as the interface conditions, are written [91]. Secondly, the set of obtained equations is space averaged on the multiphase medium [92]. In those equations, the correlation between all state variables is assumed equal to one. That allows simplifying them by neglecting the effects of cross correlation terms. For reasons of clarity, no symbol indicating that the variables are volume averaged was added.

Gas Phase
Mass Equation

$$\frac{\partial}{\partial t}\left(\alpha_g \rho_g\right)+\vec{\nabla}.\left(\alpha_g \rho_g \vec{V}_g\right)=\sum_{k=1}^{N}\left[\dot{M}\right]_{gk} \tag{22}$$

Chemical Species Equation

$$\frac{\partial}{\partial t}\left(\alpha_g \rho_g Y_g^j\right)+\vec{\nabla}.\left(\alpha_g \rho_g Y_g^j \vec{V}_g\right)+\vec{\nabla}.\left(\alpha_g \rho_g Y_g^j \vec{V}_g^j\right)-\alpha_g \rho_g \dot{\omega}_g^j=\sum_{k=1}^{N}\left[\dot{M}\right]_{gk}^{j} \tag{23}$$

Momentum Equation

$$\frac{\partial}{\partial t}\left(\alpha_g \rho_g \vec{V}_g\right)+\vec{\nabla}.\left(\alpha_g \rho_g \vec{V}_g \vec{V}_g\right)-\vec{\nabla}.\left(\alpha_g \overline{\overline{\pi}}_g\right)-\alpha_g \rho_g \vec{g}=\sum_{k=1}^{N}\left[\dot{M}\vec{V}\right]_{gk}+\sum_{k=1}^{N}\left[\vec{\Pi}\right]_{gk} \tag{24}$$

Total Energy Equation

$$\begin{aligned}&\frac{\partial}{\partial t}\left(\alpha_g \rho_g e_g\right)+\vec{\nabla}.\left(\alpha_g \rho_g e_g \vec{V}_g\right)+\vec{\nabla}.\left(\alpha_g\left(\vec{q}_g+\vec{R}_g\right)\right)-\vec{\nabla}.\left(\alpha_g \overline{\overline{\pi}}_g.\vec{V}_g\right)\\&-\alpha_g \rho_g \vec{g}.\vec{V}_g=\sum_{k=1}^{N}\left[\dot{M}e\right]_{gk}-\sum_{k=1}^{N}\left[q\right]_{gk}-\sum_{k=1}^{N}\left[R\right]_{gk}+\sum_{k=1}^{N}\left[\vec{\Pi}.\vec{V}\right]_{gk}\end{aligned} \tag{25}$$

Phase (One Equation per Phase)
Mass Equation

$$\frac{\partial}{\partial t}\left(\alpha_k \rho_k\right)=-\left[\dot{M}\right]_k^{surf}-\left[\dot{M}\right]_k^{pyr} \tag{26}$$

Chemical Species Equation

$$\frac{\partial}{\partial t}\left(\alpha_k \rho_k Y_k^j\right)=-\left[\dot{M}\right]_k^{surf,j}-\left[\Gamma\right]_k^{surf,j}-\left[\dot{M}\right]_k^{pyr,j} \tag{27}$$

Momentum Equation

$$\vec{V}_k=\vec{0} \text{ (Motionless phases assumption)}$$

Total energy equation

$$\frac{\partial}{\partial t}\left(\alpha_k \rho_k e_k\right)=-\left[\dot{M}e\right]_k^{surf}-\left[\dot{M}e\right]_k^{pyr}-\left[q\right]_k-\left[R\right]_k+\left[\vec{\Pi}.\vec{V}\right]_k^{pyr} \tag{28}$$

Interface Equations (N Equations)
Mass

$$\left[\dot{M}\right]_{gk}=\left[\dot{M}\right]_k^{surf}+\left[\dot{M}\right]_k^{pyr} \tag{29}$$

Species

$$\left[\dot{M}\right]_{gk}^{j} = \left[\dot{M}\right]_{k}^{surf,j} + \left[\Gamma\right]_{k}^{surf,j} + \left[\dot{M}\right]_{k}^{pr,j} \tag{30}$$

No Momentum Interface Equation (Motionless Solid Phases)
Energy

$$\left[\dot{M}e\right]_{gk} - \left[q\right]_{gk} - \left[R\right]_{gk} + \left[\vec{\Pi}.\vec{V}\right]_{gk} = \left[\dot{M}e\right]_{k}^{surf} + \left[\dot{M}e\right]_{k}^{pyr} + \left[q\right]_{k} + \left[R\right]_{k} - \left[\vec{\Pi}.\vec{V}\right]_{k}^{pyr} \tag{31}$$

Radiative Transfer Equation

$$\vec{e}.\vec{\nabla}\left(\alpha_g L_g^{\Omega}\right) = \alpha_g a_g \left(\frac{\sigma_B T_g^4}{\pi} - L_g^{\Omega}\right) + \sum_k \left[\frac{\alpha_k \sigma_k^S}{4}\left(\frac{\sigma_B T_k^4}{\pi} - L_g^{\Omega}\right)\right] \tag{32}$$

The radiant and conductive heat fluxes into solid phases are not neglected. We assume that solid particles are thermally thin and opaque to radiation. The radiant heat flux, in equation (25) is determined from the radiative transfer equation by using the following relation

$$\vec{R}_g = \int_{\Omega} L_g^{\Omega}\, \vec{e}\, d\Omega \tag{33}$$

This set of equations, will be reduced to improve the simple physical model of forest fire spread presented in the next section. It should be mentioned that different sub-models appear on the right side of those balance equations that need to be determined. These are not detailed here as they will not be used in the following.

Simple Physical Model of Fire Spread

Due to the amount of physical phenomena and state variables involved in fire behaviour, it is necessary to make simplifying hypotheses in order to generate a simple model to be used in a fire simulator. The thermal balance was written considering a medium equivalent to the gas and solid complex and assuming a thermal equilibrium between the two phases. We obtained the following reaction-diffusion formulation which provides the framework of the model, taking into account heat transfers mechanisms for total depth control volumes

$$\frac{\partial T}{\partial t} = -h_a(T - T_a) + K\Delta T + R - Q_{eq}\frac{\partial s_k}{\partial t} \quad \text{on the fuel complex.} \tag{34}$$

$$s_k = s_{k_0}, \quad R = \frac{a_k}{m_{eq}} \varepsilon_{fl}\, \sigma_B\, T_{fl}^4 \int_{S_{fl}} \frac{\cos\varphi_{fl}\, \cos\varphi_k}{\pi r^2} dS_{fl} \quad \text{for unburned cell ahead of the front.} \tag{35}$$

$$s_k = s_{k_0} e^{-\gamma_c (t - t_{ig})}, \quad R = 0 \qquad \text{for a burning cell.} \tag{36}$$

With the following boundary and initial conditions:

$$\vec{n}.\vec{\nabla}T = 0 \text{ at the fuel-bed boundaries.} \tag{37}$$

$$T_0 = T_a \text{ for an unignited cell at time zero.} \tag{38}$$

$$T_0 = T_{ig} \text{ for an ignited cell at time zero.} \tag{39}$$

The whole coefficients of equation (34) are called reduced coefficients as they are divided by the thermal mass per unit area of the medium equivalent, m_{eq}. Heat transfers and production processes in that thermal balance are determined as follows:

- The energy exchanged between a control volume and the surrounding air is assumed to be proportional to the difference between the temperature of this volume and the ambient one.
- The heat transferred between a control volume and its neighbouring in the burning zone located inside the litter is mainly due to embers radiation. These exchanges are represented by a single diffusion term.
- The combustion reaction is assumed to occur above a threshold temperature, T_{ig} , the value of which, for forest fuels, is commonly assumed to be 300°C. Fuel particles are dry and moisture content is not considered in this approach. All the solid degradation processes (pyrolysis and char combustion) as well as the combustion in the gaseous phase are modeled by a single simplified exothermic reaction. During reaction, the fuel mass is assumed to decrease exponentially (equation (36)). The heat generation per unit fuel mass is assumed to be constant during the combustion process.
- The flame is modelled as a radiant surface (equation (35)) as described hereafter.

The temperature T_{fl} and emissivity ε_{fl} of this radiant surface are assumed constant. T_{fl} is supposed to be equal to the temperature of the combustion zone (provided by the model of fire spread). The geometric properties of the radiant surface, namely its length and tilt angle (figure 22), are supposed to be known a priori. Their values are measured during the experiments. The amount of radiant energy impinging on the top of the litter is determined by means of the Stefan-Boltzmann law. This law makes it possible to determine the part of radiation emitted from an element of the flame front of area dS_{fl}, which impinges on unburned fuel element dS_k [70]. This amount of energy is attenuated according an inverse square law of the distance r between the two areas and also depends on the cosine of angles φ_{fl} and φ_k defined between the ray and the normal of dS_{fl} and dS_k, respectively (figure 22). The rate at which the radiant energy from the flame-front is absorbed by the fuel element dS_k is

$$\psi_{fl-dk} = a_k \, \varepsilon_{fl} \, \sigma_B \, T_{fl}^4 \int_{S_{fl}} \frac{\cos\varphi_{fl} \, \cos\varphi_k}{\pi \, r^2} dS_{fl} \, dS_k \tag{40}$$

where integral over S_{fl} is over the flame surface. Thus, the parameter R in equation (35) represents the ratio of the radiant heat flux from flames intercepted by the fuel to the thermal mass of the fuel per unit area

$$R = \frac{\psi_{fl-dk}}{m \, dS_k} \tag{41}$$

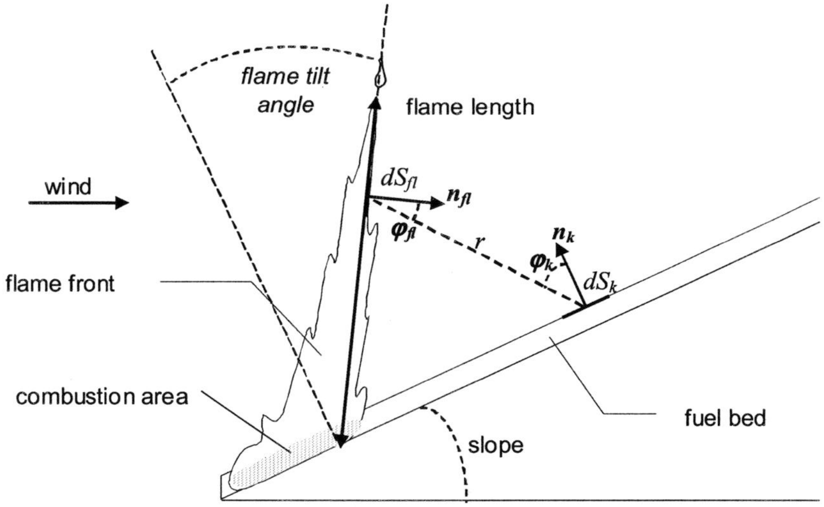

Figure 22. Radiative transfers between elementary surfaces of the flame and of the fuel bed

The model parameters h_a, K, Q_{eq} and γ_c are determined from a measured time-temperature curve obtained for a linear spread under no slope and no wind conditions [64]. They are identified once for a given fuel and fuel load and remain valid for all the experiments considered hereafter, whatever the slope and wind are.

Reduction of the Multiphase Formulation

Although the previous model was able to predict both strong slope and combined slope and low wind effects, it failed to describe fire behaviour at increasing wind velocity. The following theoretical study was conducted based on the reduction of the multiphase set of equation in order to improve it. The multiphase model reduction was performed in four steps. Firstly, as the simple model of fire spread is two-dimensional, we reduced the three dimensional multiphase set of equations to two dimensions. Secondly, since it possesses single energy conservation equation, the thermal balances of the multiphase model was reduced to a single equation. Thirdly, the resulting conservation equation of energy was written in terms of temperature. This last result showed that an advection term was missing in the simple model. It led to the last step of the reduction which concern the elaboration of a simplified flux model to determine the gas velocity trough the fuel layer $\vec{V}_g$.

Reduced Energy Equation

Equations (25) and (28) are reduced in two dimensions by averaging over the height of the fuel layer d_l thanks to the following operator

$$\overline{\overline{f}}(x,y,t)=\frac{1}{d_l}\int_0^{d_l} f(x,y,z,t)\,dz \tag{42}$$

The phenomena occurring above this layer are taken into account by way of the boundary conditions at d_l that appear when managing the average operation. The resulting equation is then simplified further by making the following assumptions:

- Pressure, stress, gravity and conduction contributions are neglected in the gas phase
- The state variable values at d_l are considered equal to their mean values in the fuel bed

We obtain the following equations of energy expressed in terms of internal energy by subtraction of the kinetic energy balance

$$\frac{\partial}{\partial t}\left(\overline{\overline{\alpha_g\,\rho_g\,u_g}}\right)+\vec{\nabla}_s\cdot\left(\overline{\overline{\alpha_g\,\rho_g\,u_g\,\vec{V}_{g,s}}}\right)+\frac{\left[\alpha_g\rho_g u_g w_g\right]_0^{d_l}}{d_l}+\vec{\nabla}_s\cdot\left(\overline{\overline{\alpha_g\,\vec{R}_{g,s}}}\right)+\frac{\left[\alpha_g R_{g,z}\right]_0^{d_l}}{d_l}= \sum_{k=1}^{N}\left[\overline{\overline{\dot{M}h}}\right]_{gk}-\sum_{k=1}^{N}\left[\overline{\overline{q}}\right]_{gk}-\sum_{k=1}^{N}\left[\overline{\overline{R}}\right]_{gk} \tag{43}$$

$$\frac{\partial}{\partial t}\left(\overline{\overline{\alpha_k\,\rho_k\,u_k}}\right)=-\left[\overline{\overline{\dot{M}h}}\right]_k^{surf}-\left[\overline{\overline{\dot{M}h}}\right]_k^{pyr}-\left[\overline{\overline{q}}\right]_k-\left[\overline{\overline{R}}\right]_k \tag{44}$$

In order to obtain a single equation of energy, equations (43) and (44) are added for all the N solid phases. Thanks to the interface relation (equation (31)), the right hand side of the resulting equation is equal to zero. For reasons of clarity, we omit the signs of average.

$$\begin{aligned}&\frac{\partial}{\partial t}\left(\alpha_g\rho_g u_g\right)+\vec{\nabla}_s\cdot\left(\alpha_g\rho_g u_g\vec{V}_{g,s}\right)+\frac{\left[\alpha_g\rho_g u_g w_g\right]_0^{d_l}}{d_l}\\&+\vec{\nabla}_s\cdot\left(\alpha_g\vec{R}_{g,s}\right)+\frac{\left[\alpha_g R_{g,z}\right]_0^{d_l}}{d_l}+\sum_{k=1}^{N}\frac{\partial}{\partial t}\left(\alpha_k\rho_k u_k\right)=0\end{aligned} \tag{45}$$

The modification of equation (45) to derive the thermal balance is achieved in two steps. First, we transform the internal energy into enthalpy. Then, we express the obtained equation in terms of temperature by using the following relation and by assuming a constant heat capacity with temperature for each chemical species j

$$h^j = C_p^j(T - T_0) + h^{j,0} \tag{46}$$

where $h^{j,0}$ represents the formation enthalpy at temperature T_0. To further simplify equation (45), we assumed thermal equilibrium between gas and solid phase inside the fuel bed. The assumption of thermal equilibrium between the two phases has already been made in several physical models [46,64]. It was verified in certain configurations as previously mentioned in the section devoted to the presentation of the experimental data. This is a very useful assumption that allows a description of propagation without describing the finest mechanisms. Finally, we obtain the following thermal balance

$$\begin{aligned}&\left(\alpha_g \rho_g C_{p,g} + \alpha_k \rho_k C_{p,k}\right)\frac{\partial T}{\partial t} + \alpha_g \rho_g C_{p,g} \vec{V}_{g,s}.\vec{\nabla}_s T + \vec{\nabla}_s.\left(\alpha_g \vec{R}_{g,s}\right)\\ &+\frac{\left[\alpha_g R_{g,z}\right]_0^{d_l}}{d_l} = -\left[\dot{M}\right]_k^{pyr} L^{pyr} - \left[\dot{M}\right]_k^{surf} \Delta H^{surf} - \sum_j \alpha_g \rho_g \dot{\omega}_i h_g^j\end{aligned} \tag{47}$$

This equation contains the mean mechanisms involved in propagation, such as convection, radiation and reactions. It is expressed in a form that enables resolution through the expression of appropriate sub-models. Equation (47) is only a part of the whole reduced multiphase model which can be derived from equations (22) to (32). However, this reduced model remains too far from a simple model to be integrated within an operating management tool. We will thus used equation (47) to improve our simple model (equations (34) to (41)).

Advective Transfers in the Simple Model

The main difference between the energy equations of our model and the reduced multiphase model one consisted in the advection contribution, which was omitted in our model:

$$\alpha_g \rho_g C_{p,g} \vec{V}_{g,s}.\vec{\nabla}_s T \tag{48}$$

Hence, we added a similar term to equation (34) leading to the improved thermal balance

$$\frac{\partial T}{\partial t} + k_w \vec{V}_g.\vec{\nabla} T = -h_a T - T_a) + K\Delta T + R - Q_{eq}\frac{\partial s_k}{\partial t} \tag{49}$$

The added term includes the gas velocity $\vec{V}_g$ and coefficient k_w, which necessitate to be determined. Coefficient k_w was derived from equation 45. We obtained

$$k_w = \frac{\alpha_g \, \rho_g \, d_l \, C_{p,g}}{m_{eq}} \tag{50}$$

To calculate k_w, we set the following supplementary hypotheses:

- The specific heat of the gas mixture is taken as constant and equal to $1360\,J\,kg^{-1}$.
- α_g, the porosity of the multiphase medium, which is 0.97 for pine needles, is assumed constant and equal to unity.
- m_{eq} is taken as constant according to [64].
- We considered that the gas mixture is a perfect gas verifying the quasi-isobaric approximation.

Finally, we obtain the following expression for k_w

$$k_w = \frac{\alpha_g \, \rho_a \, d_l \, C_{p,g}}{m_{eq}} \cdot \frac{T_a}{T} = k_w^* \cdot \frac{T_a}{T} \tag{51}$$

where k_w^* is a constant and k_w is only a function of temperature.

The Simplified Flow

A simplified model of flow was developed to calculate the gas velocity $\vec{V}_g$ (see equation (49)).

Determination of the Horizontal Velocity

We start from the mass balance equation of the gas phase (equation (22)). To simplify it, we assume as previously that the volume fraction of the gas phase remains constant ($\alpha_g \approx 1$) and that there is only one solid phase. Furthermore, we consider that the flow is quasi-static (i.e. we assume that wind variations with respect to temperature occur infinitely faster than temperature variations with respect to wind) and follows the direction of the slope. So, equation (22) becomes

$$\rho_g \frac{\partial V_{g,x}}{\partial x} + \rho_g \frac{\partial V_{g,z}}{\partial z} + V_{g,x} \frac{\partial \rho_g}{\partial x} + V_{g,z} \frac{\partial \rho_g}{\partial z} = \left[\dot{M}\right]_{gk} \tag{52}$$

Then, we apply the averaging procedure along the thickness d_l of the fuel bed (equation (40)). As the temperature profile is assumed constant throughout the height of the litter in the equation of energy, we have (omitting the average symbols).

$$\frac{\partial V_{g,x}}{\partial x} + \frac{V_{g,x}}{\rho_g} \frac{\partial \rho_g}{\partial x} = -\frac{\left[V_{g,z}\right]_0^{d_l}}{d_l} + \frac{1}{\rho_g}\left[\dot{M}\right]_{gk} \tag{53}$$

We also assume that the vertical component of the gas velocity follows a boundary layer profile (i.e. $V_{g,z}(0)=0$). We thus obtain

$$\frac{\partial V_{g,x}}{\partial x}+\frac{V_{g,x}}{\rho_g}\frac{\partial \rho_g}{\partial x}=-\frac{V_{g,z}(d_l)}{d_l}+\frac{1}{\rho_g}\left[\dot{M}\right]_{gk} \tag{54}$$

In order to solve this differential equation, the variables are determined as follow. The gas density is provided by the isobaric perfect gas law (quasi-isobaric flames)

$$\rho_g T_g = \rho_a T_a \tag{55}$$

The mass increase in the gas phase is obtained from the sub-model of fuel mass loss (equation (36)) assuming that the mass lost by the vegetal is gained by the gas phase.

$$\left[\dot{M}\right]_{gk}=-\frac{1}{d_l}\frac{\partial s_k}{\partial t}=\frac{\gamma\, s_{k0}\, e^{-\gamma(t-t_{ig})}}{d_l} \tag{56}$$

In order to close the simplified flow model, we need to determine the ascensional gas velocity at the top of the fuel bed $V_{g,z}(d)$ present in equation (53).

Determination of the Ascensional Velocity

We start from the momentum equation of the gas phase (equation (24)). By assuming, as above, that the volume fraction of the gas remains constant, that there is only one solid phase and that the flow is quasi-static, we reformulate the momentum equation in its rotational form and we subtract the mass balance (equation (22)) to obtain

$$\rho_g \vec{V}_g \wedge \left(\overrightarrow{rot}\, \vec{V}_g\right)+\frac{1}{2}\rho_g \vec{\nabla}\vec{V}_g^2=\vec{\nabla}.\overline{\overline{\pi}}_g+\rho_g\vec{g}+\left[\dot{M}\vec{V}\right]_{gk}+\left[\vec{\Pi}\right]_{gk}-\vec{V}_g\left[\dot{M}\right]_{gk} \tag{57}$$

Then, we neglect the momentum sources, the shear stresses in the gas phase and we represent the stresses at the solid/gas interface only by the drag forces. We also assume that the flow is irrotational and we set the hydrostatic pressure assumption. We obtain the following momentum equation projected onto the vertical axis

$$\frac{\partial}{\partial z}\left(V_{g,x}^2+V_{g,z}^2\right)=2\left(\frac{\rho_a}{\rho_g}-1\right)g\,\cos\phi_{sl}-C_f\,V_{g,z}^2 \tag{58}$$

This relationship implies that the vertical velocity only depends on the buoyancy and drag force effects, and that the other effects are negligible [93]. We then apply the averaging procedure along d_l, while assuming that the profile of the vertical component of velocity is linear in the fuel bed.

$$\frac{1}{d_l}\int_0^{d_l} V_{g,z}^2\, dz = \frac{1}{3}V_{g,z}^2(d_l) \tag{59}$$

We therefore have (without any average symbol)

$$\frac{1}{d_l}\left[V_{g,x}^2 + V_{g,z}^2\right]_0^{d_l} = 2\left(\frac{\rho_a}{\rho_g} - 1\right) g\,\cos\phi_{sl} - \frac{C_f}{3}V_{g,z}^2(d_l) \tag{60}$$

We set an additional hypothesis, already used in the simple model: we assume that the profile of the horizontal component of velocity is constant along the thickness of the litter. Then, by using the isobar perfect gas law we obtain the following relation

$$V_{g,z}(d_l) = \chi\sqrt{2\,d_l\left(\frac{T_g}{T_a} - 1\right) g\,\cos\phi_{sl}} \tag{61}$$

with

$$\chi = \sqrt{\frac{3}{3 + d_l\,C_f}} \tag{62}$$

C_f was estimated from the physical properties of pine needles [94]. To proceed, we used an empirical law, which considers needles as cylinders distributed in a random manner all along the fuel bed [95]. We obtained the following estimation for χ

$$\chi = 0.33 \tag{63}$$

The aim of this section was to obtain a simple description of the flow in order to improve the predictions of our model in the configuration of wind-driven fires, while keeping a short computational time. This approach is rather original because most of the models based on a thermal balance do not describe the flow and often assume a constant wind over the spreading zone [34,44]. Nevertheless, it is clear that this work remains to be extended to a two-dimensional description of the flow in the aim of obtaining a model useful for a simulator.

Reduced Physical Model of Fire Spread

By coupling equation (49) with the simplified flow, we obtain a new model that contains a thermal balance for the medium equivalent to the litter, a mass balance for the gas phase, a momentum equation along the vertical axis and the isobaric perfect gas law. This leads us to the following formulation:

$$\frac{\partial T}{\partial t} + k_w \vec{V}_g . \vec{\nabla} T = -h_a T - T_a) + K\Delta T + R - Q_{eq} \frac{\partial s_k}{\partial t} \text{ on the fuel complex.} \quad (64)$$

$$s_k = s_{k_0}, \quad R = \frac{a_k}{m_{eq}} \varepsilon_{fl} \sigma_B T_{fl}^4 \int_{S_{fl}} \frac{\cos\varphi_{fl} \cos\varphi_k}{\pi r^2} dS_{fl} \text{ for unburned cell ahead of the front} \quad (65)$$

$$s_k = s_{k_0} e^{-\gamma_c (t - t_{ig})}, \quad R = 0 \quad \text{for a burning cell.} \quad (66)$$

$$\frac{\partial V_{g,x}}{\partial x} + \frac{V_{g,x}}{\rho_g} \frac{\partial \rho_g}{\partial x} = -\frac{V_{g,z}(d_l)}{d_l} - \frac{1}{\rho_g d_l} \frac{\partial s_k}{\partial t} \text{ in the burning zone.} \quad (67)$$

$$V_{g,z}(d_l) = \chi \sqrt{2 d_l \left(\frac{T}{T_a} - 1\right) g \cos\phi_{sl}} \text{ in the burning zone.} \quad (68)$$

$$\rho_g T = \rho_a T_a \text{ in the gaseous medium.} \quad (69)$$

With the following boundary and initial conditions

$$\vec{n} . \vec{\nabla} T = 0 \text{ at the fuel-bed boundaries,.} \quad (70)$$

$$T_0 = T_a \text{ for an unignited cell at time zero.} \quad (71)$$

$$T_0 = T_{ig} \text{ for an ignited cell at time zero.} \quad (72)$$

$$\vec{V}_{g,x} = \vec{V}_\infty \text{ at the inflow of the domain.} \quad (73)$$

Gas density is obtained by using the isobaric perfect gas law. The gas temperature is equal to the temperature of the solid phase (equation (64)) since we assume the thermal equilibrium between the gas and solid in the fuel layer. Furthermore, the validity domain of the simplified equations was limited to the combustion zone. Indeed equation (68), which provides the vertical component of the velocity at the top of the fuel bed was obtained by assuming that the buoyancy effects dominate the other effects. This hypothesis is valid only in the combustion region and the calculation of the local wind can thus only be implemented in it. Outside this zone, we assumed that horizontal velocity remains constant and is equal to the velocity obtained at the interface between the burning and the unburned fuel. Indeed, in the burned and unburned regions, the influence of buoyancy onto the thermal transfers is assumed as becoming negligible and the other effects, as drag forces, are neglected.

Numerical Methods

The thermal balance equation was discretized on a uniform rectangular grid using the finite difference method with an implicit scheme. An "upwind" difference schema was used to take into account the extent of convective transfers in the wind direction. Then, the resulting system of linear algebraic equations was solved using the Jacobi iterative method. The gaseous phase equations were solved using a 4th order Runge-Kutta method. The mesh size was 0.01 m and the time step was between 0.01 s and 0.002 s for the wind velocities considered.

Validation of the Fire Spread Model at the Laboratory Scale

In the following, the model predictions are compared to two series of experiments carried out for horizontal fires in still air and for wind-driven fires across sloping beds.

Validation of the Model for Horizontal Fire Spread in Still Air

Experimental Procedure

Laboratory experiments were conducted on a dedicated combustion table in order to validate the model predictions, for horizontal fire spread in still air. The experimental device (figure 1) and the fuel preparation procedure are similar to those previously detailed. Four runs were performed across pine-needle beds (Pinus pinaster load of 0.5 kg/m^2) to measure some fire front properties. The temperature inside the fuel bed and the radiant heat flux ahead of the fire front were particularly addressed to validate the model predictions. The sampling frequency was 25 Hz. Six K-type thermocouples with 100 μm wire diameter were located inside the bed of pine needles along the spreading direction. The first thermocouple was located at 0.20 m from the edge of the combustion bench, where the ignition line was performed. The other thermocouples were spaced 0.07 m apart, along the central axis of the fuel bed. The time the fire takes to travel the distance between two thermocouples allows to measure accurately the rate of spread and to verify the establishment of the steady state. In order to measure the radiant contribution from the flame front, two heat flux sensors were located at the end of the bed of pine needles. The normal of their sensing area were vertically oriented in order to measure heat fluxes from the flame front impinging on the top of the litter. The two sensors used incorporate balanced thermoelectric panels housed between copper foils which provide response time lower than 0.5 s. Each of these hemispherical sensors is composed of different sensing areas (squares of 1cm^2) located on a heat exchanger. One sensor allows measuring the radiant heat flux and the other is capable to measure the convective and total heat fluxes. In this last case, the radiant heat flux is deduced by subtracting the total heat flux and the convective one. They can measure heat fluxes up to 10 kW/m^2. The calibration of the heat flux sensors was done by submitted them to reference heat fluxes. The accuracy of the sensors' calibration is +/- 3%.

Identification of the Model Parameters

The model parameters h_a, K, Q_{eq} and γ_c are determined from a measured time-temperature curve obtained for a linear spread under no slope and no wind conditions, following the method detailed in [64]. They are identified once for a given fuel and fuel load and remain valid whatever the environmental conditions (slope and wind). The model parameters for fire spread across a 0.5 kg/m^2 load of Pinus pinaster needles are given in table 6. These coefficients are solely fuel dependent. Thus, the classical fuel descriptors (density, surface to volume ratio, moisture content,) as well as the properties of the fuel (bed depth, fuel-element dimensions, fuel-bed porosity) are intrinsically taken into account in these coefficients.

Table 6. Model parameters for a bed of Pinus pinaster needles (fuel load of 0.5 kg.m^{-2}).

Parameter	h_a (s^{-1})	K ($m^2.s^{-1}$)	k_w	Q_{eq} ($m^2.K.kg^{-1}$)	γ (s^{-1})
value	41×10^{-3}	13×10^{-6}	11×10^{-3}	2.34×10^{3}	0.35

Fuel Bed Temperature

The temperature predictions are compared to measurements made with the thermocouples located inside the bed of fuel (figure 23). The predicted rate of spread matches the measured one (0.25 cm/s) since both temperature-time curves are synchronized. The model predictions account satisfactorily for three of the overall four steps phenomena occurring inside the fuel bed: preheating, flaming combustion, char combustion and cooling. The fuel preheating ahead of the flame front occurring during the first step is accurately represented. The temperature rise up to ignition, in this region is attributed to flame front radiation impinging on the top of the litter. During the next step, the maximum temperature predicted by the model, which occurs while the flaming, matches the measurements.

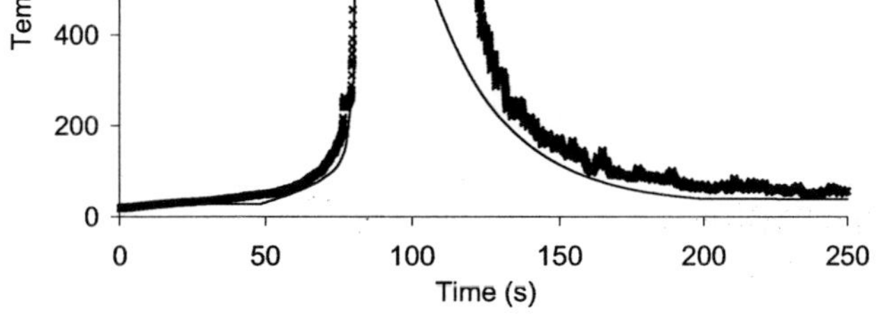

Figure 23. Comparison between predicted and measured temperature-time curves for horizontal fire spread in still air.

However, the present modelling approach does not allow distinguishing the char combustion phenomena (second peak in the experimental curve). The reason lye in the simplified formulation of the degradation processes of the solid fuel particles which considered only devolatilization. The last stage of cooling is also well predicted by the model. The model meets our objective since it is able to predict the first two stages which are the most significant for the purpose of the spreading.

Heat Fluxes

A quantitative study of the radiant heat fluxes, received ahead of the fire-front at the centreline of the bed of fuel, is provided hereafter. The comparison between model predictions and measurements is shown in figure 24. The experimental data are deduced from heat fluxes versus time measurements assuming the fire front to be a steady combustion wave. The flame front was 0.31 m high, 0.40 m wide and about 0.05 m deep inside the litter.

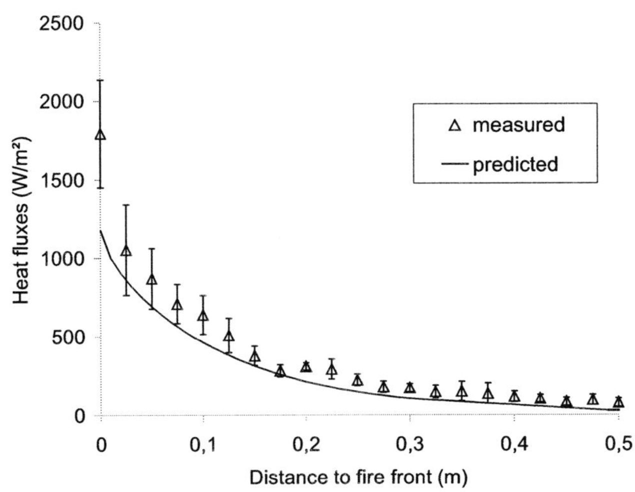

Figure 24. Comparison between the measured and predicted radiant heat fluxes upon the top of the bed of pine needles for horizontal fire spread.

The vertical bars represent a confidence interval of 95% obtained from the four experimental runs with both sensors. The order of magnitude of the radiant heat flux impinging on the top of the bed of fuel, agrees with measurements reported in the literature (fire spread conducted across bed of pine needles [96] as well as static fires [89] with similar flame dimensions). We can see that the impact of the radiation acts over a great distance from the fire-front. Preheating of the unburned fuel located ahead of the fire-front occurs over about 0.20 m, which is close to the flame length. The agreement between simulated and measured radiant heat flux ahead displayed in figure 24 explains the quality of the temperature prediction during preheating (figure 22).

Validation of the Model for Fire Spread under Wind and Slope Conditions

The model predictions will be compared here above to laboratory scale observations for fire spread across a load of 0.5 kg/m^2 of Pinus pinaster needles under wind and slope conditions.

The following configurations were studied: wind-driven up-slope fires for wind and slope ranging from 0 to 3 m/s and 0 to 15°, respectively. The experimental procedure is not detailed. It is close to that provided previously and is fully described in [75]. Only fuel temperature and rate of spread measurements were done. Unfortunately, no heat flux measurements ahead of the fire front were done.

Rate of Fire Spread

The predicted rate of spread is first provided in figure 25 for fires conducted under 10° upslope and wind velocity ranging from 0 to 3 m/s. Although the model allows describing transient effects, its predictions are only compared to the experimental rates of spread in the steady state. Predictions are in agreement with experimental data. We can notice that the fire rate of spread considerably increases with increasing wind velocity. The scattering of the measured rates of spread for a 3 m/s wind velocity is probably due to the turbulent nature of the flow. This last result put to the fore the difficulty of obtaining reproducible fire spread experiments with highly turbulent flame fronts.

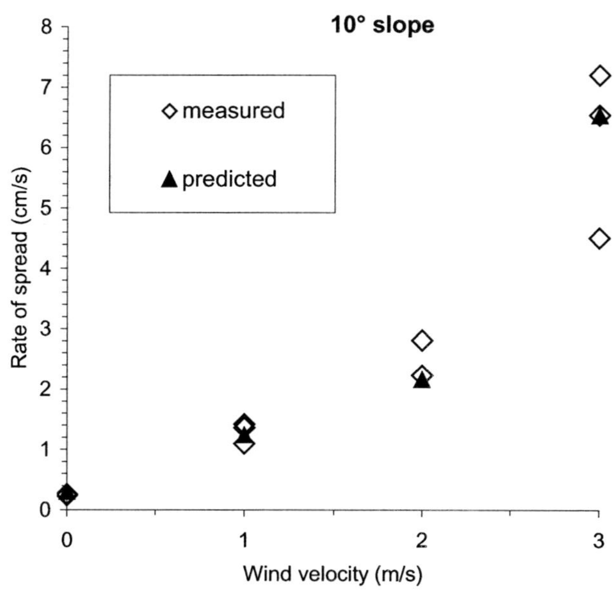

Figure 25. Comparison between predicted and measured rate of spread for 10° up slope fires under wind-blown conditions.

All the results for up-slope and wind velocity ranging from 0 to 15° and 0 to 3 m/s, respectively, are synthesised in figure 26. The model predictions account of the slope effects which further enhances the increase in rate of spread with increasing wind velocity. Overall agreement between the predicted and measured velocities is good except for the most extreme configuration (3 m/s up-wind and 15° up-slope). In that case, a mixed flow boundary layer might develop, induced by the turbulent nature of the flow above the bed of fuel. An increase in radiant and convective transfers impinging upon the fuel bed surface occurs. This effect was not considered in our modelling approach. It corresponds to the situation for which data scattering previously mentioned was marked due to random direct contact of the flame onto the top of the litter.

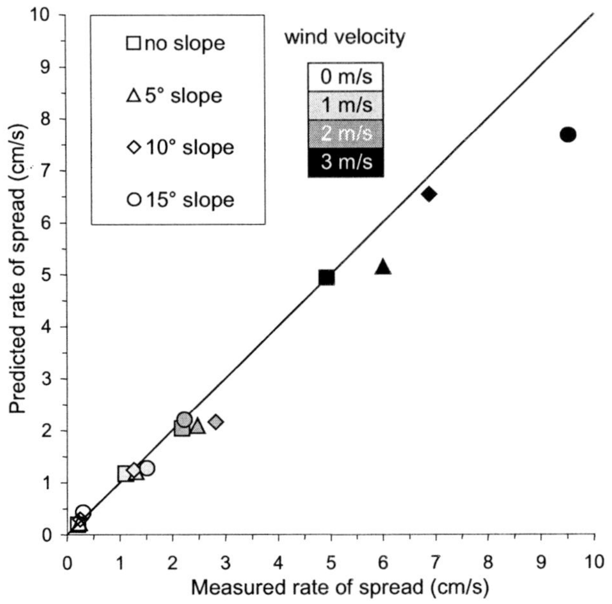

Figure 26. Predicted versus measured rate of spread for fire spread under wind and slope conditions.

Temperature-Time Curves

The predicted and measured temperature-time curves for a 10° up-slope fire spread are provided in figure 27 for 1 and 2 m/s wind velocity. It should be noted that obtaining reproducible temperature-time measurements is not easy for wind-blown fires. The differences between experimental runs are mainly due to the highly turbulent aspect of the flames under wind conditions and its geometrical and dynamic characteristics. These curves confirm that the fire travels faster when wind velocity increases. Indeed, the fire front reaches the measuring section 45 s and 105 s after the line ignition for 2 and 1 m/s wind velocity, respectively. Furthermore, the time for the unburned fuel to ignite is shorter when wind velocity increases. When the unburned fuel is preheated its temperature increases: the time for which the temperature begins to raise from ambient up to 300°C is about 30 s and 50 s, for 2 and 1 m/s wind velocity, respectively. The rate of temperature raise during the preheating is about 13 and 7 °C/s, respectively. The model predictions are in agreement with this tendency. During preheating, the long range action by radiation and convection appears to be an important feature of up-slope and wind-blown fires. It will be discussed in the following when heat fluxes will be examined. The maximum temperature zone, namely in the flame, is not accurately described by thermocouple measurements since no correction technique was applied [20] and comparison to prediction is not easy. Furthermore, the simplified combustion modelling does not take into account the wind effects on reaction kinetic bringing more oxygen into the combustion area. For instance, as shown by the temperature measurements inside the bed of fuel it appears that the increasing wind speed tends to modify the combustion processes as well as the fuel consumption ratio which greatly influence the combustion zone temperature, the flame residence time and so the fire spread behaviour.

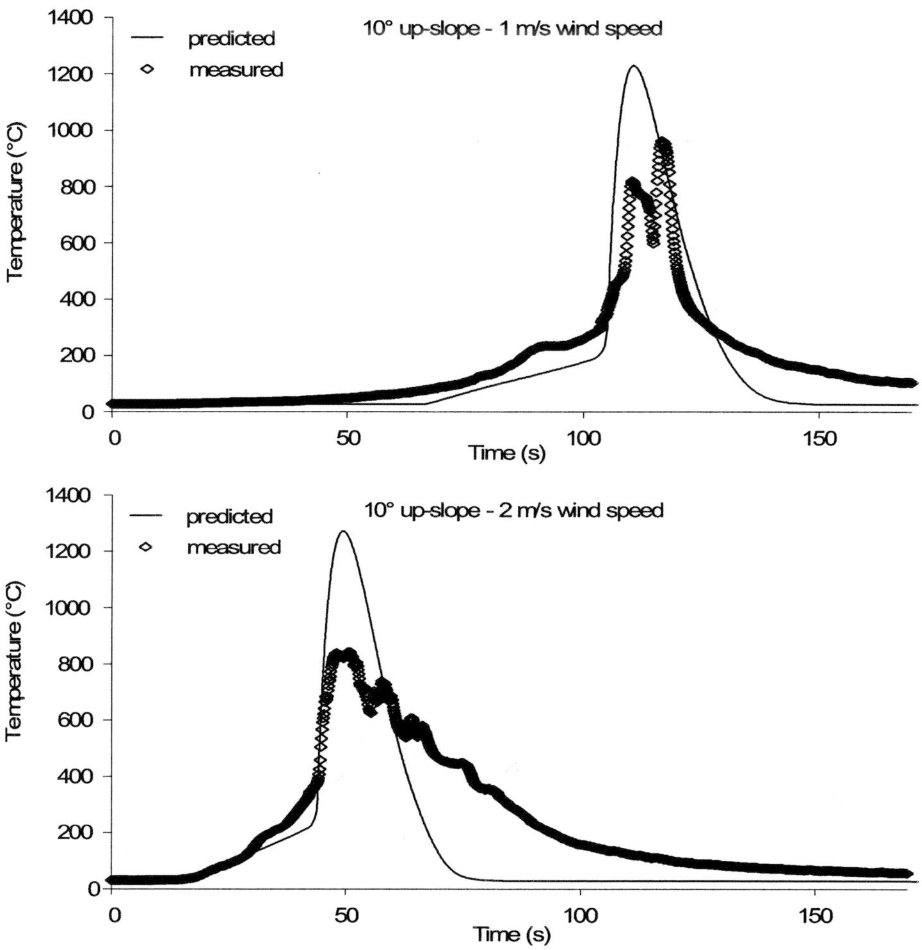

Figure 27. Comparison between predicted and measured temperature-time curves across a 10° sloping surface for wind-blown fire spread (a) 1 m/s (b) 2 m/s.

Fire-Front Shape

The comparisons between predicted and experimental fire-front shapes are not provided since wind-blown fire-fronts recorded during the experiments were not digitised. However, numerical and experimental fire front exhibit the same tendencies When, quasi-steady state is reached, fire travels faster near the centre of the bed of fuel than on the edge since the amount of heat input is greater there. The effects of wind and slope further enhanced fire-front shape distortion. Moreover, the fire-front depth increase with increasing wind velocity is also depicted.

Velocity Components Inside the Bed of Fuel

The spatial distributions of the predicted horizontal and vertical gas velocities at the centreline of the bed of fuel, for a 10° up-slope fire spread under 1 m/s wind velocity are presented in figure 28 at times 100 s. These distribution patterns occur when the fire-front has travelled about 1.5 m along the combustion table and the quasi-steady state was reached. The gas velocity components are only qualitatively examined here because of the lack of measurements in the combustion zone. The model predicts the increase of upward gas velocity in the combustion zone caused by buoyancy forces and the order of magnitude is close to the mean upward gas velocity within the flame measured in the experiments (figure 6). The horizontal gas velocity distribution exhibits a slight acceleration, before penetrating the flame zone due to the aspiration of cold gas; this phenomenon is enhanced with increasing wind velocity. Then, it decreases rapidly as it goes through the combustion zone. The horizontal component, which keeps a non nil value ahead of the fire-front, combined with the temperature gradient are responsible for the preheating by forced convection. It should be reminded that this component is calculated only in the combustion zone (grey area in figure 28).

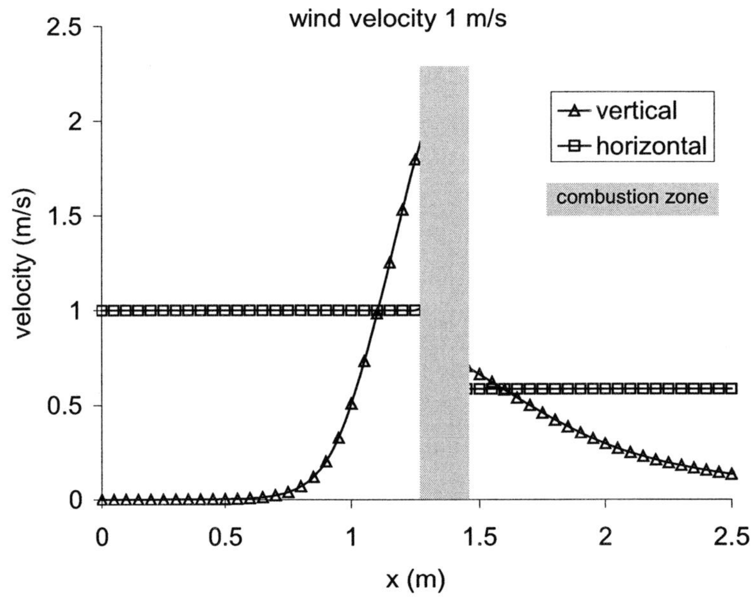

Figure 28. Predicted velocity components (at time 100 s) inside the fuel bed, for 10° up-slope and 1 m/s wind-blown fire spread.

Heat Flux Contributions

The study of the heat transfers mechanisms is of great interest for the understanding of a fire spread under wind and slope conditions. The effects of wind and slope decrease the angle between flame and unburned fuel. The flame tilt is a critical factor that determines an important part of the energy transferred towards the unburned fuel via radiation. Many authors [4,11,43,97,98] have attributed the increase in rate of spread to increased radiation impinging on the fuel due to the tilted flames. The predicted heat fluxes distribution patterns at the centreline of the bed are plotted in figure 29 for wind blown fire spread across a 10° up-slope. These distribution patterns are provided for 1 and 2 m/s at times 100 s and 55 s, respectively. Concordant fire-front shapes were provided in figure 29. The radiant and convective contributions are the most representative of fire spread under wind and slope conditions. The effect of the diffusion term is not significant under these conditions and is not discussed here. The total heat fluxes decrease with increasing distance to the fire front. As a result, about 40% of total energy is received over 0.1 m from the front. Radiation acts over a distance close to the flame length while convection is received closer to the fire front. These results are in agreement with the predictions of a detailed approach [57]. The radiant contribution to total heat transfer on the unburned fuel is greater than the convection one. It increases with increasing wind and slope. For instance, during horizontal and 10° up-slope fire spread, under 2 m/s wind velocity, the radiation represents 60% and 70% of the total heat transfer, respectively.

The predicted amount of energy received by radiation and convection ahead of the fire front are 1112 W and 486 W respectively, for a 1 m/s wind blown fire across a 10° sloping bed. For a 2 m/s wind blown fire, these values are 1679 W and 762 W, respectively. Model predictions indicate that radiation is the predominant preheating mechanism with a significant contribution from convection. These results on pine needle beds confirm the predictions of other model [43]. Forced convection is greater near the combustion region due to high temperature gradient. Nevertheless, fire spread experiments with instrumentation similar to that previously detailed will be benefit to the validation of these predictions.

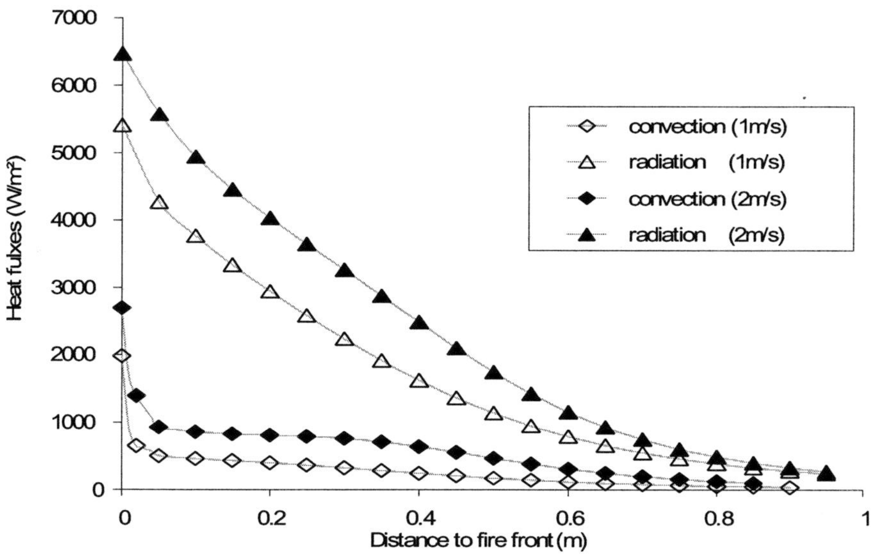

Figure 29. Heat flux contributions ahead of the fire front for 1 and 2 m/s wind aided fire across a 10° sloping surface.

Conclusion

In this chapter, we presented the contribution of our research team to experiments and modelling in the frame of forest fire research. We detailed experimental activities both at the laboratory and the field scale. Concerning the laboratory scale, we investigated the spreading of a line-fire across pine needle beds. We measured gas temperature and velocity in the fire plume thanks to a method of reconstruction of the gas temperature, by using double thermocouple-probes. This method provides results useful for the understanding of the fire behaviour and allows increasing the accuracy of measurement of the maximum temperature in the fire plume. We determined also the characteristics flame length scales from infrared image processing. These data allowed to determine the centreline temperature of flame by mean of the empirical correlation established for static fires by McCaffrey under no slope and no wind conditions. We measured also radiant heat fluxes ahead of the flame that allowed validating radiation predicted by the model and impinging ahead of the fire front.

Whether these experiments are useful to validate model assumptions, due to scale effects, they are not sufficient to study all the phenomena involved in wildland fire spread. Hence, we elaborated a protocol and a set of measuring structures to tackle the understanding of actual fires. We conducted an experimental fire that allowed providing thermodynamics data and behavioural descriptors at the field scale. Wind and vegetation effects were particularly described. We investigated also the fire front shape, its rate of spread, the amount of energy impinging ahead of it, the vertical distribution of the temperature within the fire plume as well as the wind velocity and direction during the spreading. All data presented here can be viewed as part of a necessary set dedicated to understand the fire behaviour and its main properties and this according to the weather conditions, the vegetative structure and the topography. The fire was conducted across shrubs after an exceptional rainy period for the season, which resulted in a difficult fire spread, governed by the wind. As a result, the fire intensity was not as greater as expected. This behaviour of the fire, very different from the one encountered during fire spread across fuel beds at laboratory scales, confirms the necessity to consider field scale experiments for the validation of the wildland fire spread models. We observed a change in the flame structure in comparison with the laboratory scale, as the intermittent region was larger. The temperatures measured within the flame front in the field were in the order of magnitude of the ones measured during laboratory scale experiments but the fire had a greater residence time. The heat fluxes measurements showed that radiation was an important long-range mechanism of preheating ahead of the fire front. As far as the wind is concerned, the large scale turbulence played a significant role on fire spread since it affected the flame shape, temperature and radiation emission. The whole results are encouraging but more experimental fires must be conducted in the field to further assess wind/fire coupling.

Concerning our modelling activities, we presented a simple physical model that allows describing the fire spread in two dimensions, including both flame radiation and forced convection during the fuel preheating process. We used a multiphase formulation to derive a reduced model, which has revealed both the necessity of taking into account advection in the energy balance of our model and to develop a simplified flow through the fuel layer. Thus, the reduced multiphase model has proved to be a relevant tool in the improvement of our formulation. Moreover, this reduction allows to theoretically link simple models to the more detailed models. Simple models, whose aim is to aggregate the mechanisms involved in fire behaviour in a simple manner, require a way of developing simplified equations. The method previously presented can be applied to such model. We validated the results provided by the model at the laboratory scale thanks to experiments. The effects of wind and slope on fuel-bed temperature were qualitatively predicted, even for the highest wind velocities encountered at that scale. The fire front curvature increase and fire front rate of spread increase with increasing wind and slope was also predicted. The qualitative study of the relative importance of radiation and convection heat fluxes ahead of the fire-front was presented. For the range of wind velocities considered, numerical simulations predict that fire spread is mainly controlled by the amount of radiant energy transferred from the flame to the unburned fuel with a significant contribution of convection. Nevertheless, these predictions remain to be tested against experimental data. A certain number of improvements must be made, however in the aim of including our model in a simulator. The amount of radiation impinging on the top of the bed of fuel depends on flame length, which is assumed to be known a priori. The modelling of the radiation from flame front could be improved by taking into account the temperature dependence with height using McCaffrey correlations. The simplified flow modelling can also be improved, in order to describe two-dimensional flow in the fuel layer. Furthermore, the influence of the flow on the reaction kinetic has to be included in our formulation.

The present fire spread model was relevant at laboratory scale but future research will need to describe the field-scale complex topographical conditions. The whole community of wildland fire modelling shares this consideration. Indeed, the fire description at the field scale remains a big challenge for physical modelling. Particularly as semi-empirical modelling has shown their limits in areas such as the Mediterranean one. Moreover, several phenomena are not already finely described at this scale like the influence of the wind turbulence on the flame front. We partially addressed this lack of data with our field experiment that represents a first step of conducting experimental burns in order to test and tune physical models.

References

[1] Fons, W. L. *J. of Agric. Res.* **1946**, 72, 93-121.

[2] Byram, G.M. In Forest fire control and use; Editor, Davis, K. P.; McGraw-Hill Book Company: New York, 1959; 61-89.

[3] Thomas, P. H. *Proc. 9th Int. Symp. (Int.) on Comb.* 1963, 844-859.

[4] Anderson, H. E.; Rothermel, R. C. *Proc. 10th Symp. (Int.) on Comb.* 1965, 1009-1019.

[5] Emmons, H. *Proc. 10th Symp. (Int) on Combust.* 1965, 951-964.

[6] Putnam, A. A. *Proc. 10th Symp. (Int.) Combust.* 1965, 1039-1046.

[7] McArthur, A. G. Weather and grassland fire behaviour; *Aus. For. and Timber Bureau leafter* n°100, 1966.

[8] Thomas, P. H. *Forestry* 1967, 40, 139-143.

[9] Anderson, H. E. *Fire Technology* 1968, 4, 51-58.

[10] Rios, J. *Interaction effects of wind-blown proximate flames from burning wood cribs;* Ph.D. Thesis, university of Oklahoma, 1966.

[11] Nelson, R. M.; Adkins, C.W. *Can. J. Res.* 1986, 16, 1293-1300.

[12] Wolff, M. F.; Carrier, G. F.; Fendell, F.E. *Combust. Sci. Tec*h. 1991, 77, 261-289.

[13] Dupuy, J. L. *Int. J. Wildland Fire* 1995, 5(3), 153-164

[14] Weise, D. R.; Biging, G. S. Can. *J. Forest Research* 1996, 26(10), 1849-1858.

[15] Deen Breejen, E.; Roos, M.; Shutte, K.; De Vries, J. S.; Winkel, H. *Proc. 3rd Int. Conf. Forest Fire Res.* 1998, 1, 517-532.

[16] Mendes-Lopes, J. M.; Ventura, J. M.; Amaral, J.M. *Proc. 3rd Int. Conf. Forest Fire Res.* 1998, 1, 497-511.

[17] Ventura, J. M.; Mendes-Lopes, J. M.; Ripado, L. M. *Proc. 3rd Int. Conf. Forest Fire Res.* 1998, 1, 699-711.

[18] Catchpole, W. R.; Catchpole, E. A.; Butler, B.W.; Rothermel, R. C.; Morris, G.A.; Latham, D. *J. Combust. Sci. Tech.* 1998, 131, 1-37.

[19] Viegas, D. X.; Pita, L. P.; Matos, L.; Palheiro, P. *Proc. 4th Int. Conf. Forest Fire Res.* 2002, CD ROM.
[20] Santoni, P. A.; Marcelli, T.; Leoni, E. *Combust. Flame* 2002, 131(1-2), 47-58.
[21] Marcelli, T; Santoni, P. A.; Simeoni, A; Leoni, E.; Porterie, B. *Int. J. Wildland Fire* 2004, 13, 1-12.
[22] Morandini, F.; Simeoni, A.; Santoni, P. A.; Balbi, J. H. *Combust. Sci. Tech.* 2005 (in press).
[23] Anderson, H. E. Predicting wind-driven land fire size and shape. *Res. paper int.* **305**. US Department of Agriculture, Forest Service, Intermountain Forest and Range Experimentation Station, 1983.
[24] Lim, K. H.; Kwoh L. K.; Liew, S. C.; Lim H. *Proc. 21st Asian Conf. on Remote Sensing,* 2000, Taipei, Taiwan.
[25] Cheney, N. P.; Gould, J. S.; Catchpole, W. R. *Int. J. Wildland Fire* 1993, 3(1), 31-44.
[26] Cheney, N. P.; Gould, J. S. *Int. J. Wildland Fire* 1995, 5(4), 237-247.
[27] Carrega, P. *Proc. 4th Int. Conf. Forest Fire Res.* 2002, CD ROM.
[28] Viegas, D. X.; Cruz, M. G.; Ribeiro, L. M.; Silva, A. J.; Ollero, A.; Arrue, B.; Dios, R.; Gomez-Rodriguez, F.; Merino, L.; Miranda, A. I.; Santos, P. *Proc. 4th Int. Conf. Forest Fire Res.* 2002, CD ROM.
[29] De Luis, M.; Baeza, M.; Raventos, J.; Gonzalez-Hidalgo, *J. Int. J. of Wildland Fire* 2004, 13, 79-87.
[30] Alexander, M. E.; Stocks, B. J.; Wotton, B. M.; Flannigan, M. D.; Todd, J. B.; Butler, B. W.; Lanoville, R. A. *Proc. 2nd Symp. Fire Forest Met.* 1998, 20-23.
[31] Cohen, J. D. *What is the wildland fire threat to homes?* Presented as the Thompson Memorial Lecture, School of Forestry, Northern Arizona University, Flagstaff, 2000.
[32] Pitts, W. M. *Prog. Energy Comb. Sci.* 1991, 17, 83-134.
[33] Catchpole, T.; De Mestre, N. *Aust. For.* 1986, 49 (2), 102-111.
[34] Weber, R.O. *Int. J. of Wildland Fire* 1991, 1 (4), 245-248.
[35] Weise D.R. *Modelling wind and slope-induced wildland fire behaviour*; PhD Thesis, Berkeley, 1993.
[36] Anderson, H. E. *Predicting Wind-Driven Land Fire Size and Shape. Research paper int.* **305**, US Department of Agriculture, Forest Service, Intermountain Forest and Range Experiment Station, 1983.
[37] Ridchards, G. D.; Walberg, R. *Proc. 3rd Int. Conf. Forest Fire Res.* 1998, 1, 435-440.
[38] Barros, F. J.; Ball G. L. *Proc. 3rd Int. Conf. Forest Fire Res.* 1998, 1, 879-887.
[39] Frandsen, W.H. *Combust. and Flame*, 1971, 16, 9-16.
[40] Rothermel, R.C. *A mathematical model for predicting fire spread in wildland fuels; Research INT*-**115**, USDA, Forest Service, 1972.
[41] Andrews, P. L. BEHAVE: Fire behavior prediction and fuel modeling subsystem; *Report INT*-**194**, USDA, Forest Service, Intermountain Research Station, 1986.
[42] Finney M.A. FARSITE: Fire area simulator – model development and evaluation; *Res. Pap. RMRS-RP*-**4**, USD, , Forest Service, Rocky Mountains Research Station, 1998.
[43] Pagni, P. J.; Peterson, T. G. *Proc. 14th Symp. (Int.) on Comb.* 1973, 1099-1107.
[44] Cekirge, H. M. *Comput. and Math. with Appl.* 1978, 4, 325-332.
[45] Fujii, N.; Hasegawa, J.; Pallop, L.; Sakawana, Y. *Appl. Math. Model.* 1980, 4, 176-180.

[46] Grishin, A. M.; Gruzin, A. D.; Zverev; V. G. *Fizika Goreniya i Vzryva* 1985 21 (1), 11-21.
[47] Albini, F. A. *Combust. Sci. Tech.* 1986, 45, 101-113.
[48] De Mestre, N. J.; Catchpole, E. A.; Anderson, D. H. *Combust. Sci. Tech.* 1989, 65, 231.
[49] Weber, R. O. *Combust. and Flame* 1989, 78, 398-408.
[50] Ferragut, L.; Asensio, M. I.; Montenegro, R.; Plaza, A.; Winter, G.; Seron, F. *J. Comput. Fluid Dyn.* 1996, 8, 111-116.
[51] Santoni, P. A.; Balbi, J. H. *Fire Safety J.* 1998, 31, 201-225.
[52] Lymberopoulos, N.; Tryfonopoulos, T.; Lockwood, F. C. *Proc. 3rd Int. Conf. Forest Fire Res.* 1998, 1, 405-418.
[53] Linn, R. R.; Reisner, J. M.; Winterkamp, J. L. *Proc. 4th Int. Conf. Forest Fire Res.* 2002, CD ROM.
[54] Fan, W.; Wang, J. *Proc. 3rd Int. Symp. on Fire Safety Sci.* 1991, 325-334.
[55] Grishin, A.M. *Mathematical modeling of forest fires and new methods of fighting them*; Albini (Ed.): Publishing house of the Tomsk state university, 1996.
[56] Larini, M.; Giroux, F.; Porterie, B.; Loraud, J. C. *Int. J. Heat Mass Trans.* 1997, 41 (6-7), 881-897.
[57] Porterie, B.; Morvan, D.; Loraud, J. C.; Larini, M. *Phys. Fluids* 2000, 12, 1762-1781.
[58] Sero-Guillaume, O.; Margerit, J. *Int. J. Heat Mass Trans.* 2002, 45, 1705-1737.
[59] Morvan, D.; Dupuy, J. L. *Combust. Flame* 2004, 138, 199-210.
[60] Steward, F. R. Fire spread through a fuel-bed. *Heat transfer in fires*; chap 2, 1974, 315-378.
[61] Viegas D.X., Forest fire propagation, *Phil. Trans. R .Soc. Lond. A* **356**, 1998, 2907-2928.
[62] Vaz, G. C.; André, J. C. S.; Viegas, D. X. *Proc. 4th Int. Conf. Forest Fire Res.* 2002, CD ROM.
[63] Catchpole, W. R.; Catchpole, E. A.; Tate, A. G.; Butler, B.; Rothermel R. C. *Proc. 4th Int. Conf. Forest Fire Res.* 2002, CD ROM.
[64] Balbi, J. H.; Santoni, P. A.; Dupuy, J. L. *Int. J. of Wildland Fire* 1999, 9, 275-284.
[65] Santoni, P. A.; Balbi, J. H.; Dupuy, J.L. *Int. J. of Wildland Fire* 1999, 9, 285-292.
[66] Morandini, F.; Santoni, P.A.; Balbi J.H. *Comb. Sci. Tech.* 1999, 157, 141-165.
[67] Santoni, P.A. *Num. Heat Transfer* 1998, 33, 279-298.
[68] Simeoni, A.; Santoni, P. A.; Balbi, J. H. *Int. J. Mod. Sim.* 2002, 22, 1-12.
[69] Morandini, F.; Santoni, P. A.; Balbi, J. H. *Fire Safety J.* 2001, 36, 519-543.
[70] Morandini, F.; Santoni, P. A.; Balbi, J. H. *Combust. Sci. Tech.* 2001, 166, 67-90.
[71] Morandini F.; Santoni P. A.; Balbi J.H.; Ventura, J. M.; Mendes-Lopes, J.M. *Int. J. Wildland Fire* 2002, 11, 53-64.
[72] Simeoni, A.; Santoni, P. A.; Larini, M.; Balbi, J. H. *Comb. Sci. Tech.* 2001, 162, 59-84.
[73] Simeoni, A.; Larini, M.; Santoni, P. A.; Balbi, J. H. C. R. *Mécanique* 2002, 330, 783-790.
[74] Simeoni, A.; Santoni, P. A.; Larini, M.; Balbi J.H. *Int. J. of Thermal Sci.* 2003, 42, 95-105.

[75] Mendes-Lopes, J. M. C.; Ventura, J.M.P.; Amaral, J. M. P. *Int. J. Wildland Fire* 2003, 12, 67-84.
[76] Viegas, D. X. *Int. J. Wildland Fire* 2004, 13, 101-17.
[77] Cox, G. *Combust. Flame* 1977, 28, 155-163.
[78] Tagawa, M.; Ohta, Y. *Combust. Flame* 1997, 109, 549-560.
[79] Saastamoinen, J. J.; Taipale, R.; Horttanainen, M.; Sarkomaa, P. *Combust. Flame* 2000 123, 214-226.
[80] McCaffrey, B. J. *Purely buoyant diffusion flame: Some experimental results*; *NBSIR* **79-**1910, Department of Commerce: Washington, 1979.
[81] Cox, G.; Chitty, R. *Combust. Flame* 1985, 60, 219-232.
[82] Dupuy, J. L.; Maréchal, J.; Bouvier, L.; Loïs, N. *Proc. 3rd Int. Conf. Forest Fire Res.* 1998, 1, 843-858.
[83] Audouin, L.; Kolb, G.; Torero, J. L.; Most, J.M. *Fire Safety J.* 1995, 24(2), 167-187.
[84] Zukoski, E. E. *In Combustion fundamentals of fire*; Editor, Cox G., Academic Press:London, 1995; 101-219.
[85] Zukoski, E. E.; Cetegen, B. M.; Kubota, T. *Proc. 20th Int. Symp. Comb.* 1984, 361-366.
[86] Drysdale, D. *An introduction to fire dynamics*; John Wiley and Sons: Chichester, 1985; 57-65.
[87] Beyler, C. L. *Fire Safety J.* 1986, 11, 53-75.
[88] Hasemi, H.; Nishita, M. *Proc. 2nd Int. Symp. Fire Safety Science* 1988, 275-284.
[89] Dupuy, J. L.; Maréchal, J.; Morvan, D. *Combust. Flame* 2003, 135, 65-76.
[90] Carvalho, A.; Miranda, A. I.; Borrego, C.; Santos, P.; Amorim, J. H.; Viegas, D. X. Fire as a heat source to the atmosphere: measuring and modelling. *Proc. 4th Int. Conf. Forest Fire Res.* 2002, CD ROM.
[91] Delhaye, J.M. *Proc. of NATO Advanced Study Institute*, **1**, 1976.
[92] Anderson, T. B.; Jackson, R. Ind. Engn. *Chem. Fund.* 1967, 6, 527-539.
[93] Albini, F.A. *Comb. Flame* 1981, 43, 155-174.
[94] Guijarro, M.; Hernando, C.; De Los Santos, J.A.; Diez, C. Forest fire behaviour in the wind tunnel; *Intermed. report Efaistos Proj.*, European Union, 1998.
[95] Comolet, R. *Mécanique expérimentale des fluides*; Masson: Paris, 1982; 216-230.
[96] Fang, J.B.; Steward, F. R. *Combust. Flame* 1969, 13(4), 392-398.
[97] Albini, F. A. *Combust. Sci. Tech.* 1985, 42, 229-258.
[98] Beer, T. *Boundary-Layer Meteorology* 1991, 54, 287-308.

INDEX

A

B

C

D

E

F

G

H

I

N

O

P

U

V

W

Y